T0191406

THE ROAD TO SPACE

THE ROAD TO SPACE

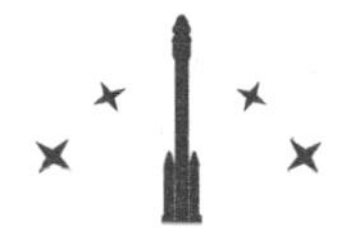

THE RECORD OF CHINA'S AEROSPACE DEVELOPMENT

LI XUANQING & LIU GANG

The Road to Space: The Record of China's Aerospace Development

Li Xuanqing and Liu Gang
Translated by Daniel McRyan

Published by NATURALOGIC PUBLISHING INC., under an exclusive license with Shanghai Jiao Tong University Press.

First English Edition 2021
ISBN: 978-1-4878-0478-7 (hardcover)
ISBN: 978-1-4878-0479-4 (paperback)

19-1235 Johnson St. Coquitlam, BC, Canada V3B 7E2

PROLOGUE

Time Evanesces, the Tune Lingers

In the divine land of China, what tune touches the deepest?

Shut your eyes and listen:

The East is Red and the sun rises …

Forty-eight years ago, accompanied by the din of radio waves, the tune, weak and strong, on and off, ethereal and mysterious, traversed from distant space.

Forty-eight years ago, it touched the entire Chinese nation, tugging at the heartstrings of every Chinese descendant, immersed in elation.

On the day when the tune sounded, history was made to remember. On the 24th of April 1970, China launched Dongfanghong 1, the first space satellite, thus becoming the fifth country after America, the Soviet Union, France, and Japan, to be capable of independently launching a space satellite.

On this very day, when the beautiful melody of *The East Is Red* sounded in the immensity of space, the whole world was in awe.

A miracle! What a miracle! Eyes widened, they couldn't help wondering how on earth China, a country that has just risen from poverty and has still bread issues to attend to, launched a satellite.

On this very day when every note of *The East is Red* grooved amid the stars, all of China was filled with elation.

As a spring breeze, the news of the successful launch of Dongfanghong 1 swept every corner of the 9.6-million-square-kilometer territory, from Tian'anmen square in Beijing to the Potala Palace in Lhasa of Tibet, from the foot of Baota Mountain in Yan'an to the bank of the Huangpu River in Shanghai, and from the riverside of the Xiangjiang River to the foot of Tianshan Mountain.

On this very day, in every city, on every street, and in every village, everybody couldn't help delightfully humming along to the melody on the radio: *The East is Red and the sun rises …*

The tune condenses Chinese wisdom.

China, as one of the ancient civilizations, propelled the development of world civilization with its four great inventions. Yet, in modern times, the ancient oriental nation has fallen. The West labeled it as ignorant and undeveloped.

A new age has come. The melody that burst from the chest of the Chinese nation reverberated around the universe, blowing minds of the biased and the proud in the West with an explicit demonstration of the infinite wisdom of China. This time, China bravely stood out in front of the wave of global technological advancement.

This tune sums up Chinese strength.

Following the detonation of the atomic bomb and the hydrogen bomb, this is another Chinese astonishing masterpiece. This great work vividly demonstrated the superiority of the socialist system. Never has there been a country in the world that united every ounce of strength and talent in the nation under extreme adversity as China did.

Pride, confidence, and strength rose in the cheeks of every Chinese person. The upbeat tune once again announced to the world that China has risen and the sick nation that suffered the plunder and ravages of western imperialism belonged to history. China can achieve everything, relying on its own strength, which is the backbone of the continuous self-improvement of the Chinese nation.

The tune encapsulates Chinese dream.

For the Chinese, space exploration has been both a folklore that has been told for thousands of years and a long craving dream. The moment when Dongfanghong 1 satellite orbited in space and the beautiful melody echoed above marked a crucial step that China took in the expedition of space exploration.

As a bugle, the tune lifted the curtain of China's space dream.

As a battle drum, the tune spurred young China as well as a considerable number of space scientists to dedicate themselves to China's aerospace undertaking with tremendous determination and a bold pledge to strive forward.

Time flies as a shooting arrow. As the tune reverberates, from scratch, China's space exploration has undergone 60 years; and it has been 48 years since the launch of Dongfanghong 1 the first artificial space satellite.

One, two, three... As the rockets soared, roaring, Chinese satellites shone brightly among the stars.

From the first return satellite to the first meteorological satellite, from the repeater satellite to Gaofen No.1 satellite... In orbit, "Chinese stars" glow splendidly, benefiting millions of Chinese. In 2016, the Chinese Plough Satellite Navigation Positioning System covered the Asian Pacific area and a global coverage is to be accomplished in the next stage. "Chinese stars" will become global, thus benefiting the world.

Step one, two three... In the spotlight of the world stage, Chinese space travel is accelerating rapidly and the space that it lays fingers on extends increasingly further.

From the first blast-off of Shenzhou Spacecraft to the first Chinese astronaut Yang Liwei and from the successful launch of Tiangong 1 to Chinese astronaut Wang Yangping's space lecturing, Chinese manned space programs are making one legend after another at an incredible speed.

From manned space flight to the Chang'e Moon Exploration project and to a Mars probe, with unprecedented energetic drive, China continues to set new records in the history of both Chinese and global space expeditions. In the vast universe there is a seat for the country that hangs Chinese knots. On the lonely Moon, the Chinese five-star red flag left its footprint.

In 60 years of glory, the Chinese space program rose from zero to one and one to infinity. Every launch radiates Chinese spirit and adds a brick to the nation's stairway to be powerful.

In 60 years of achievement, the Chinese space program laid fingers from near-Earth orbits to the Moon and to Mars. Every exploration elaborates Chinese innovation and contributes to the thousand-year-craving space dream coming true.

The space dream, the Chinese dream. In the New Year Greetings, President Xi Jinping acknowledged that by 2016 Chinese Sky Eye had been in full shape and operation, Wukong satellite orbiting for a year already, Mozi satellite flying towards the space, Shenzhou XI and Tiangong 2 traversing among the stars... In the 1402-character New Year Greeting speech, great length was dedicated to Chinese space travel, implying the significant place it occupies in the heart of President Xi and every Chinese citizen.

To an individual, 60 years is a lengthy time. Yet, to a great cause, it is a blink of an eye.

The goal stretches ahead. For Chinese space exploration, 60 years is just a new beginning.

Innovation functions as a bridge, and dreams come true. Time evanesces while the tune remains.

On the 24th of April 2016, 46 years after Dongfanghong 1 satellite took off, Chinese Space Day came into being.

On that day, Dongfanghong 1 was still operating in space. Aged as it is, it is wondered whether it can still hear the call-out from every Chinese descendant.

On the day, every verse of *The East is Red* lingered among millions of Chinese and every note ran in the hot blood in every participant of the Chinese space program.

Humming to the melody, every Chinese space program participant has cast their vision higher and further.

With the melody reverberating at ear and the road of Chinese astronomy stretching in front of the eyes, we couldn't help wondering, through thick and thin, what robust pillar sustained the great Chinese space exploration.

Let's review the epic legend.

Contents

THE ROAD TO SPACE

CHAPTER 1

The East Is Red Echoes in Space

Creak after creak.

On a serene morning, creaks from the amble on the snow echoed rhythmically. A set of footprints, two sets of footprints... Soon, footprints overlapped one another on the snow. Early risers, in tight wraps, hurried to factories and schools. Imperceptibly, the pavement covered by snow turned dark.

On the street, a tram full of passengers pulled in. The crisp ring announcing the arrival at the station echoed in the chilly wind. Somewhere further, soaring factory chimneys were spitting out thick smoke... The city of snow was gradually waking up in the early rising sun.

It was the morning of November 25, 1955 in Harbin.

The day differed nothing from the usual to the dwellers, who fared on their normal life tracks.

Whereas, for Chen Geng, a senior general of the People's Republic, it was a bit different.

He got up earlier than usual to receive an important visitor.

In martial attire, he was standing in the frosty wind in front of the office building.

Titled the deputy chief of staff of the People's Liberation Army and President of Harbin Military Engineering College, he had taken a private plane, rushing back to Harbin from Beijing one day in advance to welcome this significant guest.

Of such significance, who could he be?

At that moment, a black car pulled into the campus of Harbin Military Engineering College. The door opened and a gentle and elegant face appeared.

It was him, the prestigious scientist, Qian Xuesen.

A month earlier, on October 8, Qian just returned to the homeland. And three days ago, he officially reported to Chinese Academy of Sciences. The journey to the northeast of China was his first visiting research after his return to China.

Regarding the value of Qian, Dan A. Kimball, the U.S. Secretary of the Navy, commented in 1950 that Qian, wherever he was, was the equivalent of five Marine divisions. Back then, on the entire battlefield in North Korea, the U.S. Army possessed seven divisions as a ground force. He expressed that he would rather have him shot dead than permit his departure from the U.S. His professional insight exactly corresponded to the Chinese proverb that there is more scarcity in excellent army leaders than good armies.

Americans regarded Qian as the most capable rocket scientist. It is not hard for them to fathom what modern rockets meant for China. Therefore, efforts were made to obstruct his return to China. He underwent five years of house arrest.

Reality confirmed Kimball's insights. Yet, he still underestimated Qian's value. His calculations were far less.

The Life of Qian Xuesen, published by the Xinhua News Agency, summarized the 11 outstanding firsts Qian has made to China's National Defense Science and Technology:

China's first short-range surface-to-surface missile launch test, China's first missile and atomic bomb combination test, China's first launch of artificial satellite Dongfanghong 1, China's first nuclear powered submarine, China's first returning satellite, and China's first underwater submarine launched missile...

The firsts massively changed China and the world.

On this usual winter morning they met. One is a war-surviving general who made brilliant achievements at war, and the other is a world-class scientist who prioritized the glory of the motherland ahead of personal fame and gain as well as overcame mountains of obstructions to return to the homeland. Tightly they shook hands. The spark that was ignited by the collision of the two hearts that cared for the fate of the nation sufficed to chase away the chill of the winter.

As always, imperceptibly, history sparked a brand-new page.

On the day, Chen was deeply concerned "Mr. Qian, do you believe Chinese can make missiles?"

"Why not? Foreigners can, so can Chinese." he replied decisively.

"That is exactly what I expect to hear." uttered Chen.

The classic conversation still echoes in the ear. Since then, the name Qian Xuesen formed a close connection with the National Defense of the new China.

In his memoir, it was written that the first to address missile prospects upon his return to China was General Chen.

In a sense, their encounter accelerated the groundbreaking of China's two-bomb-one-satellite project as anything pertinent to the project was implemented faster henceforth with the coordination of General Chen.

The First to Have Sketched the Blueprint

In addition to General Chen Geng, more people were eager to meet Qian.

On December 21,1955, Qian returned to Beijing from the visit to northeast China. Hearing the news, regardless of the residential treatment, Defense Minister Peng Dehuai, commanded General Chen to have Qian invited for a discussion.

The urgency of Marshall Peng, who fought against the Americans on the Korean battlefield for three years, to meet Prof. Qian was unparalleled.

On the Korean battlefield, nuclear threats were made, outrageously and repetitively by America. President Harry Truman proclaimed in a press conference that necessary countermeasures including the detonation of atomic bomb were prepared to handle difficult military situations.

In September of 1954, Jinmen Island in Taiwan was being bombed. The Taiwan Straits echoed the roar of detonations. President Dwight David Eisenhower uttered unforgettable words that there was no reason not to deploy nuclear weapons as if there is no reason not to shoot a bullet at war.

Allegedly, before leaving the office as Commander of the Volunteer Army, Marshall Peng paid a visit to the cemetery of the John Doe volunteer soldiers. Respectfully, apologies were made for unable to take them home safely.

On the train back to China, he wrote a letter to Chairman Mao, saying "Chairman, the war ceased, and victory belonged to us. Nevertheless, we suffered a considerable loss because of inferior weapons."

Faced with western developed countries, how can the new Communist China stand straight without a killer weapon? Can Chinese develop their own killer weapon?

Meeting Qian was Marshall Peng's search for answers.

In the afternoon on of December 26, 1954, Marshall Peng and Qian finally met in a hospital ward. "Mr. Qian, as a soldier, I came to discuss fighting a war with you… I wonder whether we Chinese can make our own missile."

An answer was found in the firm eyes and meticulous analysis of Qian. Time seemed to fly in that afternoon.

In Zhongnanhai, the residential compound in Beijing housing top party leaders, Marshall Peng personally arranged a report meeting concerning the development of cutting-edge weapons by Qian to high rank ministers of the government and military, thus stirring up a missile fever among the officials.

The Second Plenary Session of the Second National Committee of the Chinese People's Political Consultative Conference was held in Beijing from January 30-February 7, 1956. As a newly recruited committee member, Qian was invited to attend and give a speech. At the evening of February 1, Chairman Mao held a feast for all members of the national committee of Chinese People's Political Consultative Conference and Qian was seated adjacent to him.

On February 4, snow covered Beijing overnight. General Chen accompanied Qian and his wife to be a guest at Marshall Ye Jianying's, during which the conversation revolved around missiles and rockets, the seemingly main course on the table.

Excitement accumulated during the meal. Subsequently, General Chen proposed seeking approval from Premier Zhou Enlai. Ye held the information that Minister Zhou happened to be attending a ball hosted by the General Office of Central Military Commission at Sanzuomen, a place in the vicinity of his home. Hence, with neither vehicle nor guards summoned, they walked straight towards Sanzuomen.

Chen entered the hall and invited Zhou out. It was the first official meeting between Zhou and Qian. In a parlor, Qian rhapsodized his scientific visions.

Having heard Qian out, Zhou solemnly asked "Mr. Qian, could you please form a written report of the visions for rocket and missile launches as soon as possible and submit it to the central government for discussion?" Qian zestfully nodded. He never imagined that the motherland appointed him a project so grand and so soon.

And there started the sketch of the grand blueprint of Chinese missiles and rockets. And Qian is the very first.

After half-month revision, on February 17, 1956, Qian officially submitted the Proposal Regarding the Establishment of Chinese National Defense Aviation Industry (for the sake of confidentiality, National Defense Aviation Industry was phrased to represent missile, rocket, and the successive space projects). The proposal detailed the organizational schemes and developmental plans for Chinese missiles and rockets. Simultaneously, a list of senior experts available to join the cause was formulated, including Ren Xinmin, Liang Shoupan, and Zhuang Fenggan, who later pioneered China's space road.

In an age when millions of tasks were waiting to be done, everything was advancing on a tight clock.

Four days later, Zhou personally examined the proposal and handed to Chairman Mao for further scrutiny. Half a month later, it was reviewed at the Expanded Meeting of the Central Military Commission.

On April 13, the State Council established a secretive department, known as Aviation Industry Commission, affiliated to National Defense. Nie Rongzhen was appointed as the director, Huang Kecheng and Zhao Erlu deputy directors, and Liu Yalou and Qian Xuesen committee members. The Aviation Industry Commission was the predecessor of the National Defense Science and Technology Commission and the General Equipment Department of the People's Liberation Army. The establishment marked a crucial step taken on China's missile and rocket path.

On April 25, at the expanded meeting of Political Bureau of the Central Committee of the Communist Party of China, Mao inspiringly articulated "We are stronger than the past and will be even stronger. We shall have more planes and cannons, and atomic bombs. Today, to avert bullying, must we have them."

On August 10, 1956, in a shabby refectory of a military camp of the western suburbs of Beijing, the Fifth Institute of National Defense was founded, with Qian as the director. There were no flowers, no colorful flags, no drumbeats at the founding ceremony but aspirations and dedications. "This is grand and promising work, to be a part of which was supreme honor" voiced Qian.

For China's missiles, rockets, and aviation, that was the day to be permanently remembered.

We Will Make Satellites Too

The history of mankind's space travel unfolded on this very day:

At the midnight of October 4, 1957, on a wasteland 2,000 kilometers away from Moscow, a rocket blasted off, its dazzling tail kindling the night sky. The first artificial satellite Sputnik 1 defied gravity, dashing towards space.

The world was stunned by Soviet Union's successful launch.

Americans were outraged. Allegedly, President Dwight David Eisenhower was playing bridge when the telegram arrived. Cards scattered on the floor when he learned that the Soviet Union satellite had prevailed. He stood up, mumbling "Unbelievable." A Washington newspaper even proclaimed 1957 to be "Pearl Harbor Year II."

Seen from today, the artificial satellite the Soviet Union launched was rather rudimental: it weighed 83.6 kilograms with merely two signal transmitters equipped; at the speed of 95 minutes per lap around the earth, it orbited on the track 215 kilometers higher from perigee transmitted radio signals relentlessly.

Regardless, it knocked open the door to the universe for the first time as the celestial envoy to space, which is a magnificent achievement that is entitled a spot in history. The US-newspaper *The Christian Science Monitor* commented that a new epoch of star trek commenced and a new promise for mankind was exhibited in the capability of the Soviet Union's satellite.

What shocked the world more was that, before the first Soviet Union satellite fever faded away, a second man-made satellite Sputnik 2 blasted off successfully on November 3rd. This trip also included a female dog called Laika, the first animal to travel to space, although she died of a lack of oxygen after four days. Back then, Soviet Union hadn't mastered the technology to recycle the air in a satellite.

Mao happened to be in Moscow in the autumn of 1957, attending the Conference of Representatives of Communist Parties and Workers' Parties of All Countries. Over the 80 years of his life, it was the only time he travelled abroad to attend a meeting.

A western journalist reported the scene:

The moment Mao stepped off the plane, he gave an enthusiastic speech at the Moscow airport. "The Soviet Union's first artificial satellite is a great achievement, marking the commencement of human's further conquering of nature." He believed that the socialist camp's possession of artificial satellite enabled contempt over everything of imperialism.

Having heard the prevailing blast-off of the second satellite, Mao delightfully expressed congratulations to Khruschev. "What a remarkable step that another satellite took off. America has been boasting yet there is not an egg tossed up there. It meant much and the superiority of socialism was exhibited. I wonder whether a common goal was chosen, 10 years for you to catch up with America and 15 for us with England."

Looking into the words of Mao, we can easily deduce that in Moscow, Mao had already begun to contemplate the prospect of Chinese satellite. The launch of the satellite astonished him greatly. If the Soviet Union could, why not China? He firmly believed China could make it happen.

During the Moscow trip, Mao paid a special visit to Chinese students abroad. In the auditorium of the University of Moscow, he addressed an enthusiastic speech to over 3,000 Chinese students abroad in a thick Hunan accent. "The world is yours, also ours. Yet down to earth, it belongs to you, energetic as the rising sun. The promise lies in you."

A young man named Sun Jiadong, sitting among the crowd, took in the words of the great leader. Titled the Chief Designer of the Dongfanghong satellites, Chief designer of China's Moon explorations, and pioneer of China's space travel, he wrote affectionately in his memoir "Chairman Mao was kind and humorous and the students abroad were immersed in massive excitement."

This is no exaggeration that Mao's speech inspired an entire generation of youth. Students abroad back then shouted out a dedication of half a century of work to the motherland.

America fidgeted when the second Soviet Union satellite succeeded. Maximum efforts were made to keep up.

On December 6, 1957, came the news that the U.S. Navy intended to launch the first American artificial satellite with Pioneer Rocket at the launching site in Cape Canaveral, in Florida.

Global attention was on America, on the launching site in Cape Canaveral. Little did anyone expect that seconds following the ignition the rushing Pioneer Rocket collapsed into a ball of fire.

Failure oft entails success. Soon, the Americans started over. On January 30, 1958, the Rocket Cupid C, carrying the satellite Explorer I, took off from Cape Canaveral.

This time, the Americans made it. Eight minutes after the takeoff, a signal was transmitted back from space. 119 days after Soviet Union's launch, America eventually positioned a cylindrically-shaped "star" above, weighing 8.22 kilograms.

The success of both the Soviet Union and the U.S. knocked open the gate to space exploration. As if in a chain reaction, more countries showed great interest, one of which was China. Numerous scientists published papers discussing the significance and usefulness of artificial satellites.

Qian made a lengthy comment on Soviet Union's first artificial satellite, pointing out that regardless of its weight and bulk, a scientist must pay attention to the carrier that took it to space, which according to the materials available was a three-stage rocket.

For a rising China, the satellite became a symbol with a special meaning, a dream for millions of Chinese.

In the afternoon of May 17, 1958, the Second Plenary Session of the Eighth Central Committee of the CPC was held, where representatives discussed artificial satellites again, they required China should also make one, because of its dual identity as the homeland of rockets and a communist country as Soviet Union.

Smoking, Mao paid great attention to the speeches of the representatives. Seeing Mao smoking heavily, Zhou Enlai realized there was a speech coming up. "Regarding satellites, let's give the floor to the Chairman" he rose and signaled the room to be silent.

The room applauded.

Peering about, Mao uttered with great calm, "Comrades, recently, you have been deeply concerned with satellites. I am no exception, of course. Last year, Soviet Union launched it to space and months ago so did America. What about us?"

He paused. Silence filled the room and all eyes were on him. Nipping out the cigarette, he shouted "We will make our own too."

There was thunderous applause and the room cheered.

Sipping on a hot tea, he continued humorously, "Certainly, we must take baby steps. Whereas, a satellite as big as the America egg, is not what we are going to shoot up. Let it be one weighing 20,000 kilograms."

The passion of every attendant present was ignited. All rose, with lasting applause.

By now, the history is deeply remembered. Chewing on every detail, a better understanding of Zhou Enlai's words show that the Chinese have a backbone and wisdom, and there is nothing so difficult as to bring China to its knees.

Just do it.

There was an interesting phenomenon in China.

It is highly likely that people named Weixing (satellite) or Xingyu (universe) were born in 1958.

Satellite was the trendiest word. Thousands of parents left the imprint of times on their newborns.

1958 witnessed the "Great Leap Forward," where utmost efforts were exerted. All walks of life were "launching their satellites" agriculture sloganeered "How bold the people, how big the productivity," industry took pride in "a 7-million steel-manufacturing army to complete a yearly task in one day"

Across the nation, "satellites" were tempestuously and increasingly being launched. However, how should the real satellite, the very first Chinese man-made satellite, be launched?

On May 29, 1958, soon after the Second Plenary Session of the Eighth Central Committee of the CPC ended, exactly 12 days after Mao's speech of the determination of satellite launch, Nie Rongzhen organized a meeting of some members of the National Defense Aviation Industry Committee. He was debriefed by Qian Xuesen and peer experts about rocket launch, artificial satellites, and continental missiles.

As Vice Premier of the State Council managing national technological work, Nie Rongzhen, the experienced veteran, shouldered the task of China's technological advancement.

Sharing similar upbringing to Deng Xiaoping, after turning 20 in 1919, he left Sichuan and headed for France on a work-and-study program. Three years later, he was admitted to study chemical engineering in the University of Labour (L'universite du Travail) in Belgium. In a letter to his parents, he wrote "What am I studying for? Never for my personal material gain, but 400 million compatriots to be dressed and fed; never for my personal happiness and contentment, but for 400 million compatriots to share. It is what I have always aspired to achieve and what I intend to pursue for a lifetime."

Lofty ambition forges greatness. Nie spent half of his life at war, where he dodged volleys of gunfire and made brilliant achievements, and the other half on technology, guiding a technological army to overcome extreme adversity and create atom bombs, hydrogen bombs, and man-made satellites at an incredible speed, therefore allowing China to walk tall again.

In August of 1958, led by Nie, the Chinese Academy of Sciences established the first satellite group, Team 581.

The name symbolizing the satellite was the priority of the Academy in 1958. For the sake of confidentiality, the team was named as numbers. Throughout China's space research history, coding major projects by the specific year is still in use as a unique tradition.

Team 581 was led by Qian Xuesen and assisted by Zhao Jiuzhang and Wei Yiqing. It consisted of three design institutes:

The first design institute, led by Guo Yonghuai and Yang Nansheng from the Mechanics Institute, oversaw the general design of satellite and carrier rockets.

The second design institute, with Lü Qiang as the director and Yang Yuanjiu, Yang Jiaxi, and Tu Shancheng in charge, was assigned to develop the system.

The third design institute, under the technological guidance of Zhao Jiuzhang and Qian Ji, conducted the manufacture of space exploration equipment and the research of space environment.

Simultaneously, Bei Shizhang from the Biophysics Institute of Chinese Academy of Sciences, and Cai Qiao from Military Medicine Institute led the research in cosmo-biology and aviation medicine; the Astronomy Institute and Math Institute also initiated calculations of the orbits.

During the visit at a science research institution, Qian inspected the use of optical devices.

For many young Chinese of today, the names of these scientists sound rather unfamiliar and distant. However, we shall not and will not forget that they, as the nation's backbone, pioneered China's space travel.

Take Guo Yonghuai as an example. He was a protégé of Qian Xuesen, and already a prestigious mechanics expert before returning to China. Regardless of personal fame and wealth, he abandoned the good living conditions in the United States and resolutely took his family back to China. He died at age 59 in an airplane crash while his body was still clinging to a folder containing top secret information.

Behind every name stands a stunning legend. Without regret, they dedicated their lifetimes to China's space exploration. Some died along the way, some worked diligently until disease stopped them. While we are living a happy life, their names shall shine brightly amid the stars.

During the "Great Leap Forward," to pay a tribute on the national day in 1958, Chinese satellite rushed the steps. No second had been spared on designing satellite and rocket models since the First Design Institute was founded.

On the edge of West Second Ring Road in Beijing stands the luxurious Xiyuan Hotel, a busy place where massive cars and crowds hurry to and fro. Guests dwelling there hardly know it was where the development of Chinese satellites and rockets commenced.

Xiyuan Hostel was the name for 60 years. The newly assembled satellite team had no office but rented several rooms at the hostel. Without computers, they utilized old-fashioned calculators; without desks, they laid prone on the concrete floor sketching graphs. Merely candles and flashlights lit the rooms, where the first satellite and rocket models were created.

On October 1, 1958, the Exhibition of Achievements of Natural Science Leap Forward by the Chinese Academy of Sciences was held in the Biology Institute at Zhongguancun in Beijing.

Two satellite models were presented, one with a detector and the other a golden puppy, as when Soviet Union launched the second satellite, a dog was included in the experiment. The puppy would bark when visitors approached the model.

The exhibition stirred up a sensation, raising the curiosity of the society and attracting the attention of Zhongnanhai, the residential compound in Beijing housing top party leaders. One after another, Liu Shaoqi, Zhou Enlai, Li Fuchun, Nie Rongzhen, and Peng Dehuai attended the exhibition.

Eventually, Mao was too curious to stay away. On the morning of October 25, with an entourage, he ambled into the exhibition hall. Lu Yuanjiu, a 38-year-old doctor of Aeronautical Engineering from MIT, shouldered the task of a commentator for Mao. He remembers every detail of the experience:

The last number Mao watched was the rocket takeoff. When the takeoff was initiated, the sitting Mao rose. Having noticed there stood a technician pulling a rubber string behind the model, he smiled "Great, let's do it. Crude yet functional. Against the westerners we shall prevail."

Upon the fading of his voice, the hall roared with cheerful applause.

Core Technology Cannot Be Bought

Under most circumstances, passion does not suffice to achieve greatness.

On October16,1958, the High-altitude Atmospheric Physics Delegation of the Chinese Academy of Sciences visited the Soviet Union, from whom they intended to learn. With their help, it was expected to throw light upon the path to the making of Chinese man-made satellite and rockets as soon as possible.

Zhao Jiuzhang was the head of the delegation.

It was a name to respect and to regret. At the difficult beginning of China's aerospace undertaking, he was significant enough to be irreplaceable. Looking up at the starry history, he shone brightly as a star, worth remembering forever.

Zhao was born in 1907 in Wuxing, Zhejiang Province. He was the confidential secretary of Dai Jitao, a veteran Kuomintang member, because he was also his nephew. A promising future awaited if he continued that path. However, disgusted by the corruption within the Kuomintang, he took another path. His intelligence and hard work helped him get admitted to the physics department of Tsinghua University and study abroad in Germany at state expense. In 1938, he obtained a doctorate in Meteorology from the Free University of Berlin. In 1939, he returned to China and became a professor at Southwest Associated University.

During his teaching years at Southwest Associated University, life was extremely harsh. He and his wife had to share only three pairs of pants. Regardless, he was fully devoted to research and teaching. When the People's Republic of China (PRC) was newly founded, the Kuomintang lured him with various temptations. Yet, his final decision was Mainland China. "Peace was hard earned after eight years of war. Leaving now would affect my work."

After the foundation of PRC, he acted as the Director of the Institute of Geophysics in the Chinese Academy of Sciences. As a geophysicist, meteorologist, and space physicist with both domestic and international prestige, he made great contributions to the development of meteorology, solid geophysics and space science in China. Upon the launching of the Soviet Union's satellite, he pointed out with sharp insight that the launch of artificial satellite was a milestone of space exploration. Subsequently, he raised the first questions to Party Group of the Chinese Academy of Sciences regarding the plan to make China's artificial satellite and corresponding institutes.

Later in Zhao's life, there was no day when he was not concerned about artificial satellites. However, he didn't survive the Cultural Revolution. On the evening of Oct. 10, 1968, unable to withstand the repeated humiliation of the "rebels," he swallowed many sleeping pills, leaving the world behind in silence. The news of his suicide reached Zhou Enlai, who was furious with tears. The famous scientist's loyalty to the nation and persistence to satellites were remarkable.

Previously on the way to Soviet Union, Zhao frowned with discomfort.

He knew the trip came with great duty. The help from Soviet Union would greatly accelerate the steps of China's making of its first man-made satellite. But, if expectations were not met, what would happen next?

After the plane landed, the delegation was housed in a Chinese hotel in Moscow, where they received the warmest welcome from the Soviet Union.

The first day, Soviet authorities arranged for them to visit the Observatory.

The second day, Soviet authorities arranged for them to visit the Institute of Space Electronics.

The third day, the Soviet authorities arranged for them to visit the biology cabin of rat experiments.

Time passed by. Soon a month was gone. Members of the delegation discovered that behind the hospitality there was the coldness to keep polite distance. Requests of other visits by China were rejected on various excuses.

Zhao could not bear it. After one dinner, he made strong demands to the receiving officials "Arrange the visits to see institute of satellite designs and the launch station as soon as you can."

"Regarding your request, superiors must be consulted. Arrangements would be made once there are clear orders" replied the officials.

Repeated strong requests worked. They were arranged to visit the rocket hall of the Central Meteorological Agency. The news lit their spirits. However, they were disappointed on site.

On the platform in the center of the hall the head of rocket was displayed, whose diameter was around one meter. Zhao and the entourages were asked to hear a Soviet expert explain the details of the rocket from three meters away.

Chinese experts could not help leaning forward to check the interior upon their first visit of the Soviet rocket. "Comrades, stop right there." Hands were reached out to prevent them and an explanation was given that the superiors commanded that the visit should not involve the interiors of the rocket.

Helpless, they retreated.

The awkward experience made the delegation realized that it was impossible to see the Soviet satellite and launch site.

The 70-day investigation came to an end. The delegation came with great expectations and left with equal disappointment. Despite that only the satellite head was exhibited, it was not a total loss at all.

After returning to China, Zhao and the delegates together submitted a detailed report to the authorities, suggesting that China should depend on itself.

Core technology that concerned the fate of a nation cannot be bought, nor loaned to by a friend. We must rely on our own and work on it. Today, reviewing the learning experience in Soviet Union allows a deeper understanding of the fact.

Eight Km Flight, the First Step to Space

At the beginning of 1959, Soviet Union presented a special and magical gift to the whole world.

At 8 p.m. Moscow time on January 2, a giant rocket took off from the launch site in the Soviet Union. A spectacular flame cut through the sky. The rocket flew high from Earth towards the Moon.

The rocket carried a red flag with the national emblem of the Soviet Union and words of pride "Union of Soviet Socialist Republics, January 1959."

The rocket, after 62 hours, passed the Moon, 597,000 kilometers away from the earth, became the first space rocket.

It was another feat in the space flight history of humans. The news of Soviet Union's successful launch astonished the world. It tremendously stimulated China. We had to make our own rockets and satellites as fast as possible, realizing the dream to dance with the stars.

However, reality was cruel.

In 1959, devastating natural disasters ravaged China. With the previous Great Leap Forward, the young Republic was struggling with the economy and mass starvation.

At the time, the weak treasury failed to finance atomic bombs, hydrogen bombs, and man-made satellites altogether. After thorough consideration, the Party Central Committee made a choice that limited resources should be allocated for the making of atomic bombs and hydrogen bombs first while the satellite program would continue to be postponed due to national economic distress.

On January 21, 1959, the Chinese Academy of Sciences conveyed the instructions from Vice Premier Deng Xiaoping that in the following two years no satellite should be launched due to the national economic situation, not a permanent halt but a delay.

Later, the Chinese Academy of Sciences made adjustments to the space technology program, proposing to shrink the goal from the launch of satellites into that of sounding rockets. A pause should happen to the study and manufacture of massive carrier rocket and artificial satellites. Instead, efforts should be transferred to a sounding rocket.

It is a rocket that could conduct environmental investigation, scientific research, and technological experience in terrestrial space. It was an indispensable experimental tool in the advancement of space technology.

In the years of famine, when the common people were struggling with basic needs, the research team of China's satellite never gave up their goals. On the contrary, they demonstrated incredible persistence on the path to space.

There was no precedent to the making of a sounding rocket. Everything started from scratch. All was dependent on their own achievements.

Wang Xiji, age 37, was the chief engineer Later, he recalled the harsh experience of tackling key obstacles, such as:

The nation was in extreme poverty.

The research team was young, averaging 20 years of age.

There was no theory nor experience.

Computers were unavailable for designing, laboratories were unavailable to conduct experiments.

The research took place in a bunker abandoned by the Japanese. In deep winter, they all acted as masons, handling bricks and stones.

Many comrades had edema due to the lack of nutrition. Yet, they still worked till late night with an empty stomach.

It is impossible for people of today to imagine how much hardship was overcome in every step taken forward. To ensure a safe initiation of the rocket engine, the milling depth tolerance of the blasting film for the start-up valve must be controlled within 0.005 mm. Two 20-year-old women bravely took up the challenge. The lack of essential equipment forced them to grind needle heads into micro graver. The required patterns were carved on printing paper first, and it was pasted on silk. Thousands of experiments were conducted to meet the requirement of design.

On February 19, 1960, the T-7M, China's first independently created liquid sounding rocket, stood on the 20-meter launching pad. The launching site was situated at an isolated place next to a river near the coast of Shanghai.

The launching conditions were unprecedentedly crude in world space-flight history.

Without an electrical speaker, the chief commander of the launching site shouted and signaled to give the command; without special filling equipment, they utilized a bicycle pump to fill the rocket with propellant; without automatic telemetry antenna, several staff rotated the antenna to track the records; and the power station was merely a borrowed 50-kilowatt generator.

Unbelievably, it was a success. The rocket soared up to 8 kilometers.

These 8 kilometers was the first step China took to into space, a pivotal step. The success paved the way for the satellite to soar and strengthened China's confidence in the launching of its own satellite.

Three months later, in the exhibition hall of cutting-edge technology at Shanghai New Technology Exhibition, Mao Zedong walked straight to the model of the sounding rocket. Pointing at it, he asked with concern,

"How high could it fly"

"Eight kilometers" answered the commentator.

Mao paused, and waved and shouted "Fantastic, 8 kilometers! This is how we should make it. Eight kilometers, 20, and 200. We shall rock the world."

According to statistics, from 1960 to 1965, T-7M sounding rockets alone conducted 9 sets of 24 high-altitude experiments, thus making essential technological preparation for the satellite launch.

The Same Proposal

Reviewing the harsh path to the construction of national defense science and technology in the country, 1964 was absolutely a year to be proud of.

On June 29, the first ballistic missile developed independently by China was successfully launched. Soon, on July 9 and 11, two more launches succeeded, implying that China had mastered the technology of missiles and carrier rockets. It marked China's space technology had achieved the transformation from zero to one, and from imitation to self-design.

What elated Chinese more was the detonation of China's first atomic bomb in Buluopo, Xinjiang on October 16, 1964. The eastern boom cheered the entire nation and astonished the entire world. From then on, the People's Republic of China stood up, becoming a big country unthreatened by any power.

Missile launch successful, atomic bombs developed; domestic economy recovering. These together reinstated man-made satellite's seat on the agenda book of national matters.

Again, history put Zhao Jiuzhang, a scientist committed to satellites, at the front line.

At the end of 1964, Zhao attended the First Session of the Third National People's Congress held in Beijing. Having heard Premier Zhou Enlai's government work brief, excitement fueled him to pull an all-nighter, drafting a proposal regarding an urgent comprehensive satellite plan. The following day, he delivered it to Premier Zhou in person.

Zhou was delighted to receive a proposal of absolute sincerity. He squeezed time between busy meetings to briefly converse with Zhao, urging a more mature plan.

On January 6, 1965, when the meetings ended, in the name as the Director of the Institute of Geophysics, Zhao submitted a more detailed report to the Chinese Academy of Sciences, which later formed an official report to the central government.

In another historical coincidence, on Jan 8, 1965, Qian Xuesen handed in a proposal concerning the development of man-made satellites to the State Commission of Science for National Defense and National Defense Industry Office, where he urged a plan at earliest convenience that should be listed as a national plan that would benefit the nation's development.

Apart from strong brains, wise scientists share sharp insights that capture strategic opportunities.

Without prior discussion, they proposed the same plan of accelerating the development of China's first man-made satellite to the central government with a mere two-day gap.

"If the carrier can be developed in 1969, I assume there is the possibility to fly a man-made satellite in 1970." Nie Rongzheng pointed that the task should only involve the transportation of the man-made satellite into orbit without further objectives. When that works, satellites for communication and meteorology are the next step.

At the beginning of May, the central government granted the proposal from the State Commission of Science for National Defense, listing satellite development as a national plan. Henceforth, full efforts were invested in China's man-made satellite. On May 31, the Chinese Academy of Sciences formed a group to develop the main body, demanding the envisaging of the first man-made satellite and a frame of a satellite series by June 10.

Zhao Jiuzhang and Qian Ji were in command of total organization. Core professionals worked day and night. Only ten days were taken to draw the initial plan of the first man-made satellite.

How should it be named?

It perplexed everyone as if parents were vexed over naming a child from a difficult labor.

Words were exchanged in the long discussion. And He Zhenghua, head of the main body group, had a whim: how about Dongfanghong 1? The name received unanimous approval.

He didn't expect that Dongfanghong, the name he gave to Chinese satellite, would become the spiritual totem of China's aerospace undertaking. Unconsciously, he made the history, and history remembered him as well.

Next, Qian Ji led a team to brief to Premier Zhou Enlai. As the most outstanding student of Zhao Jiuzhang, Qian Ji did a great amount of work at the beginning of satellite development. Decades later, when he recalled this old memory, he was taken with excitement, "Premier Zhou held my hand and quipped as the chief satellite designer also had the last name Qian (meaning money in Chinese), it appeared that Qian was indispensable in the making of atomic bombs, missiles, and satellites."

On August 2, 1965, the Central Special Committee gathered a meeting, making requirements of China's first man-made satellite: political influence must be considered; must be more advanced than the Soviet Union and the United States in the aspects of a larger weight, a more powerful launcher, a greater longevity, a newer technology, and a wider range.

The meeting also established the principle to develop Chinese satellites: from simple to complicated, from easy to difficult, from inferior to superior, a step-by-step progress. Additionally duties in the project were also clearly distributed: the State Commission of Science for National Defense would organize the entire project; the Chinese Academy of Science would take charge of satellite's main body and the ground monitor system; the Seventh Aircraft Department would be responsible for carrier rockets; and the missile experiment base would operate the launch site.

For the sake of confidentiality, the central government decided to code the program as 651 since it was in January 1965 that the proposal was made.

In the history of China's aerospace undertaking, all major decisions were made in meetings.

At present, why is there common repugnance towards meetings? There is nothing wrong with meetings, but we fail to use them properly. They are more a formality than solving problems. In fact, there is no better way than meetings in solving problems. Opinions are expressed, ideas collide, and solutions are efficiently found.

In October 1965, an argumentation meeting over the plan of China's first man-made satellite was held in the Beijing Friendship Hotel. Hosted by the Chinese Academy of Sciences, the meetings received attendance from the experts from State Commission of Science for National Defense, National Defense Industry Office, National Science and Technology Commission, General Staff Department, Navy Force, Air Force, Second Artillery, First Aircraft Department, Fourth Aircraft Department, Seventh Aircraft Department, Communications Department, Ministry of Posts and Telecommunications, 20 Test Base, and Academy of Military Sciences.

This is the first famous meeting in China's aerospace undertaking history. The meeting spanned 42 days, from October 20 to November 30. It was unprecedented in great length, great scale, and at great content. Many experts recalled it was the longest meeting of their lives and the most passionate.

Upon the ending of the meeting, Premier Zhou Enlai joined the crowd, saying "Comrades, China will have the first independent satellite. This is unprecedented greatness. So, it is significant to add a Chinese star up there and let the world witness it. Gather your ideas and argue bravely to see to it."

At the meeting, Zhao Jiuzhang made a speech concerning the general design and Qian Ji summarized the design of the satellite's main body. After 42 days of discussion, a general plan about the first man-made satellite was established. And the general requirement was summarized as to be able to fly, orbit, see, and listen.

On November 26, 1965, approaching to the end of the meeting, France shot their first national man-made satellite up to space on the independently developed "diamond" carrier rocket from a launch site in Hamaguir, in the Sahara Desert.

Following the Soviet Union, the United States, China now fell behind France. Arrival of the news urged the meeting attendants to maximize efforts to fly Chinese satellites in space.

A Series of Frustrations

For the world, 1966 was a "starry" year. The United States launched 74 satellites in that year and the number of satellites shot up by the Soviet Union was close, according to statistics.

For China, 1966 was catastrophic. The unprecedented Culture Revolution broke, sending the budding satellite project to a trough.

The launch of a satellite relies on three systems: satellite system, carrier rocket system, and ground monitor system. After the joint meeting, the three systems went into the phase of technological innovation, where rapid progress was made in all aspects. Researchers were pumped at work. Lights remained late into the night in the research buildings. A common dream was shared that the satellites would soar above.

However, the good times didn't last long. On May 16, 1966, the Cultural Revolution started. And on August 8th, the central government approved the *Decision of the Central Committee of the Communist Party of China on the Proletarian Cultural Revolution.*

The disastrous storm swiftly swept the nation, wrecking the lives of millions of Chinese, including the scientists and working personnel fighting for the satellite project.

On September 7, 1966, a gigantic and vigorous debate was held in the Chinese Academy of Sciences among thousands of people.

Sitting on the rostrum was Premier Zhou Enlai, who came to change the chaotic situation at the Chinese Academy of Sciences, protect researchers at the satellite project, and allow the nearly halted project to proceed.

In the history of the Dongfanghong satellite series, Premier Zhou was of irreplaceable significance. In the special years, he was the one who overrode all objections and obstructions at every harsh moment and ensured the successful launch of Dongfanghong.

With greatest earnestness in his speech, Premier Zhou temporarily calmed down the blind revolutionaries this time. Many researchers were able to return to their lab experiments.

Regretfully, it didn't last long. Widely spread big-character posters and big and small criticism meetings completely paralyzed satellite development.

In October 1966, Zhao Jiuzhang, Head of the Satellite Design Institute, was dismissed, criticized, and denounced, paraded through streets, and commanded to write self-criticism.

Soon, chief designer Qian Ji of the Institute was dismissed. So were an increasing number of researchers.

More devastating news kept coming. In June 1968, Yao Tongbin, a famous rocket metal expert, suffered violent criticism and died tragically at home. Four months later, stuck in desperation, Zhao ended his life.

One piece of devastating news came after another. Directors of Chinese Academy of Science, Ministry of Defense, and the central government were torn with tremendous anxiety.

On the report meeting about the development of the first man-made satellite, held on November 11, 1966, the Directors of the Chinese Academy of Sciences had to inform the central government that it was unable to continue the work on satellites but better assign new leaders or entrust the task to a new department.

To change the situation and ensure the progress of the first man-made satellite together with other science projects, the Central Committee of the Communist Party of China, the State Council and the Central Military Commission issued a decision on the implementation of military control over the Ministries of National Defense and Industry from March 1967.

Next, the Chinese Academy of Sciences turned over around 10 achievements of satellites in a decade, the prototype of the first man-made satellite, the established ground observation network, and over 5,000 researchers to State Commission of Science for National Defense, in preparation for the establishment of the Chinese Academy of Space Technology.

In February, the Academy was officially founded with Qian Xuesen as the head and Chang Yong as political commissar. Henceforth, under the guidance of the Academy, the development of the first man-made satellite was back on track.

Achieve Greatness! Eighteen Warriors Broke Through

Achieve greatness and warmly treat the extraordinary.

In 1967, the Cultural Revolution was on fire. Zhao Jiuzhang, a scientist with a problematic class background, was dismissed. Qian Xuesen took over the leadership of the development of first man-made satellite under extreme circumstances.

An urgent and tricky problem was ahead.

Who would take charge of the general design, the most core task, of the first man-made satellite?

When there was severe lack of satellite experts in the country and many prestigious scientists were dismissed as rebels, it was of extreme difficulty to find one qualified talent to shoulder the task.

The greatness of Qian was more than his extraordinary intelligence as a scientist but the sharp eyes that recognize talent.

After consideration, he recommended Sun Jiadong, a merely 38-year-old missile designer, to Nie Rongzhen,

Nobody expected that, not even Sun himself.

Years later in an interview, Sun, who was already titled an Academician and Chief Designer of China's Moon Exploration, still held a clear memory of the day when his life changed completely. He wrote in his autobiography:

In the afternoon of July 29, 1967, it was the hottest time of summer in Beijing. The sun was scorching. Bending over a table, he was in the middle of a missile design. To avoid sweat dripping from his forehead, he left a towel on his neck and sketched on the board.

Now, without notice in advance, a comrade of the State Commission of Science for National Defense drove to his office south of Beijing. Concisely, he said: "I am staff officer Wang Yongsu of the State Commission of Science for National Defense and assigned to deliver the instructions of my superiors. The country will conduct research on man-made satellites. To ensure a smooth progress, the central government has decided to establish the Chinese Academy of Space Technology, led by director Qian Xuesen and specialized in the research of man-made satellites. Qian recommended you to Marshall Nie Rongzhen. And he has decided to assign you the duty of the general design of the country's first man-made satellite."

Sun was born in April 1929 in Gaixian County in Liaoning Province. In January 1950, he enlisted in the army as a Russian translator, Fourth Air Force Academy. In

July 1951, having passed rounds of examinations and selections, Sun was sent for further study at Rukovsky Air Force Engineering College, USSR.

On March 10, 1958, he graduated with excellent grades and received the highest award, the Stalin Medal. There were only 13 medalists among all the graduates from Rukovsky Air Force Engineering College.

In April 1958, when spring blossoms began, Sun was on the train back to China. At the time, the Fifth Academy of the Ministry of National Defense was in urgent need of talents in technology. Marshall Nie commanded talent from Liu Yalou commander of the Air Force, and Sun made it on the list.

It was the first turn of his life. Despite his previous study on aviation engines, he was committed to the engine of China's missiles upon his return to the country. In the following decade, he matured together with Chinese missile technology.

The second turn was also abrupt. "I had no prior preparation, nor conditions and requirements. Right after Army Day on August 1, I packed my suitcase and reported to duty."

Henceforth, Sun began the expedition of China's aerospace undertaking, becoming one of the pioneers.

Sun often repeated the saying that a hero is nothing but a product of his time and a hero decides the course of history, which happens to be a fit summary of his legendary life.

China's satellites chose him and forged him into greatness, and vice versa. Spanning over half a century of his lifetime, he participated in the development of 33 of 100 satellites, acting as technological officer, chief designer, or chief engineer. He witnessed three milestones of China's space travel, from celestial satellites to lunar satellites. On September 18, 1999, Sun was bestowed the Medal of Atom bombs, Hydrogen Bombs, and Man-made Satellites.

In the critical times of satellite development, to initiate the work as soon as possible, Sun proposed to Qian Xuesen to select talent.

Back then, satellites were a cause everyone wanted to be a part of. In the times when a tiny matter could be magnified and over-analyzed, talent selection was difficult. A minor misstep may lead to a backfire. Young and dauntless, Sun set aside all political considerations and finalized a list of 18 talents based on various expertise. The list was quickly approved by Qian and Nie.

These are the famous Eighteen Warriors in the history of China's aerospace undertaking:

Qi Faren, Shen Zhenjin, Wei Desen, Zhang Futian, Peng Chengrong, Yin Changlong, Zhu Furong, Sun Xiangcai, Wang Zhuang, Yang Changgeng, Wang

Sun Jiadong, academician of the Chinese Academy of Sciences and chief designer of the Beidou Satellite Positioning System, giving a speech at the task meeting. (Photography Qin Xian'an)

Dali, Zhang Rongyuan, Liu Zeguang, Zheng Zhongqi, Lin Yinding, Lu Li, Wang Yifang, and Hong Yulin.

In peaceful times, how did they deserve the title of warrior?

Because in those turbulent times, where rules and order were shattered, it was not easy developing satellites. Faced with political mudslides, they had to remain calm and tackle key problems. And the path was a road no Chinese had ever walked. Every step took wisdom and tremendous courage to shoulder the risk of failure and the responsibility.

Take the insignificant electrical signal socket for example. Back then, China's basic industries were so weak that a considerable number of socket manufacturers were unable to produce it. To solve the socket problem, with letter of introduction from the Premier's office, Sun found the Fifth Factory of Shanghai Radio. After profound discussion with several experienced technicians, experiments were repeated and the special socket for the satellite was finally made.

Sun's research team made some reasonable revisions on the original plan of Dongfanghong 1. However, there was no one to approve the revised plan. Angry, Sun brought the plan to Liu Huaqing, deputy director of the State Commission of Science for National Defense. "I am Sun Jiadong, from Academy of Space

Technology. And I am in charge of the general design of Dongfanghong 1" he said, placing a thick stack of sketch sheets in front of Liu.

"This is beyond my duty and understanding," replied Liu, scanning a couple pages.

"Whatever! Approve it and we move on," answered Sun.

Liu had a deep memory of the scene. In his memoir, he wrote:

Having looked into the situation, I realized this couldn't wait. Someone must take the responsibility. So, I told him to mind only technological matters while I handle the rest. Come to think of it, besides a strong sense of responsibility, there was a silly boldness too.

At the end of 1967, the State Commission of Science for National Defense held another argumentation meeting concerning the proposal of Dongfanghong 1. Simplification was conducted. And the task was clarified as "Fly, orbit, see, and listen."

The final decision of the meeting was to broadcast the song *The East Is Red* from the satellite.

The East is Red and the sun rises…

The Tireless Stubborn Old Man

To fly, the satellite had to be carried by a rocket.

To fly is easier said than done. However, to China at the time, turning it into reality was tremendously difficult.

To explain the difficulty of satellite development, a Soviet rocket expert once joked that satellite researchers would either die early or go crazy.

On November 5, 1960, Missile 1059, China's first rocket (an imitation of P-2 of the Soviet Union) was successfully launched.

On June 29, 1964, China's first independently designed mid-short rocket lifted off in the northwestern desert. It was a new beginning of China's rocket development, a day worth remembering in China's aerospace undertaking.

Whereas, to send a man-made a satellite to the designated orbit, a carrier rocket with greater thrust was required. In December 1967, the State Commission of Science for National Defense held a meeting, setting the technical conditions of

Dongfanghong 1 as 170-kilogram weight, one-meter diameter, and 400 kilometer above perigee orbit.

Previously, China only launched one-stage rockets, which were able to break through the dense atmosphere yet unable to reach the first cosmic speed, thus unqualified as carrier rocket. Therefore, to make sure Dongfanghong 1 would fly, a brand new multiple-stage rocket had to be developed.

Qian Xuesen proposed a plan where missiles and rockets were combined: missiles power by liquid fuel act as the first and second power; the engine powered by solid fuel acts as the third; and the combination of both would make the new carrier rocket. As if in a relay race, the three-stage rocket would send the satellite into orbit. This was an innovative plan where space technology and missiles came together, and liquid and solid rockets too.

The three-stage rocket was first named Weiyun I. The official name, though, was given as Long March 1.

The heavy responsibility fell on the shoulders of rocket expert Ren Xinmin.

Originally named Dadao, Ren was born in Ningguo, Anhui Province, in 1915. The name was changed after quoted a verse from an ancient Chinese poem. As the name suggests, Ren had been a prodigy since early childhood, even named a Ningguo prodigy. When the Sino-Japanese war broke, he was pursuing knowledge at the Department of Chemical Engineering at Central University of Nanjing. In 1945, he took up advanced studies in mechanical engineering at the University of Michigan in the United States. Three years later, he obtained a master's degree in mechanical engineering and doctor's degree in engineering mechanics. In 1948, he was hired as a lecturer by the department of mechanical engineering at University of Buffalo in the US. When the People's Republic of China was newly founded, against all hardships, he returned to the country with the utmost loyalty to contribute. Later, he was fully committed to the cause of China's missile and rocket development.

In October 1992, I met Ren for the first time at the Jiuquan satellite launching center. The past of the Long March rocket series was brought up. His memory was as fresh as new. "Satellites can never fly unless the rocket Long March 1 is developed." He was determined to maximize his efforts to have the task done.

It is impossible for people of today to imagine the hardships in the research. In his memoir, Yang Nansheng, an expert on the engine of Long March 1, recorded that the inner Mongolian base offered the worst living conditions: the several old buildings of bachelor pads were where they ate, studied, and sketched; the wedded were scattered in host families near the base, whose houses were constructed by sun-dried mud bricks; meals only included steamed corn bread, potatoes, and cabbage; and the harsh weather of sand storms. Insufferable as it was, nobody backed down.

As the chief designer of Long March 1, Ren carried a yellow bag every day. He took trucks, ate at the big canteen, worked at the institute and workshop, organized discussions, coordinated plans, and checked all kinds of progress.

He had already turned 50 by then. People saw him as a tireless, stubborn old man. As an influential expert, he was popular between both parties at the rocket institute. "Which party would you side with?" asked someone. "Neither. The focus is on Long March 1. Without it, satellite launch has no promise" replied Ren. Consequently, his stubbornness caused him trouble. People put up big-character posters on his door that shouted, "Knock down the rebel."

During the Cultural Revolution, the general public's revolutionary obsession and daily class conflicts severely impeded the progress of rocket experiments. With slogans against the rebels echoing, Ren was bathed in anxiety. In the days, the first words he uttered to the designers and workers were "Comrades, whatever it is, rocket is real."

In April 1969, thanks to the efforts of numerous researchers, the development of the rocket Long March 1 finally entered the most dangerous phase: the trial run.

The trial run is a launch simulation, after the completion of rocket development, where the engine is ignited for real. Apart from the takeoff, other procedures are the same as a real launch. Only when the trial run succeeds can the actual launch be approved.

When the rocket Long March 1's trial run was about to commence, friction between the two parties of the rocket institute broke out again. Buildings were under siege and no work was to be done. Anxious as Ren was, he was helpless but reported the situation.

And it was Premier Zhou Enlai that saved the day. Hosting the Ninth National Congress of the Communist Party of China, he made calls to departments concerned upon his knowledge of the situation. "It is a matter of national glory that must not be disturbed by anyone. Both parties shall join forces and have the trail run completed on time."

At that special time, a merely technical problem could only be solved by the Premier.

On July 17, Premier Zhou gave absolute command to Qian and Ren that the trial run must be done perfectly. Following that, orders were made to over 3,000 staff to fulfill their duties with no factionalism allowed.

Shen Zhenjin, one of the trial run organizers and the eighteen warriors, recalled: "The trail was paramount. A satellite was to be put at the front of the rocket, the engine to be ignited, and both to soar and operate as required. We were to see whether the rocket could endure the adventure, whether the satellite would break. There were not enough accommodations, and over 20 researchers shared the floor of one office, squeezing one another at sleep; we would run down the hill to order a bowl of noodles or a sesame seed cake at the price of 18 cents. Monthly wage was 62 RMB, without any subsidy. Food expense during overtime hours had to be paid by ourselves."

What were these struggles amid difficulties? Shen's memory gives a vivid depiction to young researchers of today.

On September 6, 1969, three trial runs of Long March 1 were completed. Everybody huddled for hugs, and Ren had tears in his eyes.

Enthusiasm in Opening Unique Space Orbits

It was also an unprecedented challenge to keep up with the satellite.

Keeping up includes ground satellite telemetry, tracking, and command (TT&C) stations' constant control of the Dongfanghong 1's state and timely feedback of the data traced and collected after the rocket shoots it into orbit.

Before the launch of the first satellite of the Soviet Union, in the hope that China would establish a visual observation network to assist with the observation of the launch, the Astronomical Committee of the Soviet Academy of Sciences wrote to the Chinese Academy of Science, which answered the request by constructing four observation stations in Beijing, Nanjing, Shanghai, and Kunming. An optical observation group and radio observation group of artificial satellites were also founded.

After years of progress, in 1965, led by Zijishan Observatory, an observation network came into shape that consisted of 23 ground optical observation stations across the country. Nonetheless, a more accurate and stable tracking observation must focus on radio observation, assisted by the optical one. Regarding this, Chen Fangyun, a radio expert, having investigated how developed the technology was domestically and internationally, made a bold and novel proposal that the radio tracking of Dongfanghong 1 should be based on a Doppler velocimeter.

Different from overseas, it was impossible for China's ground satellite TT&C stations to spread across the globe but only within Chinese territory. And the barriers were high. "Chinese features must be represented in the TT&C technologies without copying others. Independent systems must be built. And new territories must be exploited," repeatedly highlighted Chen.

And the heavy burden of the establishment of the system fell on the shoulder of legendary Chen Fangyu, the person in charge of technology of project 701.

Chen came from Huangyan, Zhejiang province. Chen Lixin, who graduated from Baoding Military Academy and was employed by the government of the Republic of China, was his father. Having handed in his resignation, he returned to hometown because of warlordism. And then he summoned Chen Fangyun to the study room, "You must pursue knowledge and make a difference in the society. And being a high-ranking government official is not the only way." Father's words took root in Chen.

In 1931, Chen attended high school in Shanghai. A few days later, the September 18 Incident happened. He joined classmates to petition in front of the Nanjing presidential palace and was escorted back to Shanghai. After graduation, in the hope of defeating the Japanese, he applied for the Central Aviation School of the National Government. He failed because of color-blindness in his physical exam. Unexpectedly, he was admitted to Tsinghua University.

In 1938, he graduated from the Physics Department. Recommended by the school, he was prepared to work in a mechanic factory in Kunming. However, due to his unsatisfactory performance at interview, he was declined. His mentor Ye Qisun advised him to work at the radio institute of Tsinghua University, which turned out a crucial step in his life. At the age of 25, he developed a secret phone.

There had always been the intention to contribute to the fight against the Japanese. In 1941, he left the institute, came to radio factory in Chengdu affiliated to the Aviation Committee of National Government in Chengdu of Sichuan, and invented China's first radio navigation system on the plane. Seeing social corruption and political internal strife, the passionate youth refused to serve the Kuomintang. And he understood, apart from that, the alternative was to study abroad. The factory denied his departure for either the United States or the United Kingdom.

Anger eventually pushed him to part ways. He stayed in a riverside inn. In desperation, he bumped into Wang Daquan from the Aviation Committee on the street, who delivered the news of approval of his request to study in the United Kingdom as long as he agreed to return to the factory. In the spring of 1945, Chen set off to England, studying television and marine radar. He also took part in the development of the first set of British marine radar. In the summer of 1948, he returned to China. Upon his return, the Aviation Committee of the Kuomintang ordered him to the Shanghai Hongwan plane factory. His repugnance towards war resulted in an absolute disobedience and he fled back to his hometown. Consequently, he received demerits on his employment records from the Kuomintang. Nonetheless, lacking talent, the Kuomintang urgently invited him to Taiwan for his authoritative expertise. He refused again. He deliberately burned his feet so that he was hospitalized for two months, thus dodging the badgering of Kuomintang agents and staying on the mainland.

After the PRC was founded in 1949, as a positioned researcher at the physics institute and deputy director at Electronics Institute of the Chinese Academy of Sciences, Chen was engaged in radio research. In 1964, U–2 planes of the Kuomintang flew from Taiwan to investigate. Back then, the radar of mainland China failed to trace them. Later, through the State Commission of Science for National Defense, the Air Force resorted to Chen, who enhanced the anti-interference system on remote radar. Eventually, U–2 planes were shot down. And the news was brought to Chairman Mao, who left a note "Good technology, I am intrigued."

Soon after the Cultural Revolution broke out, Chen was labeled as a bourgeois academic authority and special agent of Kuomintang. Whether it was the reform through labor at workshops or enduring criticism and denouncement, never had Chen feared the obstructions. He was only bothered with one goal—to keep up with the satellite.

In 1967, the military took over Project 701 and the situation took a favorable turn for Chen. He was deployed to the monitoring base in Wei'nan, Shanxi province. The base belonged to the military establishment and was composed of soldiers inside and out. Chen was the only civilian, with no military uniform. Cadres and soldiers referred to him as a master craftsman. During daytime, he worked on the hills, and at night, he bent over in a tiny oil lamp, sketching and writing, secretly planning the monitoring network of satellites.

In the second half of 1967, the State Commission of Science for National Defense organized a large-scale survey and the sites of 18 ground satellite TT&C stations were initially decided upon. China's economy at that time made it impossible to build all 18 stations right away. After several discussions, Chen and some experts suggested, in the principle of lower cost with greater result, that half be constructed.

When the suggestion was brought up at a meeting, many doubted the considerable deduction would fail the TT&C tasks. Qian stepped forward, saying "I fully second Chen and his team. Moreover, how about two more, or at least one?"

His words left silence in the room. Difficult as Chen believed, he led the team to calculate and draw a new TT&C graph, where two stations were eliminated with 7 left.

Pleased with the result, "One less?" asked Qian.

Puzzled, "Less?" responded Chen.

Qian nodded, "Our satellite expense does not even reach one tenth of the foreign budgets. Frugal we are. The central government would grant as much funds as we demand, for our expertise is trusted. However, one more station demands millions, which could fund the education of thousands of children and feed the mouths of thousands."

The words touched Chen and whoever present deeply. New calculations were performed to eliminate one more station.

Eventually, six satellite ground TT&C stations were proposed to be built in Kashi, Xiangxi, Nanning, Kunming, and Hainan. The plan was submitted to the State Commission of Science for National Defense and approved.

Chen had a strong wish to wear military attire. Yet, under the special political circumstance, his class origin and reactionary academic authority forbade anyone to enlist him.

The famous radio expert who fulfilled the task of keeping up with Dongfanghong 1 lived and worked in military camps for a decade. It was not until 1976 when he turned 60, that the wish was granted, and he was dressed in military uniform. It was unprecedented.

It was the special soldier that soon initiated Project 863, thus becoming the father of Beidou Navigation of China's aerospace undertaking.

Chinese Stars Shone Brightly Amid the Universe

To make Dongfanghong 1 seen was more than a science task. It was also a political task that must be accomplished. It was the common wish among millions of Chinese to see the satellite soar in space. And it would greatly enhance the influence of China in the world if a Chinese satellite is seen soaring.

Whereas, Dongfanghong 1 was a globoid whose diameter was one meter. Once launched, it would orbit in the 400-kilometer space. Technically, it was impossible for celestial eyes to see a tiny lightless spherical body move rapidly from this remote distance.

Why were celestial eyes able to see Sputnik 1, the first man-made satellite of Soviet Union, which was not bulky, and a telescope too?

Scientists of different countries made careful observations, noticing two objects of different brightness moving at the orbit. How come? They were baffled. It was not until half a year later that the Tass News Agency of the Soviet Union unveiled the secret that the celestial body with lower brightness was the man-made satellite while the one with greater brightness the final-stage rocket that shot it up. To make it seen, Soviet scientists polished the final-stage rocket enough for detection from earth.

Certainly, there was the voice urging the same trick of polishing as the Soviet Union.

It would not work as it was the brand new three-stage rocket Long March 1 that launched the satellite. The last-stage rocket had a smaller diameter than the satellite. No one on earth would be able to notice anything, even if maximum polishing was performed.

So, what now?

Talents solve problems.

Where were the talents? In the throes of the Cultural Revolution, researchers with capabilities were swept to corners for nothing in that unprecedented political storm.

A dismissed talent was considered, Shi Riyao, an engineer of the Eighth Design Institute of the Seventh Machine Department.

For Shi, the harshness of life and exhaustion of work were bearable. Instead, he felt the most important part was taken away when he was forced to leave the research of space flight.

The news that he was called back to the position at the first man-made satellite together the assignment of the task that the satellite must be seen chased away his feelings of being wronged and encouraged his total commitment to the cause.

Sparks of innovation came suddenly. A simple umbrella illuminated Shi and his team. An "umbrella for observation" was to be designed to soar with the satellite. It would remain closed during the launch. Upon the arrival at the destination, it would open, and people would see it, thus the satellite too.

Trial and error. In high-speed spinning, the "umbrella" would stick together without opening, thus the expectation was unmet.

A dead end. They sought another path, altering the "umbrella" into a ball. Specifically, to design a spherical object for observation with the surface polished. Before the rocket shoots up, it would not receive inflate, thus would be light and flat. Once the rocket soared, the last-stage rocket would inflate it so that it expands into a 40-square-meter ball able to be seen. In this way, Dongfanghong 1 would be visible.

Trial and error. Success came after 11 months. Exhaustion left Shi and his team unable to celebrate then.

In the making of Dongfanghong 1, there were more researchers just as devoted as Shi. To accomplish the national mission, they swallowed humiliation and moved forward, teeth gritted.

Reflecting on this, I hold the firm belief that without their devotion, there would not have been advancement in China's aerospace undertaking, nor the prosperity in the country.

The Delicately Selected Tune to Space

When Dongfanghong 1 soared, *The East Is Red*, the most popular, most basic, and most representative tune for every Chinese, would sound in space and the world should hear this beautiful melody.

It was an exciting vision, and in fact a great challenge for technology. Given the domestic technological conditions at that time, there was no way of achieving it without extraordinary brains.

Eventually, 32-year-old Liu Chengxi, a young researcher, accepted the challenge.

Liu, from Wuxi in Jiangsu Province, was assigned to Chinese Academy of Sciences after graduating from Nanjing Institute of Technology in 1956. He was dismissed the minute the Cultural Revolution commenced. He attended meetings of criticism and denouncement, read the words of Mao, and nothing else could be done.

One day in August 1966, the directors, out of blue, summoned him for a conversation, during which he was assigned the task to have Dongfanghong 1 play the tune *The East Is Red* in space, with the assistance of some personnel. The words came as a lucky strike. "Rest assured and consider it done."

When his exhilaration faded, he calmed down and felt heavy pressure. At night, he was caught in insomnia, afraid of the fact that the failure would impact the launch of satellites, thus damage the image of the country. It was nothing if his reputation suffered, but that of the country was a big deal.

The musical device in Dongfanghong 1

Success was the only way out. There was no turning back. Liu racked his brains.

The tune would be played to the world. What musical instrument would do it best justice?

He covered basically all the musical instrument shops in Beijing, again and again comparing dozens of them one with another, such as piano, dulcimer, *liuqin*, accordion and *pipa*, *erhu*, violin, clarinet, flute and piccolo, etc. The piano was thick and deep yet too perplexing; the accordion was melodious yet not sharp enough; the flute was cheerful, yet too simple.

In the end, with the support from the Beijing Institute of Musical Instruments and the Shanghai Guoguang Harmonica Factory, Liu decided to adopt the tempo of the clock at the Beijing Station and the sound of the vibraphone, which is not only clearly melodious but applicable. They used electronic circuits to make a composite audio, ensuring that the broadcast sound of *The East is Red* was pure with a fast rhythm, and that it has a long working life and reliability.

Therefore, the team, where Liu acted as the leader, Yang Qitang was chief of technology, and Qian Xuliang and Deng Huirong were members, was in operation. Day and night, the group of young lads worked tirelessly. In the struggles to tackle key problems, they were disturbed by the revolutionary actions of the rebels; in Chongqing, they had to carry out experiments when caught between the violent criticisms between two fractions.

To ensure everything was correct, Liu checked every soldered point and selected every component.

Nonetheless, the most disturbing problem happened. Soon after electricity ran through the musical device, the tone changed. A comprehensive checkup was run over several nights and the cause was found: some fixation material had seeped into the electrical resistance.

The trial began. Liu reached out his hand and carefully turned out electricity. And melodious sounds drifted from the little box. The musical device and remote monitor device for *The East is Red* was successfully developed. Tears welled in their eyes.

To make sure the tune could be played once the satellite is launched, Liu and colleagues went on trial and error for another half a year under vacuum conditions, a simulated space environment.

It was until the first half of 1969 that the task was fully accomplished.

When the tune reached every corner and everyone was cheering for it, little did they know that Liu, the hero, was in "reform through labor" in the remote village of Zhumadian in Henan Province.

This is the perfect description of the unheralded greatness.

Remarks Were Remembered for A Lifetime

The year 1969 was a tough one for China, both domestically and internationally.

Inside the country, the Ninth Congress of the Party was convened, and the Cultural Revolution peaked; outside, the Treasure Island Incident had created unprecedented tension between China and the Soviet Union.

To walk tall before the world, on September 23, 1969, China successfully conducted the first atomic underground nuclear experiment in Luobupo; six days later, an airplane released a hydrogen bomb, a mushroom cloud erupting.

To walk tall in front of the world, Dongfanghong 1 must be launched as soon as possible without any mistakes.

In September 1969, the prototype of Dongfanghong 1 came into shape, and all systems met the requirements, and in a technologically normal status.

On July 20 of this year, Apollo 11 had landed on the Moon. That night at 22:56 and 20 seconds, Neil Armstrong made the first footprint on the Moon, announcing to the world "That's one small step for a man, one giant leap for mankind."

The news of America's landing on the Moon provoked China once again. On the road to space, China was left far behind by the United States and the Soviet Union.

More news came that Japan's first man-made satellite was in accelerated development, possibly getting ahead of China.

China or Japan, following the Soviet Union, the United States, and France, which would become the fourth country to launch a man-made satellite? It was a matter of technology as well as of international politics that involved national honor and dignity.

For well-known reasons, there was a delicate state of mind amid Chinese researchers that it was unacceptable to lose to Japan as well as other countries, such as the United States and the Soviet Union.

In October, the flight model of Dongfanghong 1 was completed. Premier Zhou demanded to hear the briefing of the satellite development himself.

In an evening of late October, Qian Xuesen and Sun Jiadong made a detailed report to Zhou Enlai, Li Xiannian, and Yu Qiu in People's Hall of Jiangsu Province.

In Sun's memory, everything that happened that night was wonderful and unforgettable.

Qian first introduced Sun to Premier Zhou, who shook his hand kindly and said "A young lad is the satellite expert, fantastic. What is your age?"

"Forty already this year," answered Sun, shy.

"Still a lad! Comrade Xuesen seems to have many young talents," quipped Zhou.

Zhou's kindness lightened the mood at the report site. Sun was relieved from intensive anxiety.

General preparations concerning the launch of Dongfanghong 1 and Long March 1 were detailed in the report by Qian. And Sun continued to give a report about the details of Dongfanghong 1.

When Sun finished the report, Zhou abruptly tossed questions at him as if in an examination. "How many cables are there on the satellite?"

Sun answered the exact number.

"How many sockets?" asked again Zhou.

It put Sun on the spot. He was lost in silence, and replied with embarrassment "Premier, I would count them the minute I return."

Zhou laughed, and kindly he said "The figures mean nothing to me, but much to you. Satellite development requires you to be careful, as surgeons who are familiar with every nerve and every blood vessel, so that no mistakes would happen."

Zhou's remarks were remembered by Sun for a lifetime.

China's first man-made satellite—Dongfanghong 1 (photography by Qin Xian'an)

At the end of the report meeting, Sun got up his courage to tell Zhou a problem that was sensitive but had to be solved. The metal emblazonments of Mao on many equipment of the satellite posed a threat to the launch.

Sun explained, "Politically, it is understandable that emblazonments are on satellite equipment given that there is the common admiration towards him. But technically, first, it creates an overweight issue, and second, unpredictable impacts may happen to the mass of the satellite once it goes up."

Taking the explanation in, Zhou stood erect and walked a few steps. No comments were made on the action. "People love Chairman Mao, naturally. But take a look around this People's Hall. Not every corner is hanging the Chairman's portrait. The goal of politics in command is to get things done, instead of vulgarization. A scientific attitude must be present in satellite development. Think it over after you return. I believe as long as the explanation is made clearly to the people, it shall not be a problem."

Tears almost rolled down his face when Zhou's words relieved him. It was the wisdom of Zhou. as the Premier; when faced with a perplexing political situation, he was always able to find a way. Without him, Dongfanghong 1 would have failed repeatedly.

Zhou's remarks served as the sword to whack away the thorns.

The Worst Failure

On November 16, 1969, on the 55-meter launch tower at Jiuquan Satellite launching base erected the first two-stage medium-and-long-range carrier rocket for the trial of Dongfanghong 1.

Trial launch as it was, it was very significant. Should the launch succeed, Dongfanghong 1 would stick to the plan and be taken to the sky by the rocket Long March 1. And the Chinese satellite beat Japan to soar ahead. Should it fail, it was a suspense whether China or Japan would make it earlier.

At 5:45 that afternoon, the rocket was ignited. As the seconds were read out, it generated a booming noise and glaring light.

Surprisingly, minutes after the take-off, near the shutdown point of the first stage rocket engine, the velocity curve ceased to rise, and target was lost.

Three days later, the debris of the rocket was found in the desert 680 kilometers away from the launching base. One program distributor malfunctioned, thus the failure to ignite the second-stage rocket and it self-destructed in the air.

It was the worst failure, caused by the poor reliability of the circuit. It was as if in a fierce football match, the most critical score slipped away in the second half. Japan was given some time to catch up, and the world was booing.

Associated Press, the Reuters News Agency, and Agence France Press published the news of Chinese rocket failure immediately. And one British media commented:

"When the beige giant was ignited, it climbed less than 3,000 feet. A glaring light shot out from the middle of the carrier rocket. With thundering boom, it exploded in pieces. In an instant, hundreds of tons of propellant splashed, and towering mushroom clouds were in horrific flames. The enormous 100-foot rocket broke into a million pieces, plummeting into flames, and scattered on the desert near the launching site. The US satellite Atlas happened to record the tragic launching experiment."

The failure didn't frighten Chinese space researchers. Having learned from the past experience, they conducted targeted improvements and began again.

On January 30, 1970, the second two-stage medium-and-long-range carrier rocket for the trial of Dongfanghong 1 stood on the launch tower at Jiuquan launching base.

This trial would determine if Dongfanghong 1 could be launched using this rocket. In deep winter, wind cut like a knife in the desert, with people watching the launch standing in full exposure to the wind, anxiously waiting for the moment.

The command was given, the rocket took off. The booming noise gradually became distant, and the rocket soared higher and higher. Eyes were fixed on the sky, anxiously waiting for the confirmation from the fall zone. Seconds later, the news came: the two-stage rocket separated successfully, and the target was hit with high precision.

Walking on air, they cheered and celebrated the successful trial. This time, Dongfanghong 1 was cleared to be able to use the Long March 1 for the launch.

Regardless, two and a half months were spent.

Twelve days after the trial, the disappointing news spread that on February 11, 1970, Ohsumi, Japan's first man-made satellite, blasted off. The Voice of America reported the event immediately. And the news spread quickly from overseas to China, and from Beijing to the Jiuquan launch base. It was a big blow to the assiduous China's aerospace team, who lost their appetite and were unable to utter a word. Some wept in solitude, and one old grey-haired expert snuck to the desert late at night, to cry disconsolately as the wind howled.

If not for the Cultural Revolution, there would not have been that many uphill struggles. It would not have been a problem for China to beat Japan first.

Nonetheless, history bears no ifs.

For China's aerospace undertaking, it was a regret never to be made up for.

Everything Is Ready

On March 26, 1970, Zhou Enlai granted permission for the rocket and satellite to leave the factory.

On April 1, 1970, in extreme confidentiality, the satellite Dongfanghong 1 and carrier rocket Long March 1 were shipped to the Jiuquan satellite launch base.

The silent desert cheered up. The entire launch site went into efficient operation, and Commander Li Fuze meticulously led the team to carry out preparations before the launch. Aerospace undertaking experts, including Qian Xuesen, Ren Xinmin, and Qi Faren flew from Beijing to give technical instructions.

It was a perplexing system engineering to launch a satellite. According to statistics, there were over 700 factories that were involved in the development of Dongfanghong 1, and 17 research institutes. A malfunction at any point in the chain would generate a catastrophic outcome.

Take communication links for example. Once the satellite takes off, thousands of kilometers of route must be guaranteed a smooth communication. Otherwise, the entire navigating zone would turn deaf and mute. To ensure the safety of every telegraph pole and the smooth communication of every link, multiple departments at the base allocated dozens of lines and hundreds of radio stations to create a giant communication network; orders were given to related provinces, cities, and counties that on the thousand-kilometer route within the navigating zone of the launch, thousands of soldiers and crowds would guard the telegraph poles day and night to prevent any damage.

To China, the launch of Dongfanghong 1 would not accept a defeat. As if a mountain, the pressure fell on the shoulders of every task participant.

From April 2–14, Premier Zhou Enlai summoned Qian and other experts twice to be posted about the front-line situation. The unusual action fully demonstrated how much material significance the central government had attached to Dongfanghong 1.

Every time when he summoned a briefing, questions about the tiniest details were asked, and he repeated his exhortation.

Zhou asked Qian "What are the locations of the first, second, and third placement of the rocket? I fear when the debris falls, poor handling would pose a threat to people's life and property. It would be worse if it falls overseas."

Qian gave a detailed answer "Once the task is complete, the first stage rocket would fall in the desert of Gansu Province, the second stage would plunge into the South China Sea, and the third would separate from the satellite in the sky of northern Guangxi, entering the orbit together without causing any celestial damage."

Zhou continued to ask Ren, "What if there is a glitch?"

Ren explained, "To cope with a glitch, there are two ways of self-destruction. One is the self-destructive system on the rocket would make a connection with the power of the detonator, thus initiating the explosion, once a malfunction is recognized; and the other is an external system would give an order from the ground to induce a self-destruction. Detonation would happen if necessary and If unnecessary, no detonation would be activated."

Another question followed: "Is the satellite reliable enough?"

Sun confirmed "Multiple experiments have proven the reliability."

On April 16, Zhou made another call instructing "When the shipment of satellite and rocket arrives at the launch site, a checkup must be done with great carefulness. Every screw has to be checked."

And on the morning of April 20, Zhou rang again, demanding the launch to be safe and reliable with nothing gone wrong, the satellite enter the orbit with great precision, and the report of the situation on time.

Zhou's demand was soon made into a banner, hanging high on the launch tower. In 2008, when I paid my visit to the Jiuquan launch base for an interview, standing at the foot of the launch tower staring at the words in red paint, I was immersed in excitement as if present at the glorious moment 38 years before, with echoes of Zhou's exhortation.

On April 23, Qian Xuesen, Li Fuze, and Li Zaishan signed the Mission Document. And the headquarters decided that at 9:30 of April 24, 1970 Dongfanghong 1's launch would take place.

It was the last night before the launch. Right after dinner, the central government rang, informing the approval by the Central Military Commission, thus giving the procedures to proceed.

It was a long wait that night. Qian was sleepless, making all potential assumptions of emergencies during the launch.

That night, trial team director Qi Faren, deputy director Shen Zhenjin, and dispatch commander Zhang Futian, were sleepless as the last comprehensive checkup was going on, eliminating all the glitches.

That night, Director of Meteorology Peng Fengshao, and his colleagues were busy working. They were concerned about the next day. "Only April 24 among these days met all launch requirements."

That night, base commander Li Fuze was sleepless, walking anxiously in the desert in the cold wind.

That night, according to Zhou's staff, he never slept a minute.

The East Is Red is Spread Across the Globe

The day finally came, April 24, 1970.

It was a sunny spring day in Beijing, but still chilly in Jiuquan satellite launch base thousands of kilometers away.

At 5:40, the rocket fueling began. Dawn broke, and the sun peeked through the clouds, as if a shy maiden. Seeing the clouds, researchers of the meteorology department were concerned. The weather was not ideal for a satellite launch.

At 13:35, fueling was complete. Next, the satellite Dongfanghong 1 and the rocket Long March 1 went into the 8-hour preparation procedures before launch. On the site, the entire engagement team were waiting, anxious and excited.

Base commander Li Fuze was also waiting, since early morning, for a call from Beijing. Once Chairman Mao granted the launch, Dongfanghong 1 would blast off as planned tonight.

The call came at 15:50. General Luo Shunchu of State Commission of Science for National Defense conveyed the message from Zhou Enlai. "Chairman has granted the launch tonight. Comrades at the site, exert utmost efforts to get the job done, succeed and make the country proud."

Once the order was made, spirits at the entire launch site were lit up. On the hall of command hung the banner with Mao's remarks: "Chinese, with ambition and abilities, must in the upcoming future catch up and go beyond the world advanced technology." Everybody was pumped, from commanders to operators at each post, preparing for the launch.

At precisely 20:00, the commander gave the one-hour preparation order. However, above the launch site, thick clouds gathered, thus low visibility. Qian was worried, so was Li Fuze.

The more pivotal the moment, the more likely the mistake. At that moment, signals of abnormality appeared in the row of equipment in the underground control room. Everyone was in high tension. After checking, it turned out to be a loss of signal on responders.

Qian ordered, "This has to be dealt with, or the accuracy of the-tracking and forecasting of the satellite would be affected."

It would take 30 to 40 minutes to conduct clearing of a fault. And it was 35 minutes from the planned launch time. Qian proposed to postpone it. Li Fuze picked up the phone and report it to Zhou Enlai. On the phone, Zhou uttered, "I agree the delay. We must fix the responders. Don't panic, stay calm."

Clearing commenced at once. Tick-tock. Everyone was anxiously waiting at the launch site and at the Beijing residential compound housing top party leaders.

"Fault found!" yelled the technicians, "It was the degradation of performance at the ground signal trigger source, not the satellite's problem." Everyone was relieved.

It was already 21:00. After discussion, Qian and Li Fuze decided the launch would take place at 21:35 and reported it to Beijing right away.

The carrier rocket Long March 1 successfully launched the man-made satellite Dongfanghong 1

Suddenly, the weather turned kind, thick clouds cracking open miraculously.

At 21:34, the highest launch commander Yang Heng ordered, "One minute!" Everyone's face froze at the launch site that moment.

"10, 9, 8, 7, 6..." Ignition. At 21:35:44, a young operator Hu Shixiang, at the launch control console, pressed the button. This young operator, 20 years later, became the commander of China's Xichang satellite launch base, and then the Vice-Minister of General Equipment Department of the Chinese People's Liberation Army.

At that second, the dark sky lit up, the desert trembling and people's hearts were pounding. In thundering roars, the carrier rocket Long March 1 blast off, spitting tangerine flames and streaking across the night sky. Rapidly, it went beyond sight.

"tata, tata..."The light of the control room detector was shimmering constantly, capturing the trajectory of rocket flight quickly.

Waiting, everybody was extremely anxious. Finally, the long 15 minutes passed. The speaker at the control room shouted, "A perfect separation between the rocket and the satellite, and the satellite entered the orbit."

We made it. The dark and cold night suddenly turned bright and warm, for everyone who contributed to China's aerospace undertaking. Hands were shaken, and hugs were given among generals, soldiers, experts, and workers at the launch site, with tears welling their eyes.

At 21:50, National Broadcasting Bureau rang, announcing that *The East is Red* was played by Dongfanghong and it had been received, loud and proud.

Zhongnanhai, the residential compound in Beijing housing top party leaders, was also waiting for front-line update.

At precisely 22:00, Premier Zhen Enlai's phone rang. General Luo Shunchu's voice came out of it, "Premier, the carrier rocket was in good operation; the separation between rocket and satellite went smoothly, and the satellite entered the orbit. In addition, the tune of *The East Is Red* had been received."

Zhou couldn't wait to reply, "Great! That's great! I'll report to Chairman Mao right away."

"Chairman, the launch succeeded." Mao received the call from Zhou. As the staff recalled, he was in great elation so that his cigarette butt was tossed away. "Great, fantastic! Premier, prepare to celebrate."

China to Walk Tall

At 6:00 a.m. on April 25, 1970, the Voice of America first broadcast the news to the world that China had successfully launched its first man-made satellite.

Later that morning, Zhou Enlai enter the hall where the Tripartite Quartet Meeting with Vietnam, Laos, and Cambodia was being held, announcing with delight:

"Friends, to celebrate the success of this meeting, I have brought you a gift from China, which is that yesterday, China successfully launched its first man-made satellite, a victory for both the country and us."

As his announcement finished, a round of applause roared at the hall.

At 6:00 p.m., Xinhua News Agency made official announcement to the world:

On April 24, 1970, China successfully launched the first man-made satellite, whose orbit perigee was 439 kilometers away from the earth and apogee 2384 kilometers. The angle between the orbital plane and the equatorial plane of the earth was 60.5 degrees, and it took 114 minutes to circle the earth. The weight was 173 kilograms. At the frequency of 20.009 megacycles, *The East is Red* was broadcast.

Then, the glad tidings rapidly spread across every city and village of China through radio, television, and newspapers. People went out to the street, drums and gongs and firecrackers, to celebrate the successful launch of Dongfanghong 1.

At 20:29, Dongfanghong 1 passed the sky above Beijing, where a sea of people gathered on Tian'anmen Square, looking up in attempt to trace the satellite under the guidance of searchlight.

Following the Soviet Union, the United States, France, and Japan, as the fifth member, China joined the club capable of independently sending a man-made satellite to space.

The foreign press commented that China's science and technology, advancing rapidly, had reached a new peak, thus fully deserving membership at the space club; the quick completion of the launch was beyond the expectation of western experts; with China in possession of atomic bomb and hydrogen bomb, the successful launch should be seen as a declaration of the capability to launch continental missiles anywhere as intended.

It had been 12 years since Chairman Mao said that we must engage in the development of a man-made satellite. Chinese kept the promise as Dongfanghong 1 met the requirements of Mao, to be "bigger than an egg," i.e. bigger than the previous ones.

Sputnik 1, the first man-made satellite by the Soviet Union, launched on October 4, 1957, weighing 83.6 kilograms.

Explorer I, the first man-made satellite by the United States, launched on January 31, 1958, weighing 8.22 kilograms.

Astérix, the first man-made satellite by France, launched on November 26, 1965, weighing 38 kilograms.

Ohsumi, the first man-made satellite by Japan, launched on February 11, 1970, weighing 9.4 kilograms.

Dongfanghong 1 weighed 173 kilograms, exceeding the sum of the weights of the satellite by the Soviet Union, the United States, France, and Japan altogether.

China's Space Day, Prosperity of the Satellites

Time zips by. In the blink of an eye, it has been 46 years since Dongfanghong 1 made it to space. In the spring of 2016, I came to the Jiuquan satellite launch base, where Dongfanghong 1 spread its wings to fly.

Time is cruel. The launch tower is already rusty. Next to it stands a marble monument that writes where Dongfanghong 1 blasted off. The steel-made tower has retired with glory, and what it witnessed allowed China to walk tall in the world.

On October 25, 1971, a year after the successful launch of Dongfanghong 1, the United Nations reinstated China's seat at the General Assembly. In 1974, Deng Xiaoping attended the Sixth Special Session of the General Assembly. It was the first time a Chinese Communist Party leader stood on the podium of the United Nations. Later, he said "Without atomic bombs, hydrogen bombs, and the satellite launch in the 60s, China would not have been a big country with major impact, nor the international status. They show the capability of a nation and symbolize its prosperity too."

April 24, 1970, the day when Dongfanghong was successfully launched, is deeply remembered by China. It was a glimmering milestone for China's aerospace undertaking and an eternal spiritual totem for every Chinese aerospace undertaking. In March 2016, it was approved by the central government and responded to by the State Council that April 24 would be China's Space Day.

The launch tower where Dongfanghong 1 blasted off on April 24, 1970. (photography by Qin Xian'an)

人民日报

毛主席语录

我们也要搞人造卫星。

毛主席提出"我们也要搞人造卫星"的伟大号召实现了！

我国第一颗人造地球卫星发射成功

卫星重一百七十三公斤，用二〇·〇〇九兆周的频率，播送《东方红》乐曲

这是我国人民在伟大领袖毛主席和以毛主席为首、林副主席为副的党中央领导下，高举"九大"团结、胜利的旗帜，坚持独立自主、自力更生方针，贯彻执行鼓足干劲，力争上游，多快好省地建设社会主义总路线，以实际行动抓革命，促生产，促工作，促战备所取得的结果。

这是我国发展空间技术的良好开端，是毛泽东思想的伟大胜利，是毛主席无产阶级革命路线的伟大胜利，是无产阶级文化大革命的又一丰硕成果。

The report of the successful launch of Dongfanghong 1 in People's Daily.

On the first Space Day on April 24, 2016, Xi Jinping, General Secretary of the Central Committee of the Communist Party of China, President of the State, and Chairman of the Central Military Commission, made important instructions to pay high tribute to the comrades who have contributed to the development of the space industry in the past 60 years, and to emphasize that a vast number of space scientists and technicians should firmly seize strategic opportunities, adhere to innovation-driven development, bravely climb the peak of science and technology, compose a new chapter of China's space industry, and make greater contributions to serving the overall situation of national development and promoting human well-being.

President Xi pointed out that exploring the vast universe, developing space industry and building a strong space power are our unremitting pursuits of the space dream.

Generations of aerospace workers have made glorious achievements represented by atomic bombs, hydrogen bombs, the Dongfanghong 1 satellite, manned space flight, and Moon exploration. A path of self-reliance and self-innovation was created. And the profound space flight spirit accumulated. The establishment of China

Space Day is to commemorate the history and inherit spirit, stimulate the enthusiasm of the whole nation, especially the young, to advocate science and explore the unknown and dare to innovate, and to unite a strong force for the realization of the great rejuvenation of the Chinese nation.

Dongfanghong 1 satellite has been long been history in our common memory. People have long been accustomed to the news that China's Long March rockets have sent satellites into space one after another.

However, 46 years later, on April 24, when people looked up to the sky, a sudden realization struck that Dongfanghong 1 satellite was never far away, as it still flies in its own space orbit. It turned out that over the years, it had always been there, to witness the rapid development of China's aerospace undertaking and the nation's prosperity. Today, aged as she is, loneliness is absent as there are about 150 Chinese satellites orbiting above, third only to the United States and Russia.

CHAPTER 2

China's Satellites Through Thick and Thin

What a familiar voice – "10, 9, 8…3, 2, 1, ignite."

What a familiar image – Thunder roars echoing, the earth trembling. The Long March rocket blasting off, and the tangerine flame igniting the night sky.

With familiar and heavy applause, in tranquility, I witnessed again a remarkable moment by Chinese aerospace workers. At 1:40 of August 16, 2016, China used the Long March 2 carrier rocket to lift off the satellite Micius, the world's first experimental satellite for quantum science at Jiuquan launch center.

Micius will enable China to achieve quantum communication between satellites and ground for the first time in the world, and build a quantum secure communication and scientific experiment system of space-earth integration. The launch was the 234th flight of the Long March carrier rocket series.

The world was amazed. The internationally prestigious journal *Nature* reported it and commented that in the special space race, China took a giant leap, getting ahead.

At present, world's technological advancement is in the second quantum revolution. Quantum mechanics, which originated from Planck's Theory in 1900, has spawned a considerable number of great inventions in the past century, such as the atomic bomb, lasers, transistors, the MRI, GPS, etc., therefore, changing the world. As experts put it, this is the first quantum revolution.

Today, quantum information technology is the latest development of quantum mechanics, representing the rising second quantum revolution. In April 2016, the European Union revealed that quantum technology would be the new flagship research project; in July 2016, the White House suggested an increase in the investment in quantum science and regarded it a national challenge and opportunity.

In the fierce global competition of quantum communication, it was not easy for Chinese aerospace workers to win the world's first title with the satellite Micius. A significant victory it was. "I believe quantum technology was of equal significance to the 21st century to the Manhattan Project to the 20th," expressed Pan Jianwei, Director of Quantum Excellence Innovation Center of Chinese Academy of Sciences and Chief Scientist of China Quantum Satellite Project in an interview. That is to say that quantum technology may change the world the way Manhattan Project spawned atomic bombs.

From the first man-made satellite Dongfanghong 1 to world's first experiment satellite for quantum science, China's aerospace undertaking has achieved the leap from a follower to a leader in space exploration.

In recently years, China's aerospace undertaking has brought countless joyful success to its people, who have therefore grown accustomed to the news of successful satellite launches. And it appears that the memory about the obstacles and failures in the exploration has faded away.

Do you know, from Dongfanghong 1 orbiting in space alone to 150 Chinese "stars" shining together, what kind of arduous journeys, unforgettable setbacks, and joyful weeping China's aerospace undertaking has endured in the past years?

Do you know, among the 234 flights from the first blast-off of Long March 1 rocket to the one that launched Micius, what sleepless nights, heartaches, happiness, and sorrow Chinese aerospace workers have gone through? Too many have dedicated their youth, and even lives.

It was a treacherous path that China's aerospace undertaking was walking.

And Chinese "stars" have gone through thick and thin.

Today, success is to be remembered, and so is failure.

The remembrance of failure prevents past mistakes to be repeated, thus an easier path to the future, and enables our hearts to grow stronger and unafraid of upcoming failures. What does not kill us makes us stronger.

Remote-sensing Satellite: Return to Earth

Unsatisfied That Only Foreign Man-made Satellites Were Flying Above

Immediately after Dongfanghong 1 blasted off, China's aerospace undertaking had the eyes on the next goal: to launch a recoverable remote-sensing satellite.

The Thirteenth Special Committee of the Central Committee clearly stated that the development of artificial satellites in China is mainly focused on application satellites, which are mainly focused on recoverable remote-sensing satellites.

As a chaser, the reason for the choice of goal is that the United States and the Soviet Union, who were in the lead, have made significant breakthroughs in the technology of recoverable remote-sensing satellites, whose practical application has posed a threat to China's national security.

As Friedrich Engels put it, "Once technological advancement can be used for military purposes, and have already been done so, they almost immediately and mandatorily alter or even revolutionize the mode of warfare operations, often against the will of commanders."

Satellites, as a result of technological advancement, have been widely used in military since their creation. In 1958, the United States began the development of the Corona satellite, a satellite that can take pictures of the earth's surface from space. Over 10 setbacks later, the United States eventually succeeded in August 1960, with films from outer space capturing the entire territory of the Soviet Union. In 1963, through returning satellites, the United States caught the Soviet Union building anti-missile weapons, which served as the evidence to the United States' accusation of the Soviet Union's violation of the bilateral anti-missile treaty. In 1964, the Corona satellite mapped all 25 long-range missile bases in the Soviet Union. In the same year, it also photographed the planned location of the first atomic bomb to be exploded in China. After the deterioration of Sino-Soviet relations, the Soviet Union used returning spy satellites to perform military surveillance and reconnaissance on China.

How can we be willing to let there only be foreign satellites above? Due to the severe situation of national security, it is not difficult to imagine the sense of urgency and mission among Chinese space researchers to conquer recoverable satellite technology.

A series of technological challenges must be overcome to launch a recoverable remote-sensing satellite. For instance, a rocket with enlarged carrying capacity must be developed, a more advanced satellite, a more perfect and reliable space TT&C network...

However, it was when the storm of the Cultural Revolution swept the nation. Class conflicts escalated, and an increasing number of aerospace workers were dismissed, some exiled to the countryside and some persecuted by the rebels...

In those special times, the fate recoverable remote-sensing satellite would never be a smooth sail.

The Initiation of the Development of Recoverable Satellites at Times of Turmoil

Cutting-edge technology is not for sale.

Americans and Soviets have already figured out the launch of recoverable remote-sensing satellites. Nonetheless, never would they share the secret of how.

In the absence of materials and experiences, regardless of the knowledge of the generation direction, every step taken was to be self-reliant and undertake a hard exploration.

The development of a recoverable remote-sensing satellite commenced at the beginning of 1966. It was listed as one of the 10-year goals by State Commission of Science for National Defense. Wang Xiji was the earliest head of technology. Later the task fell on the shoulders of Sun Jiadong.

A recoverable remote-sensing satellite must be equipped with a strong capability of earth observation, the high resolution of which was regarded as the most prior technical indicator. And the capability of earth observation was reflected in camera quality and film quantity a satellite carried. The power supply on the satellite was the zinc-silver batteries, the flying time of the satellite depended entirely on whose capacity. The realization of these technical indicators was a very difficult challenge for China's incomplete industrial system back then.

Led by Sun, in accordance with the overall technical and quality indicators, every department commenced intensive researches. From March 1970 to January 1973, the preliminary prototype development stage of the satellite was completed. In January 1973, it entered the normal prototype development stage, and noise, thermal vacuum and whole satellite vibration tests were carried out for eight months. The first recoverable remote-sensing satellite was a badminton-shaped blunt-headed vertebral body with a maximum diameter of 2.2 meters, consisting of two compartments, the instrument and the return compartments. Regarding the launch mission of the satellite, Ren Xinmin, technical director of the rocket, recalled: "To complete this task, the launch vehicle should not only have sufficient carrying capacity and the altitude and velocity of putting the satellite into orbit, but also high control precision to enable it to enter the planned orbit accurately."

Long March 2 carrier rocket took up the launch mission. After massive technicians' tackling of key problems, it adopted multiple new technologies and experienced an improvement on reliability. As a two-stage liquid carrier rocket with a takeoff mass of nearly 200 tons, it was installed with internal and external ballistic radio measuring devices able to provide parameters and information.

Ground TT&C network was also the key to the accomplishment of the mission. Returning satellites should have the ability of general satellites to fly in space, and also be controlled by the ground according to the program and return to the predetermined area safe and sound. If the ground TT&C network could not control the satellite throughout the run, as if a kite with a broken thread, the return would fail.

The main functions of the first phase of the ground TT&C network that was focused on Dongfanghong 1 were tracking measurement and orbit calculation without remote control of the satellite. Led by Chen Fangyun, technicians initiated the second phase. New TT&C stations in Changchun and Lhasa and recycle TT&C stations have been built, and a large number of TT&C equipment and various recycle measuring equipment have been added to have realized constant control of the satellite from launch to return.

At the times of turmoil during the Cultural Revolution, it was not easy for Chinese aerospace workers to tackles that many obstacles.

They were full of confidence at that moment. And the confidence mingled some blind optimism. Nobody expected how disastrous the launch turned out to be.

A Cable-triggered Incident

On November 15, 1974, at Jiuquan Satellite Launch Center, the Long March 2 carrier rocket, carrying the first recoverable remote-sensing satellite, stood on the launching platform.

"Last minute preparation!" At 11 a.m., launch director's command shouted in bitter wind.

At the launch site, all the crew held the breath, waiting for the ignition. At this point, an unexpected situation occurred – the satellite console operator suddenly found a black-out on most of the instruments on the satellite.

What now? The satellite did not switch to self-power supply as the configured program expected, in which case, the launch would only result in a two-ton useless iron bump to be shot up to the orbit.

Under normal circumstances, the issuance of the command to halt the launch requires a series of declaration and approval.

In the nick of time, Sun yelled "Stop the launch!." In that age, it took courage to make the call as it was both the technological risk and political risk that he must confront with.

The launch ceased. Most of the crew, still caught in intensive stress, had no clue what happened.

The window period was set at 11:00 to 15:30. Fueled rockets have strict parking time requirements. It is difficult to predict and control the chain reaction after a long parking time.

In a race against time, the failure must be eliminated. Sun wiped off the sweat about to roll down the forehead and organized technicians to check satellite's data log. And it was discovered that the external power plug was detached. With the failure located, technicians had to install an equivalent capacitor on the satellite circuit.

By then, the launch control room had been filled with directors of each department and relevant technicians had retreated back to the observation point 10 kilometers away. On the desert, speakers screamed to summon them back. Eventually, satellite and rocket re-entered the launch procedures.

At 15:30, on command, the rocket left the ground in thundering boom.

Six seconds later, the flight lost balance and the sway became stronger.

At the 20th second, the self-destruction system was activated, and explosion followed. And the debris of the satellite and Long March 2 plummeted to the desert less than one kilometer southeast to the launch tower.

The scene was a sea of flames. Seen from the launch site, half the sky was dyed red.

Years of efforts turned to dust in an instant. It was utmost pain for the technicians to witness the failure, and tears broke through their eyes.

The first fiasco it was.

And it was painfully carved on the hearts of Chinese aerospace workers. Decades later when Sun brought up the incident, he failed to hide the pain. "The launch on November 15, 1974 left serious damage a lifelong lesson."

How come it failed? Devastating as it is, tears were wiped, and technical personnel started to search the cause.

On the desert where rocket debris scattered, an inch-by-inch search was conducted. Back and forth, they dug the sand a foot down, sifting every tiny piece of debris out, to which the crew made claims to. The sheet copper was from this instrument, and the screw that instrument. Categorization was performed and every piece of debris was carefully checked to analyze the failure.

Soon the cause was determined, a simple matter though. A cable in the rocket was covered in intact rubber yet the copper wire inside was of poor quality. When the rocket quaked fierce at the launch, the wire disconnected and that led to a short-circuit, and further the loss of balance at the flight.

A slip with the wire triggered the loss of a satellite and a carrier rocket.

And the failure postponed the launch of China's recoverable remote-sensing satellite for an entire year.

The lesson was painful enough to be unforgettable.

The Triumph of the First Recoverable Satellite

What does not kill you makes you stronger.

Having heard of the failed launch, Ye Jianying, Vice-Chairman of the Central Military Commission, gave clear instructions "Failure breeds success. Don't be discouraged. Keep at it. We must make it happen."

The voice and determination of The Central Committee of the Communist Party of China (CCCPC) was conveyed in Ye's instructions. The fighting spirit amongst the massive aerospace workers was lit, and there came the breakthrough a heavy heart. Efforts were put onto the second launch mission.

A year later, a brand-new Long March 2 carrier rocket carrying a brand new recoverable remote-sensing satellite stood on the launch platform again at Jiuquan.

"Ignite!" At 11:30 am November 26, 1975, the rocket blasted off towards space and entered the expected orbit.

When cheers filled the air, an unexpected problem emerged.

After the satellite circled the earth, specialists of Ground TT&C noticed a sudden plunge of the gas source curve. The gas source bottle was prepared with the volume for three-day use. What if the satellite could not sustain three days as the pressure drops too quickly? The lack of nitrogen would affect the control of the satellite's position. A late order to return may result in no return at all.

Qian Xuesen, the current deputy director of State Commission of Science for National Defense rushed to the Ground TT&C Center in Xi'an. After a thorough analysis of the TT&C scheme, Chen Fangyun believed there should not occur any emergency situation; with a meticulous calculation, Yang Jiaxi considered the drop of pressure should result from the excessive coldness from external space instead of the decrease of nitrogen.

Qian's worry lingered due to the significance of the matter. He summoned Qi Siyu, Leader of Orbit Computing Group for Recovery Satellite and asked straightforward "Do you believe the satellite can sustain the flight?"

"Sure!" decisively answered Qi.

Staring, Qian tossed another question, "How can you guarantee a timely return? Depending on what?"

Qi took off his military hat, and confidently replied "On the grey hair on my head."

This is the perfect definition of taking on a responsibility.

Qian once said the success of China's aerospace undertaking never resulted from a person but a group. There are plenty of science and technology personnel brave enough to take on a responsibility as Qi in the group of Chinese aerospace undertaking. At crucial times, they would step forward and tackle obstacles.

Three day as planned, the satellite completed the orbiting task. Around 10:00 a.m. of November 29th, Ground TT&C sent out the command for return. An hour later, the satellite broke through the atmosphere and safely returned.

An old photo recorded the celebratory moment of the return.

Dragging a giant parachute of conspicuous colors, the satellite landed on a grain field filled with water. On the surrounding hillsides stood the villagers who travelled from all directions to witness the scene of bustle. However, glazing at the precious item that just came back from space, the recycling team failed to take it out of the field as a hook was left out in the design. Stuck. An elder rural man advised to clamp it with two long wooden bars. And the team lifted it to the vehicles as if a sedan chair.

The satellite traversed in space from three days and nights, during which General Zhang Aiping stayed in the shabby control room monitoring.

Having received the news of the safe return, Zhang asked about where it landed.

Staff officer answered, "Somewhere near the Guanling chain bridge in Guizhou Province."

Zhang approached the military map hanging on the wall and suddenly recalled "I have been there during the Long March in 1935."

What an interesting coincidence that this satellite should land where General Zhang fought 40 years ago as commander of the Long March's trailblazing army.

That night, Zhang was too elated to sleep, and he composed a poem:

I commanded an army of trailblazers in the Long March
It befriended a sky of megastars in the space journey
Through thick and thin we triumphed
With pride and glory, they astonished

The safe return marked China as the third country to master the technology of recoverable satellites after the United States and the Soviet Union. The technological breakthrough laid the solid foundation for China's manned space flight.

And recoverable satellite technology was considered one of the most perplexing and cutting-edge technologies back then. In 1959, America began the development of the recoverable satellite Discoverer. After 38 trials and countless failures, it was not until the 13th satellite that the recycle succeeded. The recoverable technology of space craft in the Soviet Union served to develop manned space flight, and more trials were performed. Whereas, on the first trial, China's recoverable remote-sensing satellite worked.

On December 7, 1976, China's second recoverable remote-sensing satellite successfully blasted off. After three days operation in space, it fulfilled the mission as expected and returned to the designated recycle zone on great precision.

As of 2005, five types of 22 near-earth-orbit recoverable satellites had been launched successfully. And the duration of operation increased from 3 to 27 days.

On August 29, 2005, China accomplished the launches of the 21st and 22nd recoverable satellites on the same day, marking the first-level technology of recoverable satellites in China.

Since mid-1980s, in China, the technology and data of recoverable remote-sensing satellite have been widely applied in many fields, such as petroleum exploration, railway route selection, coastal survey, geological survey and so on, and fruitful results have been achieved. Take land surveying and mapping for example: since the founding of People's Republic of China in 1949, an enormous team was organized to conduct land surveying and mapping. After 30 years of struggles, 20% was accomplished as of 1980s. Whereas, the utilization of remote-sensing satellites completed the rest 80% in 5 years, four times the result of traditional method in 30 years, with improved effect and precision.

In addition, more than 100 materials and life sciences experiments under micrograviity and space environment, and the first-class crop seed loading experiments have performed for domestic and foreign users by using the recoverable satellite platform, and gratifying results have been achieved.

Communications Satellites: the Dream of 36,000 Kilometers

The Letter Restarted a National Project

At times, civilians could also push forward the development of history.

There was a legendary story that happened in the history of China's aerospace undertaking. A letter by the general public reached Zhongnanhai, the residential compound in Beijing housing top party leaders, and restarted the communications satellite project that had been suspended for years.

At the dawn of May 19, 1974 in Zhongnanhai, lights at Premier Zhou Enlai's office were still on. Gravely ill as Zhou was, he poured over various documents and telegrams as usual.

And a letter named *Advice on the Development of Communications Satellites* caught his attention. Succinct and well-reasoned, the letter illustrated the necessity and urgency to develop China's communications satellites. The end of the letter said this:

> *China is a socialist country whose greatest advantage is to achieve maximum coordination. As long as the country gathers advantages of all sectors, it is entirely possible to send up communications satellites...*

Signed by Huang Zhongyu, Zhong Yixin, and Lin Keping.

They were no science personnel of the space flight projects but three ordinary employees at Ministry of Posts and Telecommunications.

Every man shares a duty to his country. Concerned about the undeveloped communications within the country, the three young men advocated the launch of a communications satellite as soon as possible. The letter was straightforward. When it was composed, a colleague with excellent calligraphy was invited to transcribe the content.

Their passion pushed forward the history of China's aerospace undertaking, and for that they are remembered.

It was said that Zhou was so moved that after seconds of contemplation, he took a pencil and left comments and instructions as follows:

> *Pass it to Comrades Chunqiao, Jianying, Hongwen, and Xiannian and later submit to Planning Commission and State Commission of Science for National Defense, who shall organize a meeting among relevant departments to determine the plan of the manufacture, coordination, and use of a communications satellite. And a detailed plan shall follow, and the execution be under supervision.*

This is the famous 5-19 Comment and Instruction, the significant last made by Premier Zhou Enlai when he was alive.

Staff revealed that 13 days after he finished the comment and instruction on June 1, 1974, his condition deteriorated, and Hospital 305 took him in. Henceforth, he was never able to make it back to his office, and finally he passed away.

In fact, it was not a late decision to launch a communications satellite. On June 3, 1970, one month after the success of Dongfanghong 1, Central Military Commission entrusted the development of Dongfanghong 2, a communications satellite, to State Commission of Science for National Defense.

However, it was the special times of the Cultural Revolution, in which many experts were persecuted. Work was left with nobody in charge and the science and research in total disorder. The preparation for the development and launch of a communications satellite was actually paralyzed.

Premier Zhou's 5-19 Comment and Instruction broke the deadlock and the project restarted.

The Comment and Instruction was soon implemented:

After several days, Yu qiuli, who was in charge of national planning, summoned a meeting among Planning Commission, State Commission of Science for National Defense, and Ministry of Posts and Telecommunications, to discuss about communications satellites.

On September 30th, Planning Commission and State Commission of Science for National Defense co-drafted *Report on the Development of Satellite Communications in China*. Meanwhile, the scheme validation and technical coordination of satellite communications were being executed.

On March 31, 1975, Ye Jianying, Vice-Chairman of the Central Military Commission, hosted the Eighth Member of the Standing Committee Meeting. Disputes escalated in the discussion about Report on the Development of Satellite Communications in China. Some advocated to prioritize reconnaissance satellite. Deng Xiaoping put an end to the debate, "I believe a communications satellite shall come first."

The following day, Ye signed the Report and delivered it to Chairman Mao, who was about to receive surgery due to cataracts, thus no reviewing on normal documents. The secretary inquired about the situation of his eyes. Regardless, he scrutinized the report word by word and issued approval.

Insomuch as the date of the approval of the Report by the Central Military Commission was March 31, 1975, the communications satellite project was coded Project 331.

Premier Zhou passed away on January 8, 1976. Four months later, the leading group of Project 331 was officially assembled. It was a pity that Zhou was not able witness the blast-off of the communications satellite. He was a beloved figure among whoever participated in the Project. Looking back at the history, people find it hard to forget that without Zhou's dedication, China's aerospace undertaking would not have achieved glory.

On March 3, 1977, China officially registered with International Telecommunication Union (ITU) regarding the expected launch of a communications satellite to test geostationary orbit in 1980. Five days later on March 8, ITU formally announced China's communications satellite project to the world.

36,000 Kilometer, Fantasy of a Radar Officer

Philosophers say that great dreams begin with gazing upon the stars. When human eyes and stars lock, sparks of wisdom burst.

Communications satellites are the bold predictions of Englishman Arthur C. Clarke when he gazed upon the stars.

Originally, he was an ordinary RAF radar officer. After his retirement from the army, he was fascinated by the fantasy of future radio communications. He often gazed upon the stars and thought:

A person on the horizon can look afar up to 4.7 kilometers; once he reaches the 50-meter-high roof, eyes can cover the area with a radius of 25 kilometers. On this basis, if a satellite can be launched to a certain altitude and used as a radio tandem office, the global communications problem can be solved.

After years of research and deduction, he discovered that 36,000 kilometers over the equator of the Earth, there exists an orbit that can keep the satellite stationary relative to the earth. It was later recognized as geostationary orbit by scientists.

Thrilled to the core, on May 25, 1945, he presented a memorandum to the British Star Society, elaborating on his discoveries and boldly predicting the upcoming arrival of the day of global communications. If one day human could deploy three satellites at equal distances in geosynchronous orbit, communications could be achieved anywhere on earth except Antarctica and Arctic.

The prophecy was unbelievable to people of that era. How to reach as high as 36,000 kilometers?

Strongly tempting too was the prophecy. The world would undergo drastic changes if telecommunications comes true.

None expected that it would only take 19 years for the prophecy to become reality. On August 19, 1964, the United States successfully launched Syncom 3 a geosynchronous communications satellite into a geosynchronous orbit 36,000 kilometers above the equator.

The successful launch of the first communications satellite marked the beginning of information age for the mankind. The advent shrinks the earth into a village. One hundred years ago, the news regarding the assassination of Abraham Lincoln President of the United States spread to London 12 days after the occurrence due to undeveloped communications. With communications satellites, a second suffices to disseminate it globally.

The United States used the satellite to broadcast the opening ceremony of the Olympic Games in Tokyo of Japan live for Europe and North America. When people saw the live images of the Olympics from the far east, they were left in awe and realized that communication satellites would be closely related to and indispensable to their lives.

In the spring of 1965, on the 100th anniversary of ITU, the United States launched the world's first commercial communications satellite, Early Bird, and broadcasted color TV to Europe and Americas.

The United States triumphed. Naturally, the Soviet Union refused to lag behind. Seventeen days later, Molniya (Lightning) was launched by the Soviet Union, therefore enabling regional communication and television broadcasting between the Soviet Union and Eastern Europe. On November 30, 1965, it first broadcast color TV programs by satellite, thus becoming the first country in the world to have a satellite communication network for domestic use.

At the same time, military use of communication satellites had become increasingly prominent. In addition, military satellite communication systems had been established in the United States, the Soviet Union, as well as NATO and Warsaw Pact countries.

Hereafter, communications satellites have been more heeded and widely used. As national interests, national economies, and people's livelihoods were concerned, countries began to compete on the launch of communications satellites. Vast as space is, there is only one geosynchronous orbit. To avoid collision or interference between satellites operating in this orbit, the World Radio Organization stipulated that only one satellite can be deployed at a certain distance. The orbit can only accommodate 180 satellites. Once this amount is reached, no launch is allowed.

Obviously, the geosynchronous orbit has limitations. Whoever preempts the seat holds it for good. Consequently, the quiet space turned boisterous.

Statistics show, as of 1983, the world had launched 149 satellites into geosynchronous orbit at an altitude of 36,000 kilometers above the equator.

None belonged to China.

China Deserves a Place

In 1972, President Richard Nixon visited China.

It was so significant an event that the world political pattern changed. Apart from this impact, the Chinese were able to experience the power of communications satellites.

Nixon's visit to China was accompanied by a large number of American journalists, who did a live broadcast across the Atlantic to American audiences via communications satellites. Poring over the report of the technology on newspapers, the Chinese people were amazed and hoping that China would master it as well.

To some extent, President Nixon's visit triggered a sensitive nerve of the Chinese, namely, a craving for an urgent development of communication satellites. Whether in diplomacy, military affairs, communications, or civil affairs, the country was in great need of communications satellites.

However, the Cultural Revolution made it impossible to move one step forward.

Upon the demise of the Cultural Revolution, China was faced with millions of things undone but a shortage of talents. Deng Xiaoping expected to launch a communications satellite as soon as possible and use television teaching to solve the problem of insufficient teachers. He said a teacher at the enormous Great Hall of the People was only able to educate 10,000 minds, but in a TV classroom, the number of students was infinite.

However, the progress of the development of communication satellites failed to meet the urgent need of talent nurturing. Therefore, the Party Central Committee made the decision to buy a communications satellite from the United States, whose government agreed to the purchase request.

In 1978, with Ren Xinmin appointed as the head, the delegation of the Chinese Astronautical Society visited NASA. The main purpose of this visit was to negotiate the purchase of communications satellites from the United States.

For various reasons, the deal was not made. And China was determined to make one on its own.

When the delegation returned to motherland, Ren encouraged the crew "This is for the best. We have never been more driven to develop and launch China's own communications satellite. Let's make China proud."

General Zhang Aiping added "As a member of the UN, China shall not be absent from the synchronous orbit. We will never bring shame to the country. We can buy satellites, once, twice, not infinite times. We must rely on ourselves. Develop communications satellites and launch them with our own rockets too. We can do it, and we must."

The General Accelerated the Three-Strike Conception

Upon the demise of the Cultural Revolution in 1976, General Zhang Aiping was reinstated the positions as deputy chief of staff and director of State Commission of Science for National Defense.

Again to "battle" went the general with brilliant achievements in war, founder of the People's Navy, and commander of the atom bombs, hydrogen bombs, and man-made satellites project.

At the age of 65 and in sickness, General Zhang was devoted to national defense science and technology.

"Lost time must be made up for." With a walking stick, he hobbled from one factory to another, one laboratory to another, and one institute to another. He repeatedly called on the rocket experts who were persecuted during the Cultural Revolution to have them back to institutes and launch sites.

To lift the morale, he advocated five kinds of spirit among the scientific and technological personnel: the spirit of heroic devotion, of seeking truth from science, of unity and cooperation, of self-reliance and hard struggle, and of strict discipline.

As the deputy of Marshal Nie Rongzhen, General Zhang made full a commitment to the national defense science and technology frontline in 1958. Ten years of work allowed him to understand the character of scientific researchers and the objective law of scientific research. He enhanced the technical command system that had been used for decades, which used to have Party Committee in charge, and reiterated that technologically the chief designer can override Party Committee's decisions.

Shortly afterwards, he delineated a grand blueprint for the development of national defense science and technology, namely, the strategic three-strike conception. The first was to launch the trial intercontinental carrier rocket; the second to develop the trial submarine to launch a carrier rocket underwater; and the third to launch the synchronous communications satellite.

Led by General Zhang, China's national defense science and technology welcomed a new peak.

In May 1980, the trial launch of China's first intercontinental carrier rocket passed.

In October 1982, the trial rocket launch by submarine was triumphant!

Two of the three strikes came true. And the last underwent he most twist and turns.

Conquering 36,000 Kilometers

Achieve greatness and treat the extraordinary warmly.

In September 1977, Ren Xinmin was appointed Chief Designer of China's Satellite Communications Project, codenamed Project 331.

Ren may be unheard of amid today's young people. But in China's aerospace undertaking, the title to his name is usually "pioneer," "founder," "space maestro," "authoritative expert," "national treasure." He is the idol of young scientists and technicians. The strange and familiar name can be found in every milestone and record of merits of China's space industry.

When the People's Republic of China was founded in 1949, Ren resolutely abandoned the superior living conditions abroad and returned home, becoming the first batch of high-end talents overseas who returned to build the new China. Hereafter, on the national defense front was added an indefatigable figure, who has participated in the project throughout, and has become the pioneer and leader of China's missile and aerospace undertakings.

"Project 331" was the largest, most involved, most complex, and most difficult space system engineering in China before the 1980s. It consisted of five systems: communications satellite system, with Sun Jiadong as chief designer; launch vehicle system, with Xie Guangxuan as chief designer; measurement and control system, with Chen Fangyun as chief designer; ground communication application system, with Liu Yongjun as designer; and finally the launch site system.

Ren, as chief designer of the project, was referred to as Chief of Chiefs. He had already turned 62 that year. The pressure was overwhelming for such an age.

He once recalled to me. "China's plan to develop and launch communications satellites had already been made public. There is no turning back as it concerns the national reputation. We must do everything we can to have it done."

Harder than touching the sky, a phrase people use to depict ultimate difficulty. And what Ren and his colleagues were requested to do was literally touch the sky – to conquer 36,000 kilometers and fly China's communications satellite.

It requires a more robust carrier rocket to launch a communications satellite up to 36,000 kilometers high. Therefore, the first step for Project 331 was to develop a new Long March 3 launch vehicle.

The thrust of the vehicle is determined by the engine.

Ren had eyes on the world's newest liquid-fuel rocket engine, the hydrogen-oxygen engine. It had a thrust 50% greater than that of conventional engines. At that time, the United States and France were the only capable countries, and it took them over a decade to master the technology.

As an expert in liquid rocket engines himself, Ren was well aware of the difficulty to develop hydrogen and oxygen engines. Regardless, he pressed ahead because he has a sense of urgency: foreign countries have succeeded while China started late; if dragged on, it would lag behind even more. "As long as we overcome this conundrum, the carrying capacity of China's Long March rocket will make a qualitative leap forward."

Naturally, skepticism emerged: now that the development was arduous, could we really pull it off? How about a safer plan?

At the meeting of State Commission of Science for National Defense, Ren's speech electrified the room as well as overrode all objections. "A hydrogen-oxygen engine is a mountain that we must climb. Better sooner than later. And I can guarantee that our technology is capable enough."

Difficulties had been expected, yet the reality turned out to be more cruel.

A hydrogen-oxygen engine requires low temperature propellant, liquid hydrogen, and liquid oxygen, whose boiling points are minus 183 degrees and minus 253 degrees respectively. They are easily flammable and explosive. Operators must wear cotton outfits to prevent static electricity when performing an experiment. A slight negligence would generate static electricity and lead to fire and explosion.

The engine development was hampered at the start:

In January 1978, when it was first tested, an explosion accident occurred, and 10 people were injured on the spot. Two months later, in another test, fire broke out and the flame rose straight to 10 meters high.

There were voices whispering to back down: how about slowing down the urgency? Faced with tremendous pressure, Ren firmly assured "A task is always accompanied with difficulties and risks. As long as we advance in the face of them, there is nothing we cannot conquer."

When confronted with conundrums, Ren would "sink down: be present at the front-line test site, take in the opinions of technicians, and search for breakpoints. Chief Designer as he was titled, his presence was often noticed at the test site together in the face of wind and rain, brainstorming for a solution to a technical problem.

It worked. Every time when he was stuck in a technical bottleneck, he would "sink down" and the solution would be found. Afterwards, Ren counseled his pupils "As the head of technology, judging, dealing with, and deciding on technical problems involves, first person basic knowledge and expertise together with continuous learning and renewing; second, constant practice and accumulated and concluded practical experiences together with knowledge extractions from science and technology personnel and the front-line research; and third, seeking for the truth and decisiveness."

Concerning Ren's decisiveness, there is a story to tell. Over a long disputed technical issue, he made the call. Disagreements were voiced in private. "Chief Ren's decision may not count." The words reached his ears, triggering immediate anger. He rang Tang Bangzhi, the secretary. "Call whoever spread the words, and query for his opinion on whose decision counts. I would gladly hand over the position Chief Designer of State Commission of Science for National Defense to whoever is better qualified if any."

To believe in only the reality, not books, not superiors, and not overseas. This was the work creed Ren lived by. And under his leadership, the vast number of science and technology personnel marched forward step by step against adversity.

In September 1978, the 50-second test run of the overall unit of the hydrogen-oxygen engine succeeded.

At the end of 1982, the 1400 second long-distance test run succeeded, a breakthrough in the strength of the engine.

On May 25, 1983, the hydrogen-oxygen engine's whole system completed a successful test run. Later, China became the third country to master the technology, after the United States and France.

The news that Ren passed away came when I was writing this passage. It was February 13, 2017, the day when Tianzhou-1 cargo spacecraft arrived safely at Wenchang Space Launch Site in Hainan Province. It was the beginning of the closing number of China's manned space laboratory mission.

"He was the heart of China's Long March rockets." Ren and I have met several times during interviews and reports of space flight. At nearly 80 years old, he was still climbing the launch tower; at nearly 90, when Shenzhou 5 was launched, he insisted on watching it on site; in 2009, when Qian Xuesen passed away, 94-year-old Ren Xinmin, went to offer condolences despite heavy snow; at 95 years old, he still insisted on participating in the rocket engine test run by the Institute of Sixth Academy of Astronautics. And in my humble opinion, Ren Xinmin is a great ordinary man.

On February 16, at the farewell ceremony in Babaoshan Revolutionary Cemetery, thousands of people voluntarily came to see him off, the father of atom bombs, hydrogen bombs, and man-made satellites project, and giant of space science and technology. Banners rustled in chilly wind, reading the self-summary of the 102-year-old man: "Space flight, the only thing I have done."

Ren was lucky as he was the only one of the old generation of Chinese rocket men to have witnessed the blast-off of Long March 5. On November 3, 2016, food untouched, he sat by the TV, enjoying every second of Long March 5's launch. Excitements washed over him when it succeeded. On the hospital bed, he jotted down "Congratulations to the triumph of Long March 5's first flight."

"In the last living days, he was still deeply worried about Long March 5 rocket." Ren's apprentice, the commander-in-chief of the Long March 5 rocket, praised emotionally to reporters at the onsite interview "A beacon he was when alive. And a monument he became when deceased."

A Miracle Was Made

The second hurdle to sail over for Project 331 was the establishment of a more robust Ground TT&C Network, which would enable real-time communication between the ground and the communications satellite.

The task also fell on the shoulders of a "veteran" – Chen Fangyun, chief designer of measurement and control system. That year, he had already turned 61.

There is a saying that before 60 is the prime time of a man. Chen, aged 61, proved the absolute absurdity in this statement with abundant wisdom and energy. Youth stays forever in an innovator.

Previously, the ground TT&C network built in our country could only realize launch medium and low orbit launches. Faced with the unprecedented 36,000 kilometers, it had to enhance the design to enable high power, high sensitivity and ultra-long distance. Chen Fangyun proposed a simple and practical method, the collaboration between the unified microwave system and responders on the satellite.

On July 16, 1969, the Apollo spacecraft carried astronauts to the Moon for the first time and stirred up a worldwide sensation. The inexperienced see the appearance, and the experienced appreciate to the contents. Having meticulously studied the unified microwave system the Apollo spacecraft adopted, Chen realized that the utilization would also fit China's space missions. Therefore, he designed a new unified microwave system for the measurement and control technology requirements of the first communications satellite in China.

Chen's conception was ground-breaking. It allowed our space TT&C system to develop from a single decentralized system to a comprehensive multi-functional system.

New things are accompanied with controversy. At a critical moment, Sun Jiadong, president of China Academy of Space Technology and chief designer of synchronous communications satellite Dongfanghong 2, offered full support to Chen. He believed that this new TT&C scheme was not a copy of Apollo TT&C in the United States, but a practical method in line with China's national conditions. It can transmit and receive all kinds of signals at the same carrier wave frequency, which would save a lot of equipment. A pair of satellite antennas would suffice. Weight of the satellite body and power dissipation would greatly decrease. Multiple ends were served with one strike.

According to the actual situation of Dongfanghong 2's blast-off from Xichang Satellite Launch Site and orbit entry, TT&C stations in Weinan of Shaanxi Province and West Fujian Province have been added successively to the system. However, to achieve the full-range TT&C of satellites, it was very difficult to rely solely on the existing ground TT&C stations in our country, and via the Yuanwang surveying vessel, mobile measurement and control at sea was required.

The development of the Yuanwang surveying vessel was initiated on July 18, 1967, thus the codename Project 718.

At that time, China was preparing for the flight trial of Dongfeng 5 intercontinental missile, whose maximum range was 14,000 kilometers. It was infeasible to carry out full flight test at home. The trial would have to extend to other countries and set up a landing area on the high seas of the Pacific Ocean. "Our territory is only 5,500 kilometers in length and breadth from east to west, north to south, and the range of intercontinental missiles is more than 10,000. To develop intercontinental missiles, a survey ship at sea is a must."

Data unveils that the United States and the Soviet Union have made full use of the ocean, which accounts for 70% of the earth, in the whole-course flight test of intercontinental missiles, satellites and spacecrafts. In the 1960s, the two countries successively hewed out ocean test ranges: the United States established the Pacific Range Fleet and Atlantic Range Fleet, and the Soviet Union the Pacific Range Fleet in the Far East. Based on the need of launching mission, the fleet sailed to the high seas, as close as possible to the required ballistic and landing areas, to accurately measure the relevant data.

In August 1977, Yuanwang 1 surveying vessel entered the water. Two months later, Yuanwang 2 joined and China became the fourth country to own a space survey ship after the United States, the Soviet Union and France.

A technical conundrum to be solved urgently followed the deployment of Yuanwang 1 and 2. There were inert equipment such as radar, and those for communication, navigation and meteorology in the hull, and the deck was covered with antennas of different shapes. Among them, electromagnetic interference was serious, which affected the normal operation.

Hours of immersion in thoughts, Chen came up with a clever move: segmented frequency allocation. And it eliminated the electromagnetic interference on Yuanwang vessels.

Outstanding scientists always think ahead. Yuanwang was accompanied by a large fleet of ships when launching Dongfeng 5 intercontinental missile. Nowadays, it had to sail across the ocean alone in the task of communications satellite. Can we reduce the safeguard communication vessels but still enable it to communicate with China while carrying out measurement and control? Chen originated an idea to realize solo sail of Yuanwang in the execution of the mission.

I have been to the base of Yuanwang spacecraft survey vessel, on which I have boarded for a closer look. So to speak, it was floating marine city of integration of technological essence of China's aerospace undertaking, optics, TT&C, and electronics. It not only had to sail to the Pacific Ocean to fight against storms during a launch, but also to measure and control rockets and satellites.

On July 12, 2016, officially deployed was Yuanwang 7, a new generation of ocean-going space survey ship, which was a large, independently designed and developed, and internationally advanced. Eighteen months led to the complete manufacture, achieving length of 220 meters, height of 40 meter, discharge capacity of 30,000 tons, sustainability of 12-magnitude typhoon, self-sufficiency of 100 days, and capability to carry out a mission in the Pacific, Indian Ocean and the Atlantic Ocean within 60 degrees north-south latitude. Yuanwang 7 was composed of two parts: the general platform for ships and the space TT&C communication equipment. It integrated the most cutting-edgy technologies in ship manufacture, space TT&C, navigation meteorology and ship dynamics. And it marked a new opportunity and leap forward for the development of China's offshore space TT&C undertakings, and a new improvement and breakthrough for the capability of offshore space TT&C, which was of great significance to the construction of China's space TT&C network.

Academician Li Jisheng participated in the TT&C mission of Dongfanghong 2. On the eve of the Spring Festival in 2017, I paid him a visit. He told me a thing that made him remember deeply in recalling the completion of the mission——

On February 1981, Japan successfully launched its first geosynchronous communications satellite Kiku 3 (also known as ETS IV). To introduce high-performance computers from Japan and better learn from the successful foreign experience of launching geosynchronous communication satellites, State Commission of Science for National Defense invited Japanese experts for a visit.

Having visited our computer room, a Japanese expert said, "It is impossible to complete the geosynchronous communications satellite's TT&C mission with such computers." At that time, the West imposed an embargo on high-performance computers in China. Whenever he was inquired technical questions about computers, it was the same answer that he gave, "I'm sorry, I didn't bring my notebook with me. If you want to know, please come to Japan."

How to fulfill the TT&C mission, using undeveloped computers? Colleagues brainstormed together and worked out a way: to re-mound computer storage, add computing results exchange between the two "320" computers, and optimize and integrate various software modules. Eventually, the requirements were met.

In April 1984, Dongfanghong 2 communications satellite was successfully launched. In May of that year, Li went to Japan to attend the 14th International Space Science Academic Exchange Conference. He happened to come across the Japanese expert again.

"What computers did you use to pull it off?" asked he.

"Exactly the "320" computers you saw" answered Li.

"A miracle, then" he kept asking, "Could I be honored with the details?"

The moment of arrogance played again in Li's mind. And in the Japanese expert's own words, he replied, "I'm sorry, I didn't bring my notebook with me. If you want to know, please come to China."

This is the spirit that every Chinese aerospace undertaker shared.

And it was in that spirit that they were able to overcome one unimaginable obstacle after another in the construction of TT&C network and eventually fulfilled the mission of Dongfanghong 2 using the computers Japanese experts deemed impossible.

Xichang, the Beginning of Space Expedition

The launch of Dongfanghong 2 communications satellite was set at Xichang Satellite Launch Site.

In September 1970, four months after the success of Dongfanghong 1, the construction project of Xichang Satellite Launch Site began.

And it was coded Project 7201 as the underlying construction must be completed in January 1972 according to the order of the Central Military Commission.

Sino-Soviet relations were tense and the Jiuquan launch site was too close to the border. To prepare China for stronger national defense and successive development of China's aerospace industry, the Central Military Commission decided to build a new launch site in the mainland.

Numerous investigations led to the central government's final decision on Xichang, the hinterland of Daliangshan.

Red Army's footprints covered this place during the Long March. Interestingly, it then became the new beginning of a space expedition. In the past, it was the torches of Yi compatriots that lit the night sky here. Now, Long March rockets would do the trick.

However, because of that special times, the construction of Xichang Launch Site had to strike one snag after another, yet still a suspension came. It was not until the start of Project 331 in 1977 that it was under full construction.

To accelerate the progress, three goals were set for whoever participated in the construction.

The first was to have the launching apron completed in 20 days.

The second was to erect a 77-meter launching tower at the site in 37 days. And the record of the quickest installation of a launcher was made in China back then.

The third was to finish the construction of underground command post through half a month of all-nighters. The hands were too worn out to hold chopsticks at a meal.

At last, with everything ready at Xichang Launch Site, the arrival of Long March 3 and Dongfanghong 2 was being expected.

And I stepped into the site at its 40th anniversary.

In a narrow valley it sits. Holding an old black-and-white photo, tour guide Wang Yongjun described the significant changes to me in excitement and pride, "Look, the wildness in the background is where the launch site resides."

Sky as quilt and earth as bed, camps were made on the banks of Anning River. A pot was heated when supported by three rocks. And salted edible wild herbs were the only food available." "Diarrhea happened when they drank the water here. Doctors called it Xichang disease, untreatable. Better suck it up." The hardship never faded in Wang's memory, still bitter.

Forty years have gone by. Wang and his colleagues are aged while the launch site, known as Houston in the East, has gained more energy and radiance.

On December 18, 2010, flames cut through the night sky. On the observation deck on the hillside, I witnessed that the Long March 3 carrier rocket rose with a crushing blow and lifted the seventh Beidou navigation satellite straight into the sky.

It was the closing number of China's space launch in 2010 and the 61st successful harvest of Xichang. That year, Xichang Launch Site congratulated its 40th birthday with a 100% success rate in all eight launches.

A Sudden Strike of Malfunction at the Countdown

Time flies and the "331 Project" had been under provision for more than six years.

By the beginning of August 1983, the five major systems of the Dongfanghong 2 project, namely, launch vehicle, communications satellite, launch site, TT&C, and earth station for ground communication applications, have been ready.

On August 12, the first working conference on the launch of Dongfanghong 2 was held in Beijing. Until the 27th, the meeting lasted half a month. Experts who attended the meeting conducted a detailed discussion on all aspects of technical issues concerning the launch. In the end, the Report on the Implementation of the Mission of the Dongfanghong 2 received unanimous approval.

The meeting resulted in the final decision of the launch time to be between the end of 1983 and early April 9, 1984. Three rockets and satellites were in place in case a failure should happen, and a second launch could be organized immediately.

On October 9, 1983, the Party Central Committee approved of the relevant reports, marking that the launch of Dongfanghong 2, the first geosynchronous communication satellite, has officially entered the countdown.

On October 14, 83, Long March 3 carrier rocket and Dongfanghong 2 communications satellite reached Xichang launch site.

More than one hundred departments and tens of thousands of personnel have been involved in the development and execution of the launch mission. A tiny slip of whatever link would cause unpredictable safety hazards.

For this reason, Zhang Aiping took a special trip to Xichang to ascertain the preparations. As known to all, Zhang had both identities as a general and calligrapher. Every department in Xichang arranged pens and inks to have him write an inscription during the inspection. He took it as a form of supervision. And the inscription he wrote for the launching team was "united efforts will strike thrice and stun the world," for the command team, "A crafty plan with no flubs," for Xichang Launch site, "A new epoch for national defense technology, and the faith in a victory will come true." The story about General Zhang Aiping's inscription is still on the lips of the aerospace workers in Xichang.

On January 1, 1984, Long March 3 was transported to the site and erected on the platform. Four days later, Dongfanghong 2 was hoisted on top of the rocket. Anon, all preparations fell in place. Command post set the launch time at 20:00 precisely on January 26, 1984.

Finally, as the day came, the serene valley at the launch site heard more commotion than ever.

At 15:00 on January 26, the mission entered the "five-hour preparation" procedure, and the rocket began to inject liquid oxygen, after which is completed, operators subsequently intended to inject liquid hydrogen. Prior to that, technicians carried out a second inspection on the rocket function, and suddenly detected abnormality of the rocket stabilization system.

Malfunction struck again at critical moments. Experts and technicians on the spot consulted and analyzed urgently to identify a disorder of the gyroscope platform, which called for replacement. Five hours were insufficient to eliminate the trouble and replace the platform. To secure the success of the launch, command post requested the permission of a delay from State Commission of Science for National Defense.

Receiving the call, Zhang fully demonstrated a general's demeanor. Neither furious nor critical, in a relaxing tone, he consoled "Alright, a pause then. Not necessarily a bad thing. Better a problem found sooner than later. Please tell the crew to stay calm and clear-minded. Don't panic. Be fully prepared. Buy some time while safeguarding the quality."

To front-line experts and personnel, Zhang's reply was soothing and encouraging. The trouble was eliminated, and the launch resumed.

A Good Save in Five Minutes

"Ignite!" At 20:24 on January 29, 1984, with the commander's order, Long March 3 carrier rocket lifted off Dongfanghong 2. Night sky was illuminated in tangerine flames.

In the command post hall, the curve showing the flight status on the big screen was the center of attention, and voices to report from various departments rose and fell—

The first stage, normal.

The second stage, normal.

The first ignition of the third stage, normal.

Abruptly, something unexpected happened – the second ignition of the third stage flamed out after three seconds, when the rocket had flown 940 seconds.

Thrust was lost. With the separation of satellites and arrows, it was managed to send the satellites into the near-earth orbit with the perigee altitude of 321 km and the apogee altitude of 474 km without entering the large elliptical orbit required for synchronization.

Highly elliptical orbit is the most critical step in launching a geosynchronous communications satellite. If the entry to the predetermined large elliptical orbit folds, the satellite will not be able to communicate with the ground. And the unfinished step marked the failure of the launch.

The brutal fact had tears welling the eyes of thousands of scientists and technicians.

However, grieving could wait. State Commission of Science for National Defense sent a telegram: maximize efforts to save the communications satellite.

The Xi'an Satellite TT&C Center summoned a meeting and experts from various departments urgently brainstormed. A rescue plan was quickly formed.

The key to an effective rescue was to capture it within one day. According to the design, before it entered the Large Elliptical Orbit, power was supplied by the battery on the satellite, which could only sustain one day. Once it was depleted, it would be a dead star.

But, as if a wild horse was galloping, it was circling the earth at a high speed in 90 minutes. It was no easy task to capture it.

This unprecedented difficult task fell on the shoulders of Weinan and Western Fujian TT&C stations, because at that time only the TT&C equipment of these two ground TT&C stations still held control of the satellite.

There was only one way to accomplish the save: send signals from the ground to ignite the satellite apogee engine, thus elevating its orbit.

Experts explained that this approach may lead to a horrendous situation: inappropriate attitude control. After ignition it may plummet to earth. Had the direction headed for a densely populated city, the consequence would have been catastrophic.

It was so material a matter that they had to report it to General Zhang Aiping for approval. A decisive order was given. "Go all out, if you still cannot pinpoint the satellite attitude and it should plummet overseas, I will take the blame! In a science experiment, we must be prepared for both failure and success."

"In a science experiment, we must be prepared for both failure and success." Today as we chew this sentence, we can still feel the strength! On the road of science research and exploration, and on the journey of advancing science and technology to a powerful country, only when we are able to sustain failure can we harvest success, and only when we tolerate failure can we reach the peak of cutting-edge science and technology. As commander of State Commission of Science for National Defense, he stood out for the strong heart able to sustain and tolerate failures.

As the satellite crossed national borders, there was only a five-minute tracking arc, within which, had the ground TT&C station failed to keep up, signals of command would not have reached the satellite.

What can be done in five minutes? Not even eating a simple meal. Yet, it was all the technicians of the ground TT&C stations were given.

Lap one, two, and three. Target was spotted, yet unable to be tracked.

The tenth lap it had already circled. As the orbital predications by Yuanwang became increasingly rough, there was no trace of the satellite.

The 11th lap was the last resort.

"Target spotted." The West Fujian Ground TT&C Station immediately signaled command to the satellite through microwave.

A moment later, the Station reported to the Xi'an TT&C Center: "The satellite has begun to execute the command."

Measurements implied a healthy function and the satellite's bottom was faced to the sun.

On the 13th lap, when it arrived at the apogee at the exact time of 16:09 on January 30, 1984, Weinan and West Fujian stations successively transmitted apogee engine ignition signals to the satellite.

It worked. In this second, the hall of the TT&C Center was bathed in cheers, and the scientists and technicians who fought relentlessly embraced one another tightly.

The thrust rapidly elevated the satellite's flight orbit. Soon, it rose from near-Earth orbit with a perigee altitude of 321 km and an apogee altitude of 474 km to a large elliptical orbit with a perigee altitude of 538 km and an apogee altitude of 6 480 km.

It was saved from dying and became a long-term science experiment satellite.

At 19:00 that evening, CCTV News Broadcast announced that China had successfully launched another low-orbit science experiment satellite.

Dongfanghong 2 Changed China

On January 29, 1984, the Spring Festival was only three days away.

That evening, people celebrated the successful launch of another satellite. But whoever was aware of the truth had a sinking heart.

Originally the plan was everyone would rejoice in the holidays after a triumphant launch. Yet, none was in the mood Spring Festival then? Promptly following the occurrence of the failure, Ren Xinmin, chief designer, led the expert group to analyze it. Ren motivated the group, "The urgent task is to analyze the cause and mechanism clearly and let the data speak."

On January 30, Marshal Nie Rongzhen mailed a letter of sympathy from Beijing: "As long as you take it seriously, analyze the reasons, find out the faults and draw experience, China's synchronous communications satellite will fly." Encouragement from Beijing has alleviated a little of the tremendous pressure amid the group.

Soon, regarding what caused the flop, experts initially determined that the turbofan of the third stage hydrogen and oxygen engine of Long March 3 had burned out.

What measures should be taken to avoid a second occurrence? A brainstorm fruited three enhancement measures, with which Ren was not satisfied as he believed it is no absolute elimination of the problems.

On the flight to Beijing for a meeting, inspiration struck, and Ren came up with a new solution. Subsequently, relevant research institutes and manufacturers carried out experiments and produced new parts in the shortest time. On March 23, they refitted the Long March 3 rocket that had been erected.

On April 8, 1984, two months after the first flop, the improved Long March 3 embarked on another proud expedition.

At 19:20:02, in thundering roar, Long March 3 carried Dongfanghong 2 off the ground and soared into the night sky.

"First stage rocket normal." "Second stage normal." "The first ignition of the three-stage rocket worked." "The second ignition worked."

At 19:40, following the third stage rocket's precise entry into the orbit, the satellite and the rocket separated, and the former entered the large elliptical orbit.

On April 10, the apogee engine was successfully ignited, and the satellite entered quasi-synchronous orbit.

So far, everything was on the right track with only one last step of "braking" to be taken – so that it would be in full synchronization with the earth.

Surprisingly, the battery of the satellite had a thermal runaway, or commonly a fever.

It was a threatening technical issue. The sudden rise of temperature would burn down the battery and even the entire satellite. A burned-out battery would leave the satellite no power and a satellite without power would be a piece of garbage out of control.

Data read that the temperature of the battery had exceeded the warning line by 30 degrees Celsius. Previously, in the ground test, the upper limit was a maximum of 50 degrees Celsius. As time went by, the temperature kept escalating.

It was such urgency that no hesitation was allowed. The TT&C command post quickly called the shots: to adjust the satellite's attitude in orbit at a large angle to avoid direct sunlight, so as to reduce the amount of solar energy absorbed, and then halt the heating of the battery.

However, a 25-degree adjustment of the altitude did not put a halt to the rise of temperature.

As there was the risk of control loss, the adjustment continued. It stopped when the temperature reached 45 degree Celsius, closer to the limit.

In that moment, Sun decisively commanded: another 5 degrees.

Onsite operators pulled out a paper, on which they wrote down "Sun Jiadong requested another 5-degree adjustment." Without hesitation, Sun signed it.

The signature meant full responsibility for unpredictable risks and outcomes. On the battle, the brave action set aside personal life and death for a greater cause. And on the battlefield of satellite launches, there were more times when Sun belled the cat regardless of personal gain and loss.

A miracle took place after the order was given. The temperature stopped rising, instead, dropping bit by bit. Afterwards, the satellite attitude received another adjustment from Xi'an TT&C Center.

At 18:28 on April 16, 1984, the space "brake" of the Dongfanghong 2 was accomplished, and it entered the synchronous fixed-point orbit over the equator at 125 degrees longitude. Hereafter, China possessed the first communications satellite.

At 18:00 on April 17, the communication experiment began. Beijing, Nanjing, Shijiazhuang, Kunming and Urumqi communication stations, which had already been commissioned, carried out communications, broadcasting, transmission of color TV programs, newspaper format fax and time standard frequency broadcasting. The results showed that the satellite was in normal operation and with fine transmission quality.

At 10 a.m. on April 18, on the phone, General Zhang conversed with Wang Enmao, Secretary of the Party Committee of Xinjiang Uygur Autonomous Region, far away in Urumqi, via Dongfanghong 2.

"Hello, Comrade Enmao! I'm talking to you through our own communications satellite. Can you hear me clearly?

"Perfectly clear!" said Wang Enmao at the other end of the phone thousands of kilometers away," For the first time, people of all ethnic groups in Urumqi saw the news of the day in clear images, broadcast by CCTV, just like watching TV in Beijing. Thank you for making such a great contribution to the motherland and the people!"

Elated, Zhang replied "This is the harvest of the joint efforts of all comrades participating in the research and development, and of the strong support of the people of the whole country, including the people of Xinjiang. Comrades from the base shall taste the sweetest melons there again."

After 30 years, amongst sealed documents, I discovered the conversation between Zhang and Wang. Reading it made me proud and moved. The ascent of communications satellite changed the world and that of Dongfanghong 2 did the same to China.

On April 19th, General Zhang voiced at a press conference held by Xinhua News Agency: "The successful launch of the test communications satellite validates that the technology of our launch vehicle is no inferior to that of other developed countries, and there is only a small gap, concerning satellite communication technology, between China and the world. And China fully relied on its own strength. The most important feature of this launch is that it only took one attempt to succeed while executing the test and application of communication, broadcasting and television transmission. Rarely did it happen in the history of world space technology, therefore totally demonstrating Chinese people's great talent and courage.

On May 15, Dongfanghong 2 began the live broadcast of the Second Session of the Sixth National People's Congress. Television audiences in remote cities, such as Urumqi and Kunming, were able to watch the grand event real-time.

Dongfanghong Satellite Platform Made Splendor

After the successful launch of Dongfanghong 2, State Commission of Science for National Defense decided to initiate the practical communications satellite project, also known as Project 331. And Ren Xinmin remained the chief designer.

Three Long March 3 carrier rockets and three satellites were arranged for Project 331. After two launches, there was still one rocket and one satellite. To briskly enhance modernization of communication in China, technical improvements were performed on the third satellite.

On February 1, 1986, the Long March 3 successfully lifted off China's first practical communications satellite, Dongfanghong 2A. Hereafter, there were successively four practical communications satellites launched in March 1988, December 1988, February 1990 and December 1991.

The operations of Dongfanghong 2A satellites outperformed the design target, thus the longevity beyond five years on the orbit. They were widely utilized to transmit television programs signals and broadcast programs to foreign countries. Service was provided to departments of communication, broadcasting, water conservancy, finance, education, press and publishing. Contributions were made to the development of domestic economy, national defense, the society, and technology.

Dongfanghong 3 communications satellite was launched from the Xichang Satellite Launch Center at 00:17 on May 12, 1997. Chief designer Sun described it as a new generation of medium-capacity communications satellite developed by Chinese scientists and technicians after nearly 10 years' efforts. There were 24 transponders on board, and the service longevity amounted to eight years. Compared with previous communications satellite Dongfanghong 2A, the number of transponders has increased six times, and the longevity tripled. The capacity of Dongfanghong 3 was equivalent to 12 times that of Dongfanghong 2A. It symbolized a new level that China's communications satellites have reached. Since the designated location was reached, Dongfanghong 3 have provided commercial service across the country from the beginning of 1998. It mattered in the promotion of informationization in China.

The argumentation of Dongfanghong 4 satellite platform took place in the mid-1990s. To adapt to the trend of geostationary orbit satellites toward a longer lifespan and larger capacity, the Chinese Academy of Space Technology, affiliated to Aerospace Science and Technology Group Corporation, embarked on researches.

12 breakthroughs in key technologies, such as overall design optimization, large central bearing cylinder and large capacity storage tank enabled Dongfanghong 4 satellite platform to operate for 15 years while supplying power from 6 kW to 8 kW for payload. The maximum launch weight of the whole satellite could amount to 5,200 kilograms, and most of the technical indexes proved the advancement at international level.

At 00:20 on October 29, 2006, Xinnuo 2, a self-developed new-generation high-power communications and broadcasting satellite was successfully launched by Long March 3 B at Xichang Satellite Launch Center.

Xinnuo 2 was the first to be based on the public platform of Dongfanghong 4. China Institute of Space Technology, affiliated to China Aerospace Science and Technology Corporation, was the main developer. It weighed 5.1 tons, adopted standard broadcasting satellite frequency band, and possessed 22 transponders, a long service lifespan, large capacity, safety and reliability. The design longevity was 15 years while the service longevity was 12. The satellite was utilized by Xinnuo Satellite Communications Co., Ltd. for radio and television, digital television, live television and digital broadband multimedia system in China, Hong Kong, Macao and Taiwan. The operation of Xinnuo 2 significantly benefited our country with the opening up of the international space market, improvement in the capability, security and reliability of national broadcasting, and development in live television broadcasting business.

On November 3, 2006, at the 6th Zhuhai China International Aerospace and Aerospace Exposition, Dongfanghong 4 large communications satellite platform made its first public appearance. It was the center of attention. As advanced as parallel international platforms, it was a new generation of large geosynchronous orbit public satellite platform. As it was independently developed in China, the country owned full intellectual property rights. Installing different payloads required by different users on the platform would constitute satellites for different purposes.

As Dongfanghong 4 matured, there was no deceleration at innovation amid Chinese aerospace workers, who immediately, embarked on the development of Dongfanghong 5 satellite platform.

On February 17, 2017, the China Aerospace Science and Technology Corporation disclosed China's plan to launch six communications satellites in 2017, such as Practice 13, Practice 18, Zhongxing 9A and Zhongxing 9C. Practice 18 was scheduled to be launched by Long March 5, also known as Fat Five, in June.

It was introduced that Practice 18, weighing 7 tons, had adopted quantum communications and new Hall electric propulsion systems, and would test and verify the key technologies of the Dongfanghong 5 satellite platform in orbit.

Dongfanghong 5 is a new generation of self-developed communications satellite platform with the largest carrying capacity. It is characterized by high power, high heat dissipation, long life, scalability and multi-adaptability. Compared with the existing Dongfanghong 3 and Dongfanghong 4 platforms, it has gone through significant enhancements on performance, thus as advanced as the most cutting-edgy communication satellites in the world.

Chinese "Stars" Shine Brightly

What can you not live without in daily life?

Young voices may answer cellphone, the Internet, and TV perhaps.

In fact, what they cannot live without is satellites, as without which, phone, TV, Uber, GPS, or Yelp, would immediately turn useless.

It seems far that they orbit in space. Yet, they have penetrated into every respect of national and social development. Digital age requires their constant support so that everyone's life can carry on.

Two centuries ago, German philosopher Immanuel Kant once concluded that two things profoundly astonish people's hearts; one is the sky full of stars, and the other, sublime morals.

Two centuries later, when we look up to the sky, Chinese "stars" shine brightly, illuminating the divine land of China with blessings. Now let's cast our eyes to remote space and take a closer look at the "Stars" that are crucial to our lives yet neglected.

Fengyun Satellite Series Serve Millions of Families

September 7, 1988 was an ordinary day.

As usual, after the CCTV news broadcast at 7:30 p.m., the Weather Forecast program was aired, nothing special to the general public.

But the staff at National Meteorological Administration took great pride on live air as the weather forecast was conducted with the cloud images captured by our own meteorological satellite, Fengyun 1 (FY-1).

At 5:30 a.m. dawn, China's first meteorological satellite, FY-1, took off from the Taiyuan Satellite Launch Center. At 7:09 p.m. on the video terminal of the data processing system of the Meteorological Center of the National Meteorological Administration was the first cloud picture captured and sent back from space by the FY-1. And it was presented on national television.

This day was epoch-making for meteorology in China. Before that, meteorological data was acquired from polar orbit meteorological satellites of the United States and the geostationary orbit meteorological satellites of Japan. Only when they fly over our territory could we extract data reflecting short-term weather in our country. And since that moment, it all has become the dust of history. China finally has its own meteorological satellite, and the weather forecasting will no longer depend on others.

The development of meteorological satellite in China did not have a late start.

On April 1, 1960, the United States launched the world's first meteorological satellite, TIROS 1. Henceforth, mankind has entered the era with its presence.

It is an application satellite that requires a small investment but generates great benefit. According to statistics, the annual loss resulting from bad weather in the United States in the past amounted to about 10 billion US dollars. And the utilization of meteorological satellites to improve weather forecasting has effectively reduced it by 5 billion, 25 times the annual investment of meteorological satellites.

China is one of the countries with the most menacing natural disasters, which strike in a great variety, at a high frequency, and wide range, thus resulting in massive losses. "We must build our own." On January 29, 1969, Premier Zhou Enlai set the goal with great foresight and approved the task of developing meteorological satellites in February 1970. In March 1972, the development of meteorological satellites was formally incorporated into the national plan.

However, for many reasons, the development stagnated for many years.

Owing to the reform and opening up in the 1980s, the huge demand of national economic construction and social development has accelerated the development of meteorological satellites. Ren Xinmin and Sun Jiadong had successively been in command. They led the vast number of aerospace scientists and technicians to persevere in tackling key problems and strive to catch up. With innovation, a new path was created.

As an important symbol of China's meteorological modernization, satellite FY-1 is a square box with a weight of 750 kilograms, a height of 1.2 meters, a length of 1.4 meters and a width of 1.4 meters. It circles the earth every 102 minutes, scans 2,800 kilometers wide and covers the earth every 12 hours. Following the United States and the Soviet Union China has become the third country with polar orbit meteorological satellites.

FY-2, China's second generation of meteorological application satellite, was launched at 20:01 on June 10, 1997. Unlike the two FY-1 satellites in 1988 and 1990, it is on a geostationary orbit. With a broader vision, it can cover, taking China as the center, about 100 million square kilometers of the earth's surface while observing and providing meteorological dynamics, such as cloud maps, temperature, water and gas, and wind fields in China and neighboring countries. It plays an important role in giving accurate medium- and long-term weather forecast and disaster predictions.

FY-3, a new generation of polar orbit meteorological satellite in China, was successfully launched at 11:02 a.m. on May 27, 2008. Two years later, the second FY-3 blasted off. Five years later, the third entered orbit smoothly.

FY-3 is the second generation of polar orbit meteorological satellite in China. It can carry out all-day, multi-spectral, three-dimensional and quantitative detection worldwide. It mainly provides meteorological parameters for medium-term numerical weather forecasting and monitors large-scale natural disasters and ecological environment. Meanwhile, the meteorological data it presents supports the study of global environmental change, the exploration of global climate change, and activities in aviation and navigation. provide meteorological information. After the three satellites were networked, the time to update observation data has been halved from 12 hours to 6, which massively enhance China's meteorological observation and medium-term weather forecast.

The launch of FY-4 meteorological satellite was at 00:11 on December 11, 2016. It is China's first new generation of geostationary orbit meteorological satellites. The overall performance is parallel to or beyond that of the latest generation of geostationary orbit meteorological satellites in Europe and the United States. At one stroke, FY-4 has taken a significant leap from "running behind" to "running in the lead."

How impressive is FY-4?

Five hundred pictures of lightning can be taken in one second. Detection of frequency and intensity of lightning in a region became achievable, and the first issuance of lightning warning was realized in China; at the altitude of 36,000 kilometers from the earth, the observation of weather change is effortlessly done, and so is a precise vision of any spot that it wants to see; zero errors would occur in merging images; the life span is designed to reach seven years; 24 hours gazing at the earth can be implemented and the inspection of the typhoon area takes place every 3 minutes.

Yu Xinwen, General Commander of FY-4 project, rhapsodized: "FY-401 has penetrated through a world conundrum, the mutual interference among multiple instruments working simultaneously. And there was the effective installment of four kinds of loads on the same satellite platform, and the realization of the first comprehensive observation of the atmosphere by multiple means in the world."

On February 27, 2017, the National Defense Science and Technology Bureau and China Meteorological Administration jointly released the first batch of images and data acquired by FY-4. It served as solid proof to the normal operation and proper collaboration between FY-4 and ground application system, thus marking that China's geostationary orbit meteorological satellite has been successfully upgraded and replaced.

Tian Yulong, chief engineer of the National Defense Science and Technology Administration and secretary-general of the National Space Administration, introduced that FY-4, as a new achievement of independent innovation in China's space science and technology, has made a series of major technological breakthroughs. Compared with the first generation of satellite observation system, the time resolution of observation has doubled, the space resolution up by six times, and the ability to observe atmospheric temperature and humidity up by thousands of times, the amount of whole satellite observation data up by 160 times, and the number of observation products up by three times.

Today, China has successfully sent up 15 meteorological satellites (8 in orbit), becoming the third country in the world to have both polar and geostationary orbit meteorological satellites in operation in space after the United States and Russia. And the engagement pattern is formed as a combination of multi-satellites in orbit, integrated operation, mutual backup, and timely encryption.

Over the past decades, through the unremitting efforts of aerospace scientists and meteorological scientists, Fengyun series satellites experienced from 0 to1, miniature to mammoth, and feeble to robust. Today, they provide services to more than 2500 users in more than 70 countries and regions around the world, and support applications in many industries and fields.

What are FY series capable of?

The staff of the China Meteorological Administration explains that they undertake four tasks. First involves daily weather forecasting, such as temperature, humidity and precipitation; secondly, it monitors large-scale meteorology and its derivative natural disasters and changes in ecological environment, such as snow and ice coverage, fog coverage, and drought scope; next, global environmental change is under monitor and geophysical parameters are provided to support the study of the law of global climate change, climate diagnosis and prediction. The fourth is to provide global and regional meteorological information for aviation, navigation, agriculture, forestry, ocean, hydrology, and other areas concerning national economy.

The depiction may fail to leave a deep impression. Another way may do the trick.

Since 1988, the National Satellite Meteorological Center has fed tons of fire information to forest and grassland fire prevention departments annually and monitored all major fires throughout the process. On February 28, 2011, FY-3A and 3B jointly spotted the fire in Lijiang, Yunnan Province, at 12:00 and 14:00, respectively. And the in-time report of the situation offered a hand in the putting-out.

As of 2015, since FY-2 was put into operation in 1998, 168 typhoons landing or affecting China have been monitored with full coverage without missing any, which contributed greatly to reducing casualties and property losses.

Over the years, when natural disaster struck, such as the flood in the Huaihe River in 2003, the forest fire in the Great Hinggan Mountains in2006, extremely freezing weather in early 2008, and the severe drought in southern China in 2009, the high-hanging meteorological satellites have been a helpful friend.

From the Beijing Olympic Games to the 60th anniversary of the founding of the People's Republic of China, the Shanghai World Expo to the Guangzhou Asian Games, the Wenchuan and Yushu Earthquakes to that of Zhouqu Debris Flow ... The meteorological support of the Fengyun Series has contributed a lot to the accomplishment of the important tasks in the center of worldwide attention. Experts reveal that Fengyun series are the most effective and widely used civil remote-sensing satellites in China. Preliminary calculation estimates that the input-output benefit ratio of meteorological satellites exceeds 1:40.

Over recent years, in addition to serving the motherland, the benefits of Fengyun series have transcended national boundaries to all mankind. Nearly 100 countries and regions receive meteorological data, part of which is extracted from Chinese meteorological satellites and used by meteorological, environmental departments, universities and research institutes in the United States, Europe, Australia, Japan, Singapore, Malaysia, Vietnam and other countries and regions.

A situation of tripartite confrontation came into form among China's Fengyun series and American and European meteorological satellites. At present, the meteorological forecast in the Eastern Hemisphere mainly relies on China's meteorological satellites for relevant information. The World Meteorological Organization (WMO) has incorporated the Fengyun series into the meteorological satellite sequence of global operational application. They have gained the membership in Global Integrated Earth Observing System (GEOSS). Together with European and American meteorological satellites, an observation network has been created for the earth's atmosphere, ocean, and surface environment.

On July 4, 2016, the National Bureau of Defense, Science, Technology and Industry and China Meteorological Administration jointly convened the symposium on the development of FY satellites. At the press conference, Wu Yanhua, deputy director of the Bureau, announced that by 2025, China has planned 14 atmospheric observation satellites: one FY-2, four FY-3, three FY-4, two precipitation measurement radar satellites, one solar terminator orbit satellite, one geostationary orbit microwave detection satellite, one high precision greenhouse gas integrated detection satellite, and one atmospheric environment monitoring satellite.

Knowledge of global meteorology benefits millions. By 2030, China's FY meteorological satellites will have achieved network observation as the third generation of solar synchronization and geostationary meteorological satellites, therefore forming a global observation system.

In the starry sky, FY satellites are the space eyes of Chinese. And they will weaken the mystery in meteorology in both China and the world in the upcoming future.

"How's the weather tomorrow?' "What's the index of PM 2.5?" "Good for outdoor activities?"

When we check the weather trends with great ease on the phones, let's not forget the FY series, and more generations of devotees who sacrificed their youth for the FY projects over the 4 decades.

Ocean Satellites Appreciate the Blue

On September 11, 2012, CCTV broadcasted the marine environmental forecasting near Diaoyu Island at 13:57. And after the news program at 19:30, the weather forecast for the island and waters in the perimeter was released.

From then on, changes of the natural world of Diaoyu Island and its adjacent waters have been exposed to public view.

Several foreign presses commented that this action was more than an ordinary marine environment forecast, but another act of declaring sovereignty after China declared the territorial sea baseline of Diaoyu Island."

This broadcast was the masterpiece by China's ocean satellite as well as a highlight of its capability.

In childhood, we were asked: what is the size of our motherland?

There is the common answer blurt out: 9.6 million square kilometers. Growing up, we realize inaccuracy in the answer as the blue territory is neglected.

China has 18,000 kilometers of coastline, over 2 million square kilometers of shallow sea continental shelf, more than 6,500 islands, and 3 million square kilometers of jurisdictional sea area. There is abundance in biological resources, port resources, mineral resources, oil and gas resources, and tourism resources.

For ages, however, the oceans have been neglected. Naturally, this results in the neglecting of the development and launch of ocean satellites.

In 1978, the United States launched the world's first ocean satellite. Afterwards, Japan, Canada, Russia and EU countries have successively joined the club.

The oceans cover 71% of the earth's surface. To develop, utilize and protect independent blue territory more effectively, countries have vigorously developed and applied ocean satellites as a space technology. And exploration and patrol are better conducted.

Remote-sensing of ocean satellites cannot be replaced by other means of ocean survey and observation. During the 105-day effective operation, the first experimental ocean satellite in the United States obtained data of wind direction and speed of global sea surface equivalent to the sum of all ship observations since the 20th century.

Compared with the world, the development of China's ocean satellite had a late start, but the starting point was high and the pace fast.

On May 15, 2002, China's first ocean satellite, Haiyang 1A (HY 1A), was launched from the Taiyuan Satellite Launch Center. Henceforth, the absence of ocean satellites became history, marking that China has gained a seat in the club of advanced ocean satellite remote-sensing.

On April 12, 2007, HY 1B was successfully launched. Bai Zhaoguang, chief commander and chief designer introduced that compared with HY 1A, it provided more than three times more information due to longer life span and excellent performance. The exploration scope could cover the whole globe while the utility increased exponentially. The effective operation of HY 1B indicates that China's ocean satellites and its marine applications have made great progress towards serialization and scale. And based on warships, buoys, coastal stations and satellites, China's ocean stereoscopic monitoring system has stepped up to a new level.

On August 16, 2011, HY 2, China's first ocean dynamic environment observation satellite, was launched from the Taiyuan Satellite Launch Center. According to personnel in charge of the National Defense Science and Technology Bureau, it is capable of a relentless detection of ocean dynamic environmental data, including wind, wave, current, tide and temperature in all weather, all-time and global environments. It will further perfect China's three-dimensional marine monitoring system and propel the development of space and marine undertakings. In the meantime, it ends the monopoly of Western countries on marine dynamic remote-sensing data and strengthens China's international competitiveness.

The technological breakthrough of HY 2 has astonished the world: the precision of its satellite orbit determination has reached 2-3 centimeters, and technique precision for tracking and orbit determination of China's ocean satellites has evolved from tens of meters to centimeters. In one swoop, China achieved what was achieved internationally after 3-4 decades using only one tenth the time.

And a bigger surprise is that the team who accomplished a series of technical breakthroughs on HY 2 turned out to be so youthful.

Chief designer Zhang Qingjun was only 38 the project began. The average age of the chief designers of each subsystem was 36. The chief designer of comprehensive test was 29 years old and that of load 31 years old. The "post-80s" account for 43% of the total research team.

Ocean satellites appreciate the blue from above. The person in charge of the National Satellite Ocean Application Center revealed that by 2020, China will have launched eight ocean satellites to enable environmental remote-sensing monitoring of the marine color environment and dynamic environment in all jurisdictional areas of China and even of the globe.

The *Business Development Plan of Land and Sea Observation Satellites* details that the eight ocean satellites include four ocean color satellites, two ocean dynamic environment satellites and two land and sea radar satellites. While enabling remote-sensing monitoring of the marine color environment and dynamic environment in all jurisdictional areas of China and even of the globe, it will also consolidate the monitoring of the waters near Huangyan Island, Diaoyu Island, and all islands of Xisha, Zhongsha and Nansha archipelago.

Gaofen-1 Satellite: Smart eyes

Grim was the day of March 8, 2014.

Airliner MH370 vanished in the dark night. It was gnawing pain to have the life or death of 239 passengers, including 154 Chinese, undetermined.

The world united to search for the airplane that had lost contact. In the unprecedented international scour, China allocated 21 satellites, including Gaofen-1, to forage about the suspicious marine space. The result was notified to 25 countries.

Australia expressed at a press conference on the progress of MH370's search and rescue that the images captured by Gaofen-1 were "very valuable and helpful."

A month later on April 17, China's Gaofen-1 and FY-3C officially were officially appointed for charter duties at the 31st International Charter on Space and Major Disasters.

Therefore, Gaofen-1 stood in the center stage of world attention.

On August 3, 2014, an earthquake struck Yunnan province. In the relief work, Gaofen-1 worked its mojo again. An urgent observation was given to help the rescue.

Regardless of high exposure of Gaofen-1 on television and the Internet, little did the public know the Gaofen series.

On April 26, 2013, it blasted off at Jiuquan Launch Center, as the country's first low-earth-orbit remote-sensing satellite whose estimated lifespan exceeded five years. Thus, a new epoch of ground observation unfolded. As the space photographer, Gaofen-1 can lay eyes on the bikes on the ground, and within 4 days, scan the whole earth.

It is also the first satellite of the ground observation system with high resolution of China's significant earmarked science project, one of the significant earmarked projects validated in the *Outline of the National Medium-and Long-Term Science and Technology Development Plan (2006–2020)*. The projects consist of space-based observation system, near-space observation system, aerial observation system, ground system, and application system. The plan involves the launch of 5–6 observation satellites during the implementation of the 12th Five-year Plan in the purpose of the establishment of a ground observation system with high space resolution, high time resolution, and high spectrum resolution. As of 2020, in tandem with other observation means, a stable operating system shall have come into shape. It shall be capable of the observation throughout all day, all weather, and all the globe.

When praises were still being shouted at the Gaofen-1, Gaofen-2 took off.

On August 20, 2014, the Long March 4B carried it to the intended orbit.

The eyes are sharper. Resolution precision reaches one meter. During in-orbit probation, effective support was provided to Yunnan Ludian Earthquake, Jinggu Earthquake, Sichuan Kangding Earthquake, Chile Earthquake, Indian Debris Flow, and Beijing APEC Conference.

On December 29, 2015, Gaofen-4, as the closing number of space launches of the 12th Five-year Plan, blast off at the Xichang Satellite Launch Center.

By far, it has been the high-earth-orbit remote-sensing satellite with the world's best eyesight able to acquire 50-meter resolution visible light and 400-meter resolution mid-wave infrared remote-sensing data. Commonly, in geosynchronous orbit about 36,000 kilometers away from the earth, its big eyes can see tankers sailing in the sea. At present, mostly hundreds of meters, or even thousands is the spatial resolution of high-earth-orbit remote-sensing satellites around the world.

What team created the perplexed eye in the sky?

Unbelievably, the average age of the research team was merely 31. In four years, they built Gaofen-4.

In tandem, three Gaofen satellites form a supplementary observation system with great efficiency. Yu Dengyun, Deputy Director of Science and Technology Committee of China Aerospace Science and Technology Group and Chief Engineer of Gaofen-4 explained "As Gaofen 1 and 2 circled on low orbits to patrol the earth for a closer look, Gaofen-4 operates at a high orbit to gaze below. It can observe a designated area continuously with great time resolution. Unlike earth-circling satellites, it does not take it every a few hours or days to observe the same area again.

On August 10, 2016, Gaofen-3 was successfully launched at the Taiyuan Satellite Launch Center. As the only radar satellite in the special Sky Eye Project, its ability gets even better.

First of all, it is a "bat satellite" able to penetrate through clouds at dark nights. Compared with other optical remote-sensing satellites in the Gaofen series, it was mostly characterized by all-day and all-weather imaging. Day or night, sunny or stormy, imaging works. Statistics tell about 70% of disasters in China are earthquakes, floods, and debris flows. When typhoons, storm surges, earthquakes, landslides, floods, and other natural disasters strike, bad weather is inevitable, therefore the difficulty for optical remote-sensing satellites to conduct any detection. The sharpness of the eye is often reduced drastically. "Gaofen-3 can stab through clouds, fog, rain, snow, and smog, to conduct microwave imaging. And first-hand information is quickly provided" affirmed Jiang Xingwei, director of the National Satellite Ocean Application Center of the State Oceanic Administration and deputy director of high-score special application system.

Next, it is a "diverse satellite." Gaofen-3 is a synthetic aperture radar satellite with the most imaging modes, 12.

It not only contains the traditional imaging modes of strip and scan, but also switches freely in the modes of spotlight, strip, scan, wave, global observation, and high and low incidence angle. Earth and ocean observation are both achievable. One satellite for multiple uses. Yu Weidong, deputy director of satellite system payload of Gaofen-3, Institute of Electronics of Chinese Academy of Sciences, introduced the spatial resolution ranges from 1 m to 500 m, and the width from 10 km to 650 km. A large scope of observation can be conducted, maximum seeing images in the radius of 650 km at a time, and roads, general buildings and ships on the sea can be distinguished.

Thirdly, it is a "long-life satellite." The designed lifespan is eight years, longer than the usual three-five of previous Chinese satellites, and six-seven of the remote-sensing satellites from other countries.

Sky eye in the sky overlook the divine land of China. Gaofen satellites have a wide range of applications, including land survey, environmental protection, emergency relief, resource development, agricultural output estimation, and urban planning. Nowadays, the they are quietly changing our lives:

—an enhancement in accuracy and efficiency of land surveys, timely and effective tracking and monitoring land resources, and simultaneous monitoring the environment in a large area to identify the smog in many cities.

—in the process of rapid urbanization, Gaofen provides data to support a feasible plan. Every year, the Ministry of Land and Resources demands data concerning 19 million square kilometers from Gaofen satellites. It has gradually replaced nearly 80% of the data from foreign satellites with the same resolution.

—an acute sense to discover where mining in violation of regulations is undertaken. In 2015, 192 key mining areas were dynamically monitored by Gaofen satellites data, and remarkable results were achieved. The real-time observation "gaze" of Gaofen-4 is expected to play an important role in early warning of forest fires, detection, and extinguishing thanks to its relentless monitoring.

At present, the promotion of Gaofen satellites' special application has been totally initiated. Twenty-one provincial administrative regions have established Gaofen data and application centers, and achievements have been harvested in land, environment, surveying and mapping, agriculture, forestry and other industries.

Quantum Satellites Rewrite the Future

January 18, 2017 was another memorable day for China's space industry.

On this day, Micius, the world's first quantum satellite, was officially delivered for use after completing the four-month on-orbit test mission.

The project began in 2011. In 2016, the launch succeeded. And it was put into operation in early 2017. The fact that the transformation from design drawings to Micius satellite was completed in six year left the world in awe.

"Compared with western countries, our science has been the chaser for so long that many formed the idea that this is the normal state. And it feels awful. So, we named it Micius," said academician Pan Jianwei, chief scientist of the Quantum Satellite Project. Micius first proved that light travels along a straight line in an experiment and put forward the concept of particles and the rudiment of Newton's law of inertia, for which he was titled a Saint of Science. "We would like to remind people of the fact that Chinese are also good at science. We were, are, and will still be."

The great prospect of quantum communication and computing lures every major country to throw millions in the development of related technologies. The European Union announced in April 2016 the investment of 1 billion euros in support of quantum technology, the third flagship research project in the EU after graphene and brain science. Thomas Carrarko, one of the project leaders and head of the German Centre for Integrated Quantum Science and Technology, revealed that when the EU determined the first two flagship projects, quantum technology was not as mature as now because that was before Edward Snowden disclosed the large-scale surveillance activities in the United States. Now Europe demands more secure communications technology.

The White House also tweeted at the end of July 2016, citing the July report *Promoting Quantum Information Science: Challenges and Opportunities for the Nation* of the National Science and Technology Commission. An increase in the investment in quantum information science was suggested. In fact, several US governmental departments, such as National Defense, NASA, National Institute of Standards and Technology (NIST), have long made massive investments in quantum information technology. Over the past decade, NIST has produced five Nobel Prize winners of Physics in quantum regulation and quantum information.

High-tech companies are also throwing millions at quantum technology. For example, Google, together with NASA and the University of California, Santa Barbara, established the Quantum Artificial Intelligence Laboratory and launched the Quantum Computer Development Program; in 2014, IBM announced an investment of 3 billion US dollars in some areas focusing on quantum computing over the next five years; and some relatively small technology companies are budding, such as Battele, which is running the first commercialized quantum cryptographic communication network in the United States.

"I believe the importance of quantum technology in the 21st century is paralleled to the Manhattan Plan of the 20th" commented academician Pan Jianwei, director of the Center for Quantum Excellence and Innovation of the Chinese Academy of Sciences and chief scientist of the China Quantum Satellite Project. Quantum technology could change the world pattern just as the Manhattan Plan did with atomic bombs.

In the new round of the world science and technology race, China took the lead. At the end of 2016, the well-known academic journal Nature labeled Micius Quantum Satellite and human beings' first detection of gravitational waves as the international science event of the year.

By far, Micius has set up the experimental platform of space-earth integrated quantum communication first in the world. It enables the conducting of space quantum experiments. Academician Pan introduced that the research team is currently carrying out experiments in an orderly manner and has obtained preliminary data. The team is confident of successfully completing all the preset experiments, including satellite-ground quantum key distribution, wide-area quantum key application demonstration, satellite-ground two-way quantum entanglement and satellite-ground quantum teleportation.

According to China's plan of quantum communication, after the launch of quantum satellite and the establishment of Beijing-Shanghai trunk line of quantum communication at the end of 2017, a wide-area quantum communication system was initially formed in China. By 2030, China will have become the first country to build a global quantum communication network.

In June 2017, the National Development and Reform Commission issued Development Plan of Urban Agglomeration in Yangtze River Delta. There is the outline of the blueprint for the practicality of quantum communication and the proposal to speed up the construction of quantum communication network in major urban areas of urban agglomeration and build a wide-area quantum communication network there. The proposal also advices to actively engage in Project Beijing-Shanghai Trunk Line of Quantum Communication, promote the use of quantum communication technology in Shanghai, Hefei, Wuhu, and other cities, and popularize the application of quantum communication technology in governmental departments, military, and financial institutions.

"With the rapid development of science and technology in China, I believe that quantum communication satellites will benefit millions of households in about 10 years. I hope that in my lifetime, I will witness the birth of quantum internet with quantum computing as the terminal and quantum communication as the security guarantee." Academician Pan Jianwei confidently said, "I believe Chinese scientists can do it."

CHAPTER 3

The Long March Went Global

On the bank of Yudu River, moonlight poured down. In serenity stood carrier rocket Long March 5.

The night of October 17, 1934 was also a night when moonlight resembled water. The Red Army, in torches, crossed Yudu River and traveled far. And the journey lasted three years.

It never occurred to the soldiers of the Red Army that in straw shoes, their feet would walk a path of faith and victory, known as the Long March.

To their great surprise, decades later, the name Long March was used on a domestic carrier rocket,

"Ignite." On the 80th anniversary of the Long March on November 3, 2016, China's new carrier rocket Long March 5 blasted off on the Wenchang Space Launch Site, in Hainan. It was a critical step in the upgrade from a country able to undertake aerospace missions to a powerful country adept at them.

Fifty-seven years ago, a Long March rocket took off from the middle of nowhere. From the altitude of 8 kilometers, each flight soars higher and higher. Eventually, it stands in the center of the world stage.

The success of Long March 5 is the 238th flight of the entire series. The number, often neglected at celebration, is the most convincing proof to the fact that Long March is one of the best carrier rockets in space flight around the globe.

The Long March by the Red Army has become history.

Yet the Long March in China's aerospace undertakings continues.

Long March rockets keep soaring.

Long March 1, 2, 3, 5, 6, 7... The series of rockets are flying towards further space.

On February 28, 2017, at the Australian International Air Show, the joint appearance of China's new carrier rockets Long March 5 and 7 became the center of attention.

Today, Long March rockets have extended to the world, becoming well-known "gold medal" rockets at home and abroad as well as a "beautiful business card" of China's aerospace industry. Praise is given at every mention of Long March rockets.

There is no smooth sailing in the achievements at science and technology. The success of Long March rockets is no exception. Yet, the failures along the way are either little known or totally forgotten.

Time heals the "scars" on Chinese aerospace workers. However, the remembrance of failure is key to success. This is very important for the vigorous development of China's space industry and for the new generation of young workers who are creating a new glory today.

Right now, let's open these dust-laden memories and go through the ups and downs of the Long March rockets.

Love at First Sight

April 8, 1984 is a glimmering milestone inf China's space exploration.

On this very day, with great thrust, carrier rocket Long March 3 sent China's first communications satellite Dongfanghong 2 into a 36,000-kilometer geosynchronous orbit.

It set the Thames on fire. Following Dongfanghong 1, it astonished the world.

Immediately, the spotlight was cast on Long March 3 as the world's most advanced, hydrogen-oxygen engine, was adopted the launch. China was the only country, in addition to the United States and France, to have harnessed the technology.

The praise lasted for quite a while from mainstream foreign press that the successful launch of Long March 3, as a milestone in China's space flight history, which demonstrated the rise to prominence and power of Chinese rockets.

Faced with repeated flattery, China was not lost in wild joy. Instead, a bold idea burst forth after a scrutiny of the words.

Now that the Long March 3 had received high recognition globally, how about using it to send up foreign satellites in the expansion of international market of satellite launches?

To some extent, the public opinion around the world concerning Long March 3's triumph pushed its step forward to the global stage.

Another push came from the profound changes in the domestic situation.

In 1984, reform and opening up picked up the pace and the prosperity of economy became the core goal of the nation. In that summer, at the enlarged meeting of the Central Military Commission held in the northern coastal city of Beidaihe, Deng Xiaoping pointed out "No war shall break out in five years, thus no need to be prepared against war and natural disasters."

To this end, the army's strategic policy had been greatly adjusted from the state of imminent combat to the track of peacetime construction. At once, the State Council and the Central Military Commission put forward new ideas for the transformation of military products into civilian products for the National Defense Science and Technology Commission and the Ministry of Space. In addition to the projects listed as national key projects, the country will no longer allocate special funds to finance the costly army of space flight. The secure jobs for the Chinese aerospace workers are no longer secure.

Following the instructions from the Central Committee, to adapt to reform and opening up, the Ministry of Space has put forward the policy of a "transition," that is, from planned economy to market economy, from trial to commodity, and from military products to civilian products.

Simply put, we are on our own.

Now, a huge opportunity lies ahead for Chinese aerospace workers – to break the Long March rocket into the international market. Fame and profit are to be earned.

The Ministry of Space soon took action. They recruited people from different units and formed a group of ten, led by Vice Minister Liu Jiyuan, to explore the international market. They were responsible for the preparations for early market development.

At the end of 1984, a formal report on the Long March 3 entering the international market was presented to the leadership of the National Defense Science and Technology Commission.

General Zhang Aiping was excited about the news. It is recalled that on that day he got up from the rattan chair with a rattan cane, crying "Great! It is a matter of building an international reputation and military influence to have Chinese rockets in the international market. We must have it done. Now that we are equipped with the skills, we shall dare to compete, to challenge, and to strive forward and break free from restrictions."

On April 22, 1985, at the International Symposium on Space Technology, China released its first exploratory "balloon" to the West, titled "China's Space Activities and the Possibilities of Providing International Services." This speech aroused great interest amid all attendants.

In June 1985, the Chinese delegation participated in the International Aerospace Exposition held in Paris. Wu Keli, head of the delegation, delivered a message to journalists from all over the world that China's Long March rockets planned to share the international space market. Meanwhile, at the International Space Conference held in Geneva, Chen Shouchun, head of the delegation of China's aerospace industry, also announced that Long March rockets would be available in the market to undertake foreign satellite launches.

Chen made a tempting offer that Long March 3 would charge 50% less for the launch service than peers. There was no intention to gain enormous profit. And the materials and human labor were much cheaper in China.

On October 27, 1985, Li Xu'e, Minister of Space Affairs, on behalf of the Chinese Government, officially announced to the world through Xinhua News Agency that China's self-developed Long March 3 and Long March 2 will be available for contracting foreign satellite launching at preferential prices.

Hereafter, the door to China's space flight was officially wide open.

Unexpectedly, the first order soon came. In November 1985, the Swedish Space Company visited China and requested a tour of the rocket assembly workshop. And it was love at first sight. After brief talks, the two parties signed a letter of intent to launch a Swedish postal satellite with the Long March 2. On January 23, 1986, China and Sweden signed a launch booking agreement in Stockholm.

It was the first foreign launch booking agreement. Immediately, *People's Daily* published the news from Stockholm.

Long March Rockets Against Adversity

1986, the year of disaster in human space flight, yet the year of opportunity for Long March series to enter international market. On January 28, the US space shuttle Challenger exploded 73 seconds after liftoff. At 10,000 meters, it turned into a fireball, killing all seven astronauts, including a female astronaut Christa McAuliffe, who used to be a teacher. The world was in deep grief.

The crash of Challenger seemed to have opened Pandora's box. Failure tailed.

On February 22, 1986, the French Ariane rocket launched a Swedish satellite. It exploded in the air, causing a loss of more than $100 million. Prior to that, in 1985, the Ariana rocket had already suffered a defeat.

On April 18, 1986, rocket Titan was launched at California Air Force Base. A huge explosion happened shortly after liftoff, destroying launch platform and the hundreds-of-millions-of-dollar communications satellite and also injuring 58 people.

On May 3, 1986, rocket Delta of McDonnell Douglas Company took off with a meteorological satellite worth more than 50 million US dollars. After 91 seconds, the flight attitude lost control. Ground controllers were forced to execute self-destruction orders. It burst into a ball of flames.

The failing rockets were the main forces. The series of accidents left satellites launches of those countries in total chaos.

A long investigation was required to locate the problem in the crash of Challenger. In the meantime, all space shuttles were suspended. NASA disclosed that it was not until the first quarter of 1988, the suspension would end.

Titan would not be cleared for flight in two years at least. McDonnell Douglas Company believed that it would take no less than a year and half for the recovery to complete. The French Ariane had a credibility crisis after two consecutive fiascos.

Scores of satellites awaited launches in a great shortage of rockets. Supply failed to meet demand. Under this circumstance, the spotlight was naturally cast on China, whose Long March 3 delivered perfect performance.

On February 1, 1986, it sent up another communications satellite to space. The contrast between foreign failure and Chinese success was so obvious that Long March 3 won unprecedented favor.

In this case, China struck the iron while it was hot:

At the 23rd Scientific and Technological Subcommittee of the United Nations Committee on Outer Space, the Chinese delegation stressed in particular China's position. Outer space, as the common wealth of mankind, should be used exclusively for peaceful purposes and for the benefit of all mankind. China is willing to put Long March 2 and 3 into the international market and make positive contributions to the development of international space cooperation and the development and utilization of outer space.

Next, China's Long March 3 made an official appearance at the International Space Conference held in Washington, USA. Representatives from dozens of countries, such as the United States, France, Britain and Australia, showed great interest. Shortly after, 17 space companies in the United States, Canada and other countries have established business contacts with Great Wall Corporation of China.

In July 1986, the "Request for Certain Issues Concerning Launching Foreign Satellites" report drafted by the National Defense Science, Technology and Industry Commission and the Ministry of Space was submitted to the State Council and the Central Military Commission.

On September 4, official approval was issued, agreeing to list the launching of foreign satellites as a national key project. As the original report was drafted in July 1986, it was named "867 Project," which was organized and implemented by the National Defense Science and Technology Commission, and required the support and cooperation of relevant departments throughout the country.

At the conference, General Zhang Aiping shouted: "Launching foreign satellites is a national honor and a matter of national prestige and interests. No failure is accepted."

This was not only General Zhang's personal words, but also the voice and oath of Chinese aerospace workers. Nobody expected the road to the world stage would be of tremendous adversity.

AsiaSat 1: Triumph at One Stroke

Opportunity does not wait.

The positive situation for Long March Rockets to go global took a turn for the worse in 1988.

Flight plans of American space shuttles resumed earlier than expected. They picked up commercial launches again. America's Titan rockets and Delta rockets undergone technical enhancement. French Ariane rockets recovered to the best state and had seven consecutive launches.

The excellent performance empowered Ariane rockets to occupy 50% of the international market of satellite launches. Three years of booking contracts had been signed ahead with many countries. And the commercial profit amounted to 15 billion francs.

Tempted by the huge profits in the international market, Japan proactively developed a large H-2 launch vehicle with a 2-ton launching capability in an attempt to get a piece of the cake. They also set up a satellite launch service office in Washington, D.C., at an alarming rate.

The Soviet Union's Proton Rocket has also taken action, striving to enter the international market.

As if in a track and field race, many competitors suddenly accelerated forward and left behind the Chinese Long March rockets, which had been running side by side.

The sudden changes caught China off guard. Bad news flooded in.

The bankruptcy of Trey Satellite Company spread rapidly and annulled the launch contracted signed with China.

Next came the bankruptcy of West Union Satellite Corporation and the annulment of the draft agreement of collaboration.

Successively, all launch contracts and booking agreements between China and foreign countries were cancelled.

Cruel was the market. Foreign satellite companies were still skeptical about China's Long March.

For a while, it was a dilemma for China's space flight.

What now? In great adversity, the stable performance of Long March 3 made an effective self-promotion:

On March 7, 1988, Long March 3 successfully launched the newly developed communications satellite Dongfanghong 2A 01 into space. On December 22, it sent up Dongfanghong 2A 02.

Every cloud has a silver lining. After meticulous investigation, Asia Satellite Telecommunications Holdings Limited, which was co-invested by Cable and Wireless of Britain, China International Trust and Investment Corporation and Hutchison Whampoa of Hong Kong and registered in Hong Kong, finally chose to launch AsiaSat 1 with Long March 3 to provide better communication services for the Asia.

A launch contract was signed between two parties. Yet, it was merely a small step taken.

It was Hughes Global Services an American company that manufactured AsiaSat 1. According to international law, the launch of satellites developed by the United States requires an exit permit issued by the US Government. Without that, regardless the contract signed, China is prohibited from launching it.

The obtainment of the exit permit became a barrier that must be overcome. And China took the initiative.

At a press conference, Lin Zongtang, Minister of Aerospace Industry, reiterated the basic policy of the Chinese government on foreign launches: when foreign satellites are sent to China for launch, their technical security can be guaranteed, and China does not seek any technical confidentiality; China's foreign launch service is only a supplement to the world launch service market, and will never pose a threat to European and American launch service providers.

The US government held an indifferent attitude as usual. On September 2, 1988, Winston Lord, the U.S. ambassador to China formally expressed to China that the premise for the issue of an exit permit was that the two governments must first reach an agreement and understanding in negotiations on launching U.S. satellites.

Therefore, a tough negotiation commenced between two parties. The head of the Chinese negotiating delegation was Sun Jiadong, a satellite expert. That of the U.S. negotiating delegation is McAllister, Assistant Secretary of State, a diplomat who had expertise in politics and professional negotiation. However, Sun's wisdom on the negotiating table amazed his rival McAllister. He accused that China's satellite launch price was not normal, but a market dumping subsidized by the government. Sun rebutted it was because Chinese labor force was much cheaper; the monthly income of an ordinary worker in the States should be $3,000–4,000, while that of one in China RMB 100; when the hundred-of-times difference was applied in labor-intensive rocket manufacturers, the cheaper charge would be no wonder at all.

The bargain came back and forth. Sun recalled that the negotiations, as national sovereign interests were involved, were very difficult.

The first negotiation took place in Diaoyutai of Beijing from October 18–22 1988.

The second was held in Washington D.C. from November 26–30 that year.

The negotiations fruited three agreements signed between both parties: Memorandum of Agreement on Satellite Technical Security, Memorandum of

Agreement on Responsibility for Satellite Launch and Memorandum of Agreement on International Trade in Commercial Launch Services. It was another big step China took to enter the international market.

To shield China's space activities under international law, the Standing Committee of the National People's Congress approved China's accession to three outer space treaties, namely, the Convention on International Liability for Damage Caused by Space Objects, the Agreement on Rescue of Astronauts and Objects Launched back into Outer Space and the Agreement on Registration of Objects Launched Outside Space in November 1988. China gained membership at the international space flight club, a legal assurance to the entry to international market.

Hardly had one wave subsided when another rose.

At the turn of spring and summer in 1989, the U.S. government suspended the issuance of satellite exit permits to China. A ban put the launch of AsiaSat-1 in jeopardy. On the one hand, China proactively attempted to make negotiations. On the other hand, preparations were performed according to plan: the newly built comprehensive satellite testing plant in Xichang Launch Center was about to complete; Long March 3 was demanding extra hours on the production line to catch up with the progress.

Eventually, on Christmas Eve 1989, President Bush nodded to the launch of AsiaSat-1 with Long March 3. At 12:40 on February 12th, 1990, a special plane loaded with AsiaSat-1 landed at Xichang airport. The final stage came, and the closer to the end, the bigger the pressure.

From February 12th, when the rocket entered the launch site to the launch day, Long March 3 underwent a series of quality checks, such as unit test, horizontal test, vertical test, satellite-rocket joint test, etc. Over 200 on-board instruments in the unit test were detected any significant defects. To Xie Guangxuan, chief designer of Long March 3, chief scientist of the Hughes Global Services and chief designer of AsiaSat 1 said: "My satellite is reliable and so is your rocket. The excellence of Chinese expert put me at ease. We will make it."

During the preparations, disturbing news spread one after another:

On February 22, Ariane rocket failed at launching a Japanese communications satellite; on the 28, a reconnaissance satellite sent by an American space shuttle exploded; on March 14, a Titan rocket disappointed an international communications satellite.

Meteorology stood in the way of success. Stormy weather was also frequent at Xichang Launch Site. China, America, and Asia Satellite Telecommunications Holdings Limited eventually set the date of launch as April 7 after serious considerations. Yet, luck did not side with us that day. In the afternoon, clouds gathered, and rain poured when it was four hours away from the launch.

A postponing to 21:30 was approved. By 21:00, thunder had quieted down and the rain stopped. The command post was crowded with over 200 people, including aerospace experts with different skin colors and languages, presidents and chairmen of the consortiums, and celebrities from Hong Kong and Taiwan.

On the hills six thousand kilometers away from the launch site gathered a large crowd of bystanders. In the number of hundreds of thousands, they came from Xichang and Chengdu to show support.

Ignition, the moment highly anticipated, was ordered at 21:30. In the shout of command, tangerine flame boosted Long March 3 off the launch tower.

The serene canyon boiled in cheers.

About 20 minutes later, Long March 3 made an accurate delivery of AsiaSat 1 to large elliptical orbit. The remote sensing data depicted the high precision of entry of Long March 3. The allowable deviation of orbital inclination was 0.2 degrees while the actual was 0.001 degrees; the allowable deviation of perigee height was (+18 km) while the actual was only –1 km; and the allowable deviation of apogee height was (+280 km) while the actual was only –46.8 km.

The following day, major press around the world covered the news of the successful launch of an American satellite by a Chinese rocket.

The Washington Post and *The New York Times* commented that the launch of a foreign satellite announced China's official entry to international market and status in technology.

The Christian Science Monitor considered it both a technical and diplomatic achievement for China.

Thames believed it meant that China has become a major rival in commerce.

China Daily in Thailand evaluated that it was a delight; China's satellite technology was qualified and less expensive, thus the stronger rivalry.

Long March made the name. The world accepted it with embrace. And China's space flight therefore became internationally commercialized.

Three Sketches and A Major Deal

After a taste of success with AsiaSat 1, the ambition extended to an Australian satellite.

No other satellite was as bittersweet nor unforgettable as Optus B1 in the memory of the older generation.

In Rock the Universe, a Chinese documentary of space flight, when asked which launch left the deepest impression, silver-haired grey men answered unanimously that it was the Australian satellite.

Why Optus B1? The answer lies in the intricate experience of the launch.

Before the explanation, let's take a look at Long March 2E.

It was Huang Zuoyi who sketched the first design. On a cold winter night, inspiration struck. He realized the installment of four more propellers on Long March 2C, which had never failed, would create a rocket with enormous thrust.

The following day, Huang took the sketch to have Yu Menglun, an orbit expert at Rocket Institute, analyze the calculations. A conclusion was reached that theoretically it could reach the capacity of 1/3 that of an American space shuttle.

The words cheered Huang up massively. He took the sketch to seek approval of Wang Yongzhi, then deputy director of Rocket Institute, whose satisfaction was reflected in the immediate answer of yes.

Huang had a sharp sense of market. Having seen the failure of Challenger an American space shuttle, he predicted a temporary crisis of launch vehicle in international market, which was a rare opportunity for China. As long as a rocket with 1/3 the capability of the American creation could be made, China could quickly win the orders of satellite launches globally.

Then, Huang and colleagues performed a further investigation of the market.

First, key technology of international commercial satellites was in the possession of America. Hughes Global Services, as the strongest, dominated over 60% of the share in the market.

Next, international communications satellites in the 80s were epitomized by AsiaSat-1. The common weight was around 1.4 tons. However, a new generation with big capacity, multiple functions, and over 3-ton weight, would gradually replace the smaller ones.

Long March 2E
(Photographed by Qi Xian'an)

Third, satellites capable of launching satellites over 3 tons were only American space shuttles and rocket Ariane of France. The over-qualification and high cost allowed rocket Ariane to dominate 40% of the market.

Last, there was still a great potential in carrier rocket as the global communications satellite market would experience an increase in demand.

The conclusion was explicit that regardless the success of AsiaSat-1, the 1.4-ton capacity of Long March 3 limited the prospect. To hold the market position, Long March 3 did not suffice. To meet the international demand, there was necessity to develop a new rocket with thrust as great as Long March 2E.

With the consensus of an urgent mission, Rocket Institute immediately summoned Wang Dechen, Huang Zuoyi, and others to further validate and perfect the plan of Long March 2E. Three sketches were completed, upgrading the capacity of low-orbit rockets from 2.5 tons to 9.2.

To promote Chinese rockets, a special agency operated in Los Angeles, the China Launch Service Office. There were only two employees, Huang and his assistant.

Potential emancipates in great pressure.

Never had Huang imagined that a bookworm like himself would turn into one of the best rocket salesmen in China before the experience of harsh commercial competitions in America.

Rocket Ariane charged zero for the first two services at the early entry to the market.

The promotion of Long March 2E was a huge trouble in the absence of a physical rocket, of a success rate. Only three sketches were the promotional material. In the adversity, Huang managed to persuade Hughes Global Services to take a look at the Long March 2E program.

In October 1986, a delegation was sent to Beijing to investigate the details of the program. Wang Yongzhi elaborated to the visitors, who turned out rather content.

On November 26, 1986, Hughes Global Services mailed a letter to China's Rocket Institute and expressed the willingness to have a satellite launched by Long March 2E between September 1989 and September 1991. A US$100,000 deposit was paid.

On January 16, 1988, Australia Communications made public announcement that Long March 2E would shoulder the duty of launching the communications satellite purchased from Hughes Global Service. The possible denial of permit from the U.S. government would lead to Australia's reconsideration of the contracting agreement signed with Hughes Global Services.

On September 9, 1988, the Sino-American negotiation fruited the official approval of the State Department.

On October 31, the last negotiation between two parties began. When it fell into a deadlock, Huang untied the knot.

China charged 25 million dollars for a launch service, so cheap that it was even lower than half of the international price. He harnessed the advantage to manipulate the minds at the negotiation table. "Unfortunately, little time do we have. No contract signed today would open doors to another company for business collaboration tomorrow." At the end, America compromised. In the morning of November 1, Wu Keli, assistant manager of Great Wall Company and Dorfman, vice president of Hughes Global services, signed the contract.

This was the first time to even happen in world space launches. Before the manufacture, three sketches landed a major deal among China, Australia and the United States.

The 18-month and 14-month Results

There was no finished product of Long March 2E yet. Did Americans fully trust the quality of Chinese rockets?

The answer was negative. To ensure a good quality, Americans listed a series of strict requirements in the contract:

1) Around 1990, America would perform a detailed review on launch service preparations including Long March 2 and the launch site. If sufficient evidence of incapability to launch the satellite or to have it done on time was found, America was entitled to end the contract with a one-million-dollar fine on China.
2) Launch March 2E must complete a trial launch in Xichang Launch Center before June 30, 1990. Should the trial fail, the contract would end with China fined with one million dollars.

These provisions served America assurance yet China a great challenge.

Only 20 months were given to China to have the first trial done since the effectiveness of the contract on November 1, of 1988.

To develop a prototype of a new rocket would take normally 4 to 5 years in China back then. The time for America was at least 2 to 3 years. Tick tock. The pressure was suffocating.

It was a win or die battle.

On December 14, 1988, Li Peng, Premier of the State Council chaired a meeting of the State Council and adopted a resolution on launching Long March 2E. Afterwards, the mission of launching the Australian satellite was added to the list of national key projects.

And there were only 18 months left.

On February 16, 1989, the National Defense Science, Technology and Industry Commission officially appointed Wang Yongzhi as chief commander, Yu Longhuai as deputy commander, and Wang Dechen as chief designer.

At the initiation meeting, Wang Yongzhi made a promise that the resignation letter would be in place if he failed to erect Long March 2E on the launch tower before June 30, 1990.

Launch Tower of Long March 2E at Xichang Launch Center (Photographed by Qin Xian'an)

A normally unacceptable schedule was laid in front of the personnel of each system.

The task required solutions to 20 major theoretical problems such as stability of rocket body, bundling connection, longitudinal coupling and separation of booster rocket and the completion of over 24 sets of 440,000 drawings of the comprehensive rocket design within 100 days; the production of over 5,000 special sets of tools and hundreds of thousands of components within 400 days; and the implementation of nearly 300 grounds tests, such as simulation, integrated matching and booster, fairing, separation of satellite and rocket, and full-rocket vibration.

Deadline cornered every unit to work at an extraordinarily rapid pace.

Meetings with over 300 factories in 74 cities of 25 provinces were held to negotiate purchase contracts by raw material preparation center of Long March 2E within a month.

Over 5,000 deals were made in one and a half days during the national urgent meeting of special materials for Long March 2E, and 90% of the urgently needed materials were made certain. They demanded 7445 pieces of metal materials, over 1,100 mechanical and electrical products and 580,000 electronic components in short notice, and 98% arrived in Beijing at a rarely fast speed.

To make the deadline, efforts were maximized.

It was common that employees at the general assembly plant pulled all-nighters. The special team led by Zhaoming, which was responsible for the processing of pipe conduits, had five all-nighters at the first work shift. Only naps were taken when the pipes went through acid pickling. Directors of the plant handed over a cigarette to Zhao the day mission was complete. "No, thanks." Right after the reply, he fell into deep sleep.

A component on the rocket required extremely high machining precision, at which only an old craftsman, named Xu Qingsong was adept. Unfortunately, Xu suffered from bladder cancer and had just had a surgery. What then? Having found out the details, weak as he was, he dragged himself to the workshop. It was high advised that he should rest. Yet, he shook head to that, saying "China can afford monetary loss in failing the contract but a disdained reputation."

"China cannot afford a disdained reputation." It was also the voice of all the frontline personnel who had been working day and night.

Long March 2E's dynamic analysis was a world-class cutting-edgy technology. Predictions were made that China's incapability of the calculations would lead to a total handover of parameters with a US$20 million payment. Researcher Zhu Liwen and his colleagues spent 150 days in the computer room programming over 10,000 entries with less developed computing devices. A modal synthesis method was adopted. Draft papers stacked up to 1.5 meters were used. And the fruit was a mathematical model that proved the prediction wrong. When Zhu took it to Hughes Global Services in Los Angeles for review, American experts gave it full validation after a quick browse.

In the canyon of Xichang, another silent battle started on March 7, 1989: the construction of the first giant launch tower. The contract left it a short time: 14 months.

The architecture of the same scale took Americans 19 months, and French 29.

At the end of 1988, an inspection team from Hughes Global Services flew across the ocean to the canyon. Stenhauer, head of the American Astronautical Association, and Smith, chief science adviser to Hughes Global Services, widened their eyes searching for something. At last, they stopped and asked Qu Congzhi, head of the launch center, "Where is your launch station to be?"

"Right under your feet. We will build a large launching station in 14 months.

"Here? 14 months?" They shook their heads as they glimpsed at the rocky ground. "Impossible. Two to three years is what it will take in the West, where modern construction operates."

Smith spoke more aggressively. "14 months sounds like the words after ecstasy." Having noticed the offense in the statement, he added at once "Nothing is certain though as miracles happen sometimes."

14 months involved the constructions of a mobile service tower with the height of 97 meters and weight of 4500 tons, a fixed umbilical tower with the height of 74 meters and weight of 1050 tons, three lightning arresters at the height of 170 meters, diversion grooves, power supply rooms and control rooms. It was as difficult as reaching for the stars to build an enormous launch station ranking the third in the world, and first in Asia.

The canyon in west Sichuan opened arms to a special installation team that once built the experiment tower of atomic bombs at tall as 104 meters, and experts of installations of all kinds across the country.

A flood rarely seen in a century roared in the early dawn of September 4, 1989. The huge debris flow cut off all the railways, highways, transmission lines and communication lines leading to the canyon site. 100 trucks of steel components urgently needed for station construction were blocked outside. Four hours after the outbreak, thousands of flood warriors organized by Xichang Satellite Launch Center set up a front line and rushed around the clock for one whole month to restore the railway, highway, transmission line and communication line. And to recover the time delayed by natural disasters, officers and soldiers of the special installation brigade worked excessively for 420 days and nights, there were 1,200 overtime hours and 3 kilograms of weight loss per person.

"My love, when you read this letter, I am gone for good. Darling, don't be sad. I have been well prepared for a sacrifice the moment I chose this sacred cause of peril..." It was a quote from the letter of farewell to the wife by Li Lianlin, director of launch and test department of Xichang Launch Center during the mission. Luckily the success prevented the letter's delivery. Yet, the message of full commitment was delivered in the simple words.

The 97-metre-high mobile service tower required two 32-meter and 120-ton cross beams be installed on the top. However, there was the lack of large vehicles or hoisting machinery to move the two giants. With bare hands, not only did they unload and transport them to the site, but also install them on the tall towers with a precision of a height gap less than 2 mm between two beams.

On April 20th, 1990, the launch tower erected in sweat and blood stood tall in the canyon of Xichang. In July, American experts Stenhauer and Smith were invited to witness the first test launch of Long March 2E. A picture together in front of the soaring launch tower was suggested. Just as the photographer was about to press the shutter, Smith signaled a pause, "A moment, please. I sincerely apologize for my foolish words years ago."

There was little faith amid western experts of space technology in China's manufacture of Long March 2E before deadline. Rumor has it that Ariane was ready to replace China after the overdue on June 30, 1990 as one contract of launch service had already been drafted.

Yet, it was merely a wishful thinking of Ariane. On June 26, 1990, one day ahead of the date stipulated in the contract, China's newly developed Long March 2E stood tall on the new No. 2 station launch tower of Xichang Launch Center.

Words spread that it was Wang Yongzhi's deliberate arrangement to have it there one day earlier, so that a proof of capability would be presented.

The arrival of thunderstorm season was perilous to the launch. Danger occurred during the fueling. Wei Wenju and Lu Ahong led a crew of workers to eliminate the threat. Wei's lung was fully burned. He sacrificed life to the cause.

Details of the accidents were highly confidential. There was little knowledge of the demise when the celebration of the successful launch of Long March 2E carried on. And I could not help wetting my eyes when I found out the secret at the 40th anniversary event of Xichang Center.

On the frontier, there are millions of ordinary folks who dedicate their lifetime to the development of national space mission the way Wei did. In a sense, the road to greatness is laid on their sweats.

In July 17, 1990, the preliminary trail flight of Long March 2E went through. Apart from the model Optus B1 (technically a piece of iron with the same weight), it carried a Pakistan test satellite.

The successful trial flight implemented the terms as agreed in the contract. Enormous reputation in commerce was the harvest. And it raised international attention, especially France's. *Le Monde*, a French daily newspaper wrote that it was an intimidating rocket, whose success injected more confidence to China's carrier rockets in the competition of world's commercial launches, and the victory against America and Europe would continue.

And the pace picked up.

At midnight of January 30, 1992, Optus B1 traversed across Pacific Ocean and safely reached the satellite plant in Xichang.

Afterwards, an official announcement of the launch time was delivered: March 22.

0.15 mg of Aluminum Redundancy Took the Blame

It was finally March 22 after 4 months of ups and downs.

Carrying Optus B1, Long March 2E made the appearance.

The night, Party and state leaders, as swamped with the affairs in the Fifth Session of the Seventh National People's Congress, squeezed time to witness the launch, in the command hall of Beijing National Defense Science and Technology Commission; hundreds of domestic and foreign dignitaries who specially made the trip sat anxiously on the observation deck of Xichang Launch Site; and folks from all directions left no empty space on the hills in the vicinity.

Statistics tells that over 600 million people in mainland China watched the launch on television. In addition, people from Hong Kong, Macau, Taiwan, America, Australia, and Europe had their eyes focused on the event.

The launch was broadcast live by China Central Television to the world. Domestic media flooded to the site. Amid the journalists, I was there in great anticipation.

It was too paramount a launch.

Apart from the inscription of the label China Space Flight, Long March 2E carried the imprint of the three national flags of China, America, and Australia. It was the symbol of collaboration and sharing of risk. None wanted any problems. A simple glitch would result in enormous monetary loss and a damaged reputation for China and America, and a break in domestic communication for Australia.

No failure was affordable. All the personnel of the launch had the firm belief that no failure would happen. Preparation of a celebration was complete.

A feast had already been reserved at Hotel Tengxunlou hundreds of kilometers from the launch site; speeches of congratulations had been folded and put in the pockets of delegates from three parties; firecrackers and drums were in place at the Rocket Institute; the script for live broadcast was only written to praise a success at China Central Television.

Instead, expectations failed.

At 18:40, ignition was initiated.

It was absolute silence among the people as every breath was held.

The rocket bottom spat brownish yellow smoke and the body began to strongly tremble.

Eyes were locked on the sky for a glorious blast-off.

The flight should have happened four seconds after ignition. Yet, seven seconds passed. Long March 2E sat still on the launch tower.

An order of emergency shutdown came from the Command Hall.

It was total chaos, at the site and in their minds.

On television, the audience received the live image of the announcement of the fiasco in clouds of smoke from Tang Jin'an, the then-general manager of the Great Wall Company.

There was no time to grieve.

At first notice, technicians rushed to the site, regardless of peril, to save the rocket.

The body has tilted, and three of the four bolts of fixture have been dislocated. The rocket would detonate a massive explosion should it collapse as it was loaded with over 400 tons of fuel, equivalent to 10,000 tons of explosives. In the radius of 2.5 kilometers from the launch site, a sea of fire would rise.

Operators, technicians, experts, and soldiers immediately started the save.

39 hours of relentless work prevented the demise of Optus B1 and the launch site.

Calm reactions of the Chinese to danger won the respect from America and Australia. The survival of Optus B1 kept the collaboration alive and the confidence intact.

On March 24, Bob Mansfield, President of Optus Communications, Australia, and Steve Dorfman, chairman of Hughes Global Services, USA issued a joint statement in Sydney that despite the failed launch of a communications satellite for Australia on the evening of 22, the plan would remain unchanged. They held full confidence in China's ability to locate the cause and execute a re-launch as soon as possible.

Black March 22 broke the heart of every Chinese aerospace undertaker. It was a pity for the country and its people. And the result was the pressure as heavy as a mountain.

AFP first reported the fiasco minutes after the occurrence. Soon the news spread across the planet.

The fiasco was a huge humiliation to the self-esteem of millions of Chinese. After an hour, telegrams and phone calls bombarded Xichang from all over the country. Words of disappointments and anger were made clear to the workers: a disgrace to the Great Wall, an utmost upset to the Chinese people, a huge shame, and never forget the 1.1 billion eyes...

In contrast to the misunderstandings from the general public, the Party Central Committee and authoritative specialists offered consolation and encouragement, which eased the pain among front liners.

The central leadership consoled that failure breeds success; stay humble at a victory and stay encouraged at a loss; problems will be located and solved; high morale is significant.

As a veteran, Qian Xuesen came forward for motivation.

"As I have always said, loss is not always a bad thing. In fact, it keeps companion to a scientist. Flowers, applauds, and praises would fade. And loss is not to be feared as millions of losses are the bricks that made mankind's space flight. The high risk inevitably involves losses and barriers, something we must confront with."

Also, General Zhang Aiping wrote to cheer them up.

"Nobody makes perfect. What matters is to figure out what goes wrong. Only when the culprit is found and properly dealt with can success arrive."

The greatest urgency was to identify the cause.

No seconds were to be spared as China itself, America, and Australia awaited the result.

On April 9, the fault analysis review committee consisting of Tu Shou'e, Wang Yongzhi, Liang Shoupan, Liang Sili, and Xie Guangxuan disclosed the findings:

It was the unwanted aluminum redundancy weighing 0.15 mg between the fourth and fifth electric shocks of the program distributor that led to the blast-off.

None expected an answer as simple as a tiny piece of aluminum scrap.

It was an unforgettable lesson learned.

On April 13, the launch was officially defined a total fiasco by China's Ministry of Space Flight. Yet, Hughes Global Services, USA, advised a change to "abortion" as China's definition would trigger a massive compensation from an international insurance company, which would prevent another flight.

Consensus on the definition of "abortion" was reached between both sides. The insurance company did not halt the plan, and another chance of launch was permitted.

It was a plan that served every side's interest.

100 Days Made A Miracle

The winner does not take any blame while the loser does not get to explain.

Chinese aerospace workers had to swallow the loss and keep moving forward as actions spoke louder than words.

And there was common agreement among them that the second chance could not afford to blow.

The clock was tighter: only 100 days left.

Under any circumstance a new Long March 2E had to be built before June 30th.

At midnight of March 30, a special train pulled over into a station in Beijing carrying the physically damaged rocket engine and the emotionally damaged rocket developers. The door slid open. An image was presented to the welcoming party as several technicians were carrying the deeply ill Yu Longhuai chief director of Long March 2E to the platform.

He embarked on the work the earliest. And it had been over 100 consecutive days of work at the front line. He was too exhausted and stressed out that a fever of 39 degrees Celsius took him on the train ride.

Three days of rest and fluids helped the recovery. Immediately, he returned to the command post. Wang Dechen, chief designer of Long March 2E, Shen Xinsun, director of Chinese Rocket Institute, and many crew members also embarked on the work in labs, design rooms, and workshops soon after they walked off the train.

The improvement of Long March 2E went on at an alarming rate.

Scientists and technicians not only put forward solutions to the problems detected, but also completed the review and approval of 24 sets of 440,000 drawings in a short period of a dozen days. Based on the problems they have discovered, they added 22 modifications to lightning protection, sweating prevention and rain prevention during rainy season.

There was no fancy talk but diligent work. The employees worked almost to the gross neglect of their health. Many moved their bedding rolls to workshops for quick naps. The lathe was rotating, the welder whistling, and the chemical milling fluid boiling. Pu Jinming, a veteran worker, contended for every minute to process key components. His hands became so red and swollen that he could not even hold his lunch box. Yet, he still refused to take a rest. Craftsman Jiang Min worked overtime at night and knocked out two teeth when riding a bike home. When the workshop leader advised him to rest, he replied eagerly, "There is no way to sit idle while comrades are fighting at the front line."

During the golden week of Labor Day, over 3000 staff kept working at the frontline. None demanded payment for extra hours. There was one common goal: a capable rocket that would make China proud.

On May 14, 1992 when diligence was given in the preparation of Long March 2E, a sad news came that Marshal Nie Rongzhen, the pioneer and founder of China's defense science and technology, passed away.

In the days of the first launch of the Optus B1, Nie was deeply concerned. Having heard the setback, he did not sleep well for days. And he urged the comrades of the National Defense Science and Technology Commission that on the one hand, we should conscientiously draw lessons from prior experience, and on the other hand, we should locate the cause for better control. It was a great pity that he did not live to witness the launch of the Optus B1.

With grief channeled into strength, Chinese managed to build a new Long March 2E within 100 days, and a new record was born.

At 12:30 on June 30, 1992, a special train with the new Long March 2 departed from Beijing. After five days and four nights of long-distance travel, it arrived at Xichang Satellite Launch Center.

The launch plan resumed. And rounds of discussions at the command post eventually set the date as August 14.

At this point, a tricky decision had to be made urgently – would the launch be conducted quietly behind closed doors, or broadcast live as last time? The failure still haunted all personnel involved, which was reflected on their discouraged facial expressions.

Some suggest a live broadcast, arguing that without the guts to go live, there was no going global. Some suggest otherwise, explaining the risk of another humiliation.

As the dispute was not settled, the decision was made by the Central Committee that a live broadcast won.

On the evening of August 12, China Central Television delivered the news that China would launch Optus B1 again at Xichang Satellite Launch Center at 7:00 p.m. on August 14, and there would be a live broadcast.

It was a brave decision. As news spread quickly, domestic and international attention was attracted.

It was another chance on the world stage. In the glare of millions of eyes, 7:00 p.m. on August 14, 1992 came. "Ignition!" A flame burst out from the bottom as if a roaring dragon. The rocket was lifted off.

Soon it went beyond a view from the ground. And there was suffocating atmosphere of anxious waiting in the command post.

At the 159th second, the first and second stage rockets separated as planned; at the 200th, with the fairing falling off, the rocket flew out of the atmosphere; at the 572nd, it entered the orbit and adjusted its flight attitude; at the 675th, the rocket and the satellite separated successfully, and Optus B1 entered the orbit.

Chinese cheered, and the world was in awe. Against adversity, Chinese, who bore the expectation of the nation, made a miracle happen. They accurately identified the cause in 18 days, produced new rockets in 100 days, and re-launched the satellite in 145 days, thus a new leap forward in China's space industry.

Immediately, Tang Jinan, President of the Great Wall Industrial Corporation of China, gave an address. He deeply thanked the people of the whole country and the US-Australia partners for their support after the fiasco of "3.22" and welcomed more collaborations. On behalf of the US side, Jiang Sen made the evaluation that it was a significant milestone: "This is the most important page in the history of Hughes Global Services." "It is a great event," commented Gordon Pike, the representative of the Australian Satellite Corporation, "China's Long March launch vehicle series are of high quality. I am sure that in the world launch service market, they will continue to shine."

Yu Longhuai, the chief director was spotted weeping at the moment of the successful entry to the orbit. After the "3.22" setback, he also shed tears. He made a thorough investigation at the accident site for several days and nights until the cause was found. When the Optus B1 launch succeeded, a reporter asked Yu Longhuai, "Now that the reputation is made, would you consider resigning?" No direct answer was given. But later, he composed a poem to express the ambition:

Crescent Moon together with a sky of stars,
shine upon the boundless desert;
the roaring rocket aimed up at cloud nine,
and the flames dyed the horizon;
firecrackers shouted and tears welled;
and here we go again for another climb.

Optus B2: An Unraveled Mystery

The setback was gone. The triumph is not the end.

The launch of the Optus B1 is a crucial step for China's space flight to go global. But as agreed in the contract, only when the Optus B2 goes up would the mission be complete.

The launch time was August 1992 as stipulated in the terms. Yet, the loss the first launch resulted in a postponement to December.

Three days after the shoot-up, Chinese resumed to the preparations for another launch.

Everything was going smoothly.

On October 16, new Long March 2E was manufactured, and on the 31, the second Optus B1 arrived at Xichang Launch Center.

The only twist was to set the launch date. It was postponed from the previous date of December 9-21, as requested by America.

At 7:20 a.m., ignition was made as scheduled. The Long March 2E soared into the dawn sky.

Thirty minutes later, a report came from the Xi'an Satellite TT&C Center that the satellite had been sent to the elliptical orbit of 200 km to perigee and 1050 km to apogee. Meanwhile, America also received feedback from Los Angeles and Australia that it had entered the predetermined orbit. At 19:51, the heads of China and the United States signed the launch documents. Next, Chinese Commander Hu Shixiang made an official announcement of the successful launch.

The following day, the foreign press gave it wide news coverage.

AP praised the successful launch as a huge encouragement to China's ambition to join the international competition of satellite launches. Reuters called it a victory that gave the world a peek of Beijing's capability.

However, on the third day, emergency happened. Australia declared the failed connection to the satellite since 3:00. No signal was received, and there was no trace of the satellite. There was no confirmation of the cause between the rocket and the satellite itself. Yet, most the blame was cast on China's rocket.

What went wrong? Three parties held an emergency meeting to summon experts to quickly go through various data in order to identify the cause. And the preliminary conclusion was that the flight was normal, and the satellite made the entry to orbit as planned.

But why there was no response?

Soon after, burnt satellite debris and fairing debris were found in the mountains near the Xichang launch site. After analyzing them, China and America inferred that the fireball that appeared 47 seconds after the take-off might be the satellite explosion.

China and the America formed a joint investigation team to dig deeper at the truth. After seven months of joint investigation and analysis, Hughes Global Services and Great Wall Industrial Corporation issued a joint statement on 14 August 1993:

Based on the analysis of telemetry data of the rocket and the inspection and special test of the fairing debris, China Great Wall Industrial Corporation and China Launch Vehicle Technology Research Institute confirmed that there was no defect in the design, manufacture and assembly of the rocket nor fairing. Hughes Global Services accepted this opinion. At the same time, Hughes Global Services determined that no satellite designer or manufacturer had defects that led to the failure. China Great Wall Industrial Corporation and China Launch Vehicle Technology Research Institute took this opinion as well.

The joint statement simply made clear that neither China nor the United States was responsible for the accident. So, the culprit for the disconnection became a mystery, which remains unsolved today.

Long March Series Hits Rock Bottom

Since the 1990s, a market economy in China bloomed beautifully. An unprecedented storm of reform swept across the country. Enterprises and factories proactively responded to the trend, and so did China's space industry. On April 17, 1993, the National Aerospace Corporation and National Space Administration were established, marking that market economy has fully taken over the control on China's space industry from a planned economy. National Aerospace Corporation issued the general policy of "developing space flight, enhancing domestic products, improving efficiency, and going global." International cooperation was actively made with the United States, France, Germany, Russia, Ukraine, Brazil, and Japan to accelerate the pace of China's space flight to go global.

In 1996, the 47th International Aerospace Flight Convention was held in Beijing. In the speech at the opening ceremony, Chairman Jiang Zemin stressed that Chinese government and people were willing to make greater contributions to the peaceful use of space resources and the expansion of space applications, and to the progress of human civilization.

In 1994, the Long March rocket had a stable performance of all five successful launches. On August 28, the Long March 2E sent up the third Australian satellite made in the United States at Xichang Satellite Launch Center.

In 1995, space was particularly entertaining – the United States and Russia collaborated, and space shuttles and the space station had a "space kiss" twice, firing up the enthusiasm in space flight.

However, 1995 was a tough year for China's space industry. On January 26, in the midst of a bitter cold wind, Long March 2E took off with the US Apstar 2 satellite. Unfortunately, the rocket exploded in the air, destroying the satellite too. The explosion triggered a chain reaction. Afterwards, the performance of the Long March rocket has been alarmingly unstable.

In December 1995, after 11 months of preparations, Long March 2E made another attempt to send up Apstar 2 and the efforts paid off. Before the joy healed the wound, a more tragic failure ensued.

The day of adversity was February 15, 1996. At 8 p.m., China launched the large-capacity telecommunications satellite Intelsat 708 developed by the United States using the Long March 3B, which was then the most capable rocket in high orbit developed by China It possessed the world's top technology of second self-ignition under vacuum conditions. Intelsat 708 was an American creation with a fine reputation in the world. Together they made a strong team. And the expectations were high. However, 20 seconds after the blast-off, the rocket veered off course and plummeted. Huge explosion ensued, leaving the valley in roaring flames and smoke.

One of my colleagues experienced the peril. "I was at the launch site" he recalled, "When I saw the rocket that just took off exploded into pieces, my brain buzzed as if an explosion incurred there." And he was lucky to hide in a nearby air-raid shelter, or the consequences would have been unthinkable.

Over the years, the scene of the explosion February 15, 1996 replayed in my mind every time I hosted an interview concerning a launch and signed the risk letter. After the accident, Liu Jiyuan, general manager of National Aerospace Corporation, was drowning in suffocating pressure. Rumor had it that his hair turned grey overnight. He recalled: "Two happenings that have had the deepest impact and the greatest pressure on me. One is the "Cultural Revolution" and the other the "215" explosion. I have endured countless failures, among which none was as devastating."

The explosion sank China's space industry to rock bottom. The business of foreign launch service was hanging by a thread. Space News of the United States predicted successive failures would make it more difficult for China to obtain insurance and scare off more potential customers.

The prediction came true soon. Trust on the Long March 2E started to collapse and contracts were withdrawn. Hughes Global Service, the partner with the longest cooperation, halted the next business negotiation and signed the contract with Japan instead.

As the credibility of Long March rockets fell, the launch of Apstar 1A became the key to turning adversity into success. Therefore, the mission was more than a simple rocket launch but the duty to rebuild China's image. "We don't fear failure, but we can't afford another" said Huang Chunping, leader of the trial team, "There was a mountain-sized burden on our shoulders." At 18:47 on July 3, 1996, the Long March 3 successfully sent Apstar 1A into orbit. And the burdensome mountain was gone. Whoever participated in the mission breathed with great relief, and Long March's reputation was saved before too late.

But, as if a new scar is torn up sooner than the old wound heals, another failure that incurred 46 days later had China's space flight hit total rock bottom. On August 18, 1996, another Long March 3 carrier rocket attempted to shoot up a ChinaSat 7, an American communications satellite, and ended up with a disappointment. The rocket did not explode but the engine shut down 48 seconds earlier than programmed and the satellite ended up on a useless low orbit.

Foreign presses jeered that Chinese ought to know their limitations and quit the international launch market.

It is common knowledge that going global is never smooth sailing, during which setbacks and barriers are frequent visitors. Whereas, the frequency of their appearances is beyond expectation.

Regaining Trust

It is a thorny path to success.

Between giving up and striving on, the Chinese picked the latter.

Failure after failure led to deep soul-searching among the Chinese aerospace workers. They could not be more aware of the fact that one more failure would end the future of China's space flight.

"Tighter quality control, more careful work, and more cautious thinking." The Central Committee of the Party, the State Council and the Central Military Commission have given clear general requirements for China's space launch.

On December 27, 1996, the National Aerospace Corporation held the Third International Commercial Launch Conference in Beijing. Experts, core technicians, and managers from the front line of commercial launch services were invited to attend. The conference aimed at freshening the minds of the front-liners. And the result was a new Long March development plan, namely the Life Project, whose core purpose was to enhance the reliability of launch vehicles.

Bai Bai'er, assistant manager of National Aerospace Corporation, elaborated that the name "life" should serve as a constant reminder of careful work to avert deadly accidents.

Lessons of blood gave birth to the idea that quality is life. Thereafter, the National Aerospace Corporation established strict quality control standards on the basis of the summary of prior experience in quality control in the past decades. The standard of "Accurate positioning, clear mechanism, recurrence of problems, effective measures, and drawing inferences" were implemented it in every link of the space system.

Immediately after New Year's Day 1997, the Ministry of Satellite Launch Measurement and Control (TT&C) of China embarked on three checks, namely problems check, gaps check and hidden dangers check, from bottom to top in all launch centers to better guarantee a successful launch. Xichang Satellite Launch Center hung the slogan that read: "Mobilize all to search for problems, gaps and hidden danger to eliminate the tiniest possibility of failure."

The Taiyuan Satellite Launch Center further improved the whole process quality monitoring system for the first international commercial launch case. Since the day the Iridium satellite entered the site, a weekly business analysis report meeting was held, and special consultation was conducted on tackling dozens of dilemmas.

The Center also strengthened a series of rules and regulations, thus forming an atmosphere where quality was the focus anywhere, anytime, and for anyone.

1997 was a remarkable year for China, because the country would once again exercise sovereignty over the newly-returned Hong Kong. The year was also meaningful for China's space flight, as a goal was set that the six launch missions scheduled that year should only succeed.

1997 was a tough time for China's space flight as well as a turning point. It was in that period that ideas "quality is life" and "every screw matters" took root in the nation. It was also in that period that China established a series of standards to ensure successful launches.

Huge pressure failed to impede China to deliver six triumphant launches out of six attempts. The beautiful Long March rockets won over hearts again.

On May 12, 1997, the first attempt succeeded. The Long March 3A lifted off from Xichang Satellite Launch Center, and in great precision, delivered the new generation domestically developed communications satellite Dongfanghong 3 into orbit.

On June 10, the launch of Fengyun 2, a new generation of meteorological satellite, by the Long March 3 also came out satisfactorily. The 160-day in-orbit test cleared it to be deliverable for use.

On August 20, the newly developed Long March 3B, a bundled launch vehicle, carried the Philippine Mahuhay satellite up. And its fine performance demonstrated the ability to shoot up a load of 5,000 kilograms into a high orbit.

On September 1, the modified Long March 2C was effectively launched at the Taiyuan Satellite Launch Center at the first attempt, and two Iridium model satellites from Motorola Inc.(US) were delivered into orbit with fine accuracy.

On October 17, the Long March 3B, the one with biggest thrust in the series, effectively sent up the Apstar 2R a communications satellite manufactured by Space Systems Loral (US).

On December 8, the modified Long March 2C successfully launched Iridium. This was the first commercial launch of an Iridium satellite under the contract of the Great Wall Industry Corporation of China and the first international commercial launch of the Taiyuan Satellite Launch Center.

Success tastes sweeter in great adversity. Chinese space flight rose again in 1997, stronger and more mature.

The Golden Age of Long March Rockets: 1998–2017

The most unforgettable Chinese memory in 1998 was the joint efforts of soldiers and civilians against the huge Yangtze River summer deluge. Also, impressing was the Long March 3B's launch of Xinnuo 1 a European-made communications satellite that left the deepest mark on China's space flight.

Xinnuo 1 was manufactured by Aérospatiale, a French company. It was installed with 24 C-band transponders and 14 KU-band transponders on board. The service that it provides includes telephone, fax, radio and television, data transmission and other for users in China and Asia-Pacific region. The launch was China's first time with a communications satellite made in Europe.

The mentioning of the Long March 3B would naturally remind us of the 2.15 explosion. And this launch was a chance or Long March 3B to return to international commercial launch stage. Should it succeed, the reliability would be trusted again; should it fail, it would be the last bow the rocket takes on the global stage. China's space flight would have no credibility left and the confidence of Chinese aerospace workers would be entirely shattered.

Therefore, maximum importance was attached to the mission from the officer at the highest position to the worker with the lowest pay grade. Party and state leaders made trips to Xichang Satellite Launch Center to supervise the launch. Suffocating tension permeated Xichang Satellite Launch Site. Many described the same feeling of an intangible rock pressing the heart.

All preparation could not have been more scrupulous. Liu Jiyuan, general manager of National Aerospace Corporation looked as deep as into tiny screws. He found it difficult to rest assured without checking the fueling of rocket on site.

At 17:20 that evening, the command of ignition came, and Long March 3B, carrying Xinnuo 1 communications satellite, took off and headed into space. And applauds and cheers bombarded the command post.

The 100% victory rate (6 out of 6) in 1997 lifted Long March rockets off rock bottom while the launch in 1998 allowed them to break through the dark clouds to kiss the sunshine.

Confidence was built greater than ever. And the bitter taste of failure rarely happened in China's space journey ever since. Instead, the pace was picked up and Long March rockets entered its golden age.

On June 1, 2007, China's Long March launch vehicle series welcomed the 100th flight:at 00:08, Long March 3A effectively sent Xinnuo 3, a communications satellite, into space at the Xichang Satellite Launch Center.

Since Dongfanghong 1, it took 28 years for the Long March rocket series to complete the first 50 launches, while only 9 years to complete the second 50. The frequency increased from one launch every few years to a few in a year, or even an annual dozen. The development kept accelerating.

Reviewing the history of Long March rockets, Zhu Yilin, academician of the International Academy of Astronautics, summarized, "Today, China has undoubtedly become a major space power in the world."

On December 8, 2014, China's Long March rocket welcomed the 200th flight. At 11:26 p.m., China used the Long March 4 to deliver the CBERS-04 into orbit from the Taiyuan Satellite Launch Center.

The first 100 launches by the Long March series spanned from 1970 to 2007 with a 93% rate of success; and the time gap shrank from 37 years to 7 for the second 100 launches to complete at a 98% rate of success.

Liu Jiyuan, the winner of the most prestigious award in aerospace conferred by the International Academy of Astronautics (IAA) and former general manager of National Aerospace Corporation, elaborated that 13 types of Long March launch vehicles had been put into use since 1970 They went from zero to one, from the tandem type to strap-on type, from normal temperature propellant to low temperature propellant, from one rocket one satellite to one rocket several satellites, from launching satellites to launching manned spacecraft and lunar probes. Gradually, the ability to launch different types of satellites and manned spacecraft in low, medium and high earth orbit has been developed.

Ignition after ignition left the world in awe. And behind every glorious moment lie the brilliance and selfless dedication, perseverance and diligence, and independent innovation and brave attempts of the Chinese space team.

Apart from the physical achievements, the virtues reflected in the splendor of atom bombs, hydrogen bombs, man-made satellites, and manned space flight have become a legacy of spiritual treasures for the Chinese nation.

Liang Sili, expert on rocket control, recalled that in the years of pioneering attempts, lights were never off in the office; there was enthusiasm in extra work or study; the words echoing in the corridors were "work till the last breath." The space team, whose stomach was rarely half full, made miracles of the missiles with the equipment that was deemed impossible.

Wu Yansheng, director of the Chinese Carrier Rocket Institute analyzed that despite the popularity of rocket studies, they put less cash to the pockets than a breakfast restaurant even could; so, we never hesitated about the decision; diligence, and perseverance enabled us to master core technology and key technology in space flight when developed countries implemented the policy to permit little high technology export; generations of talents of space flight were nurtured that harnessed the world's cutting-edge technologies.

It was a team with an average age of 35 that shouldered the responsibility of designing Long March rockets. And the situation of youth in charge was common in basically every space flight frontier.

Lei Fanpei, current director of the Long March Rocket Unit and chairman of Aerospace Science and Technology Group predicted that it was highly likely that a Long March rocket could complete the 300th launch in the next seven years.

When the hundredth launch had been reached, there was another milestone for China's space flight. On May 15, 2007, a Long March 3B effectively launched a telecommunication satellite for Nigeria from the Xichang Satellite Launch Center, a breakthrough in the whole export of China-made satellites. It served as an announcement of China's ambitious entry to the global market of commercial satellites. This was another significant progress after China's entry into the world commercial launch market.

On January 16, 2016, a communications satellite from Belarus was effectively launched at the Xichang Satellite Launch Center in the early morning. It was the opening for China's whole satellite export to the European market. The development of China's space flight was speeding up.

China has made 43 international commercial launches and 49 satellites for 20 countries, regions and international satellite organizations since the AsiaSat-1 of Hughes Global Services in April 7, 1990. In addition, 10 international commercial launch services have been provided.

The number of satellite exported from China broke through from zero in 2007, and increased to nine communication satellites to Nigeria, Venezuela, Pakistan, Laos, Belarus, and other countries.

Data from the Great Wall Industrial Corporation shows that China's share in the international market is expected to increase to 15% by 2020. The person in charge of the company promised that in the future, more efforts would be made to help Chinese communications satellites to enter the international mainstream market and promote remote sensing satellite exports.

According to the Thirteenth Five-Year Development Comprehensive Planning Outline of the China Aerospace Science and Technology Group, by 2020, the number of spacecrafts in orbit in China will have exceeded 200; the number of annual launches will have reached 30; and the membership to the advanced space flight club shall have been obtained.

New Adventure for Chinese Rockets

The stage of China's space flight will be as big as the capability of launch vehicles allows.

Standing at the new starting point of 200 launches, the Long March launch vehicle series embraces new adventures. In the future, China will need a new generation of rockets to carry out deep space exploration and interstellar travel.

It is a clear fact that in spite of the many achievements Long March Rockets have made, there is still a gap between our carrying capacity and the first-class international level.

The carrying capacity of the United States, Russia, and Europe on low-Earth orbit has gone beyond 20 tons, and that of Japan has exceeded 15 tons. The carrying capacity of the United States on geosynchronous orbit has reached 13 tons, that of Europe over 12 tons, that of Russia more than 6 tons and that of Japan beyond 8 tons. China's active launch vehicle's carrying capacity on low-Earth orbit spanned from 0.3 tons to 9.5, and the geosynchronous orbit capacity from 1.5 tons to 5.5.

The gap was vast.

The presence of the gap punctures the bubbles of past achievements. Since 2015, Chinese aerospace workers have kept the diligence going on the development of Long March rocket series. The goal was to achieve a new breakthrough from the quantity of launches to a greater carrying capacity.

At 7:01 on September 20, 2015, the first of China's new generation launch vehicles, Long March 6 soared high from the Taiyuan Satellite Launch Center. It carried 20 micro-satellites, such as the Xiwang 2, Tiantuo 3, Naxing 2, ZDPS 2, and LilacSat 2 developed by National Aerospace Corporation, National Defense University of Science and Technology, Tsinghua University, Zhejiang University and Harbin University of Technology.

Long March 6's triumph has set a new record of one rocket carrying multiple satellites and filled a gap in the capability of a quick entry of multiple satellites into orbit.

A new milestone was made.

A mere five days later, China's new four-stage solid launch vehicle, the Long March 11, was ignited at the Jiuquan Satellite Launch Center and effectively launched four micro-satellites into space. The success marked a major breakthrough in key technologies in solid launch vehicles for China. It was of great significance to improve the type spectrum of launch vehicles and strengthen the capability to enter space.

Robust rockets are the key to growing into a powerful country with space flight. In 2016, China delivered a piece that condensed years of work and the world bowed.

On May 25, 2016, China's new generation medium-sized launch vehicle, the Long March 7, made its first successful flight at the newly built Wenchang Space Launch Site in Hainan. The event initiated the space laboratory mission of China's manned space engineering.

Characterized by high reliability, safety, non-toxic and pollution-free, the Long March 7 has a low-Earth-orbit carrying capacity of 13.5 tons, nearly 60% higher than the well-known Long March 2F. The Long March 7's development suffices to serve the needs of launching cargo spacecraft in China's manned space station project and the long-term needs of renewal and replacement of manned launch vehicles in the future.

It has adopted three-dimensional technology. From design to production, an all-three-dimensional digital platform has been used. As China's first "all-digital" rocket, it digitalized the whole life cycle of Chinese launch vehicles.

Fan Ruixiang, chief designer of the Long March 7, is a perfectionist. He has strict technical requirements to the extreme: "Quality comes first. The rocket must first ensure high reliability." Long March 7 adhered to the concept of environmentally friendly: "green energy, green materials, and green technology." Nine independent evaluation meetings, 31 design reviews, 146 thematic reviews, and retrospective reviews have been carried out from prototype completion to sample production. It was an eight-year marathon.

The prospect for applications is rather wide. It can be simply summarized as cargo flight for the short term and manned flight for the long; and various modifications are feasible for comprehensive mission coverage. Chief designer Fan Ruixiang explained that the ability to launch high-, medium-, and low-orbit application satellites could be equipped in a short time through simple adaptability transformation. Therefore, urgent demand in the current mainstream satellite launching market at home and abroad could be met and China's space industry would be able to forge major market-oriented and internationalized rockets and gold medal rockets.

When discussion still revolved around the Long March 7, a new sensation came – the Long March 5.

The Wenchang Space Launch Site in Hainan has once again stood in the center of global attention. At 20:43 on November 3, 2016, the Long March 5, China's new rocket with the greatest thrust by far, took off.

At the launch site, Long Lehao, an academician of the Chinese Academy of Engineering, could not hide the pride on the face. "I waited 30 years for this moment to happen. Long March 5 did not fail my expectations" he said. "As a typical representative of the new generation in the Long March rocket family, it took a great leap while laying the foundation."

At the celebration, Luan Enjie, former director of the National Space Administration and academician of the Chinese Academy of Engineering, spoke in excitement, "The success of the first flight of the Long March 5 marks the upgrading of our launch vehicle and the entry of its carrying capacity into the club of international advancement. It is a key step to transform from a country able to undertake aerospace missions to a powerful country adept at them."

As well-known to all, since the project was established in 2006, Long March 5 has gone through 10 years. However, it was little known that the proposal was made 30 years ago. It was three decades of hard work that realized the ambition.

A set of data reveals the difficulty of the Long March 5. Statistics tells the sum of research and development projects numbers more than 800; scientists have carried out 1,289 general tests and more than 20,000 ground tests, and produced more than 18,000 machine units of various types; the electrical system alone contains 110,000 components, 15 times the total number of components of the Long March 2 and 3. "The dynamic characteristic parameters to be analyzed for a booster of the Long March 5 are almost the workload of a rocket before." explained Wang Penghui, head of a test Institute of the National Aerospace Corporation.

As the epoch-making foundation model of the Long March rocket family, the Long March 5 was enormous: nearly 57 meters in height, as tall as 20 stories; the body had a diameter up to 5 meters, bundled with 4 boosters with 3.35-meter diameter; the first use of two 50-ton hydrogen-oxygen engines and four boosters with two 120-ton liquid oxygen kerosene engines each; 10 engines were ignited simultaneously. The maximum pressure created by one engine was 500 atmospheres, equivalent to pumping water from the Huangpu River in Shanghai to the Qinghai-Tibet Plateau as high as 5000 meters.

For the largest thrust, the Long March 5 was nicknamed Hercules. the total take-off weight of the rocket was about 870 tons, the take-off thrust exceeded 1,000 tons; the low-Earth-orbit carrying capacity reached 25 tons, the geosynchronous transfer orbit carrying capacity 14 tons, 2.5 times better than the Long March 3; the whole body adopted 247 core key technologies with fully independent intellectual property rights. "In terms of comprehensive performance, the Long March 5 has the capability paralleled to the mainstream carrier rockets in the world, which will greatly enhance China's independent entry to space" said Li Dong, chief designer of the Long March 5.

As the pioneer of launch vehicles with a large carrying capacity and the main force of deep space exploration, the Long March 5 will be used for the missions of the third phase of future lunar exploration projects, manned space station and the first Mars exploration. A new adventure is to be taken.

In 2017, the Long March rocket family welcomed a new member.

At 12:11 on January 9, 2017, China successfully launched the Jilin-1 Video-03 with a small carrier rocket, Fast Boat-1A, at the Jiuquan Satellite Launch Center, carrying two cube satellites, XY S1 and Kaidun-1.

The mission was organized in the form of a pure commercial launch contract and operated in full accordance with market behavior. It was the first commercial order of Fast Boat rockets, a new step for China in commercial space flight.

The Fast Boat 1A was quickly ready, as there was no need to fix the launching tower. The use of solid rocket motors avoided fueling. Two days and a staff of ten sufficed to pull it off.

"As the name suggests, the feature of Fast Boat 1A is fast." Liang Jiqiu, chief designer of Fast Boat at the Fourth Research Institute of National Acrospace Corporation, compared that the launch took only eight months from the order made by the client to the final delivery of launch services, while it would take more than a year for general rockets to have the work done. A new record was set.

Zhang Fu, vice president of the Fourth Research Institute of the National Aerospace Science and Corporation and chairman of Aerospace Science and Technology Rocket Technology Co., Ltd., said: "For commercial launches, the lack of competitiveness gets no orders. Fast Boat rockets charges about $10,000 per kilogram of load, an attractive price."

Recently, the world's space powers have started the development of heavy-lift launch vehicles: in October 2010, the United States clearly proposed to accelerate the development of new heavy-lift launch vehicles; at the end of 2013, the Russian Federation Space Agency planned to start the development of heavy-lift launch vehicles, with the expected carrying capacity ranging from 16.6 tons to 130.

A competition for heavy-lift launch vehicles has begun.

This time, the Chinese will not lag behind. The white book *China's Space Flight 2016* wrote:

> *In the next five years, China will develop and launch non-toxic and pollution-free medium-sized launch vehicles, improve the spectrum of new launch vehicles, and further enhance reliability; carry out key technical research of heavy-lift launch vehicles, break through the key technologies, such as overall rocket body, liquid oxygen kerosene engine and hydrogen and oxygen engine with a large thrust; initiate the implementation of heavy-lift launch vehicle projects; carry out research on low-cost launch vehicles, new space-to-Earth reusable transportation systems and other technologies.*

According to Wu Yanhua, deputy director of the National Space Administration, China's super heavy-lift launch vehicle is proposed to be named the Long March 9, which is expected to have the first flight in 2030.

The development plan of China's heavy-lift launch vehicle plans that in the future, the diameter of the rocket body will reach 9 meters, the total length will be nearly 100 meters, the take-off mass will be 3,000 tons, and the carrying capacity on low-earth orbit will be 100 tons. Through the reorganization of the existing modules, the carrying capacity will reach 125–130 tons, which is the basic requirement for manned lunar landing and large-scale deep manned space exploration missions in the future.

The Best Fuel

Long March rockets have made glamorous achievements over the years. Regardless, the journey continues as the Chinese always keep it in mind that Long March rocket carries the spirit of the Long March of the Red Army. They dare not disappoint the great name of the Long March!

Today, the China Academy of Launch Vehicle Technology stores a special national flag in its collection. The bright red flag bears the signatures of 232 veteran Red Army soldiers who participated in the Long March. Every signature sends sincere trust and encouragement.

The day that the flag was handed over was the 70th anniversary of the victory of the Long March of the Red Army and the 50th anniversary of the founding of the aerospace industry. Nearly 1,000 employees of the China Academy of Launch Vehicle Technology stood solemnly in the rain. President Wu Yansheng took the flag from Li Zhongquan, an old Red Army soldier who had climbed over snow mountains and grasslands.

This is a mission handover, a spiritual continuation. The spirit of the Long March was, is, and will still be the best fuel to thrust forward China's space flight.

The Long March of the Red Army and China's space flight have been tightly intertwined in a special way. Marshal Nie Rongzhen, who served as the political commissar of the First Red Army during the Long March, was the pioneer of national defense and space flight in China. Sun Jixian, who led 17 warriors in crossing the Dadu River, served as the first commander of the predecessor of the Jiuquan Satellite Launch Center, China's first integrated missile test range, 22 years after the victory of the Long March. There are more who were enlisted in the Long March and later became the pioneers of China's space industry.

Today, Long March rockets have become the spiritual link between the Red Army's Long March and China's space flight. "It is no coincidence that our launch vehicle was named after the Long March. The spirit of space flight and the spirit of the Long March run in a single line. Moreover, the Red Army's indomitable and fearless style in the Long March has always inspired the national rejuvenation. A leader of the China Academy of Launch Vehicle Technology spoke in excitement on the spot, "China's rocket industry should overcome all difficulties and obstacles and be victorious as the Long March did."

A high-quality-talent team of a new generation of launch vehicle's technical innovation and test has risen, with nearly 80% of the research core under 35 years old.

A young team can bear hardships, tackle obstacles, and dedicate wisdom and energy. They are the ones who forge the golden reputation of Long March rocket.

"What motivates diligence and boldness in scaling heights? The belief that we are the creators of the most important tools of our country and the escorts of the big rockets." Zhang Shu, born in 1989, and the technical director of the software of the Long March 5 TT&D control system, expressed the aspiration of this young team in his own poems.

The young team frees concerns over the future of the Long March rockets.

CHAPTER 4

The Lofty Ideals of Manned Space Flight

Astronaut Yang Liwei gazed up at the gate of the Jiuquan Satellite Launch Center to catch a glimpse of the sky. The sun was about to rise and the desert to wake up from a sound sleep. Twilight sketched the beauty of the horizon.

The scene reminds me of the words of Konstantin Tsiolkovsky, a Russian and Soviet rocket scientist and pioneer of the astronaut quote: "Earth is the cradle of humanity, but one cannot live in a cradle forever."

It was in 1911 when Tsiolkovsky wrote the words. And it is them that inspired the dream of space flight across the planet.

Russians, Americans, and Europeans have successively crawled out of the cradle over the past century. And the Chinese closely followed.

It was 5:28 a.m. on October 15, 2003, the moment of a lifetime for Astronaut Yang. To make it this far, he had devoted relentless hard work and sweat since he joined the Chinese People's Liberation Army astronaut team in 1998.

It was a historic leap for China's manned space flight. To make it this far, it took Chinese aerospace undertakers 36 years of bittersweet explorations since the initiation of Dawn, the first manned spacecraft program came out in 1967.

Both sides of the corridor outside the launch center were filled with people coming to see him off. They were silver-haired aerospace experts, teenagers holding flowers, people of all ethnicities dressed in festival costumes, comrades-in-arms with Yang, and reporters including myself ...

With keen attention, astronaut Yang took the first step.

As the poet Liu Xiang of the Han Dynasty wrote, "As the wind howled and the water froze, the hero left and never returned." Tension was in the air. After all, it was the first time that China actually sent astronauts into space; there were too many unknown challenges and too many uncontrollable risks. And in humanity's conquest of space, 22 astronauts have already sacrificed their precious lives.

"With the five-star red flag flying in the wind, how thunderous the song of victory, singing our great motherland ..." Heroism permeated the air – the warriors' expedition represents the country, 56 ethnicities and hundreds of millions of compatriots.

Waving colorful flags and flowers, those seeing him offs shouted, "Yang Liwei, we wish for your safe return."

Every word radiated a ray of warmth. Astronaut Yang was aware of that his wife and children, his parents, national leaders and common folks, were all gazing upon and cheering for him. Standing a few meters away, I noticed that every step he took was firm and strong.

"Commander in-chief, ordered to carry out the first manned space flight mission, I am ready to go on standby, please give instructions! Yang Liwei, astronaut of the Chinese People's Liberation Army Astronauts Brigade!"

"Off you go! The motherland and the people wait for you to triumph!" The commander-in-chief issued a solemn order.

"Yes Sir!" Yang returned a standard military salute and boarded the bus to the launch site.

On that day, I had the privilege to witness this glorious moment of manned space flight in China. It was a great expedition of the Chinese nation: to realize the dream of flying, something the country has craved for over a thousand years.

Looking back from the historic moment of Shenzhou 5 expedition, we can clearly see Chinese arduous exploration and tortuous tackling of key problems to make the dream come true; Looking forward, we can clearly see the accelerated development and lofty ideals of China's manned space flight.

Whoever Controls Space Controls the Earth

Reviewing the development of world space flight, every step of manned space flight is eye-popping.

Scientists usually define manned space flight as the most complex, sophisticated, profound, grandest, challenging, and risky system engineering of mankind to date.

This feat is parallel to the achievements of the Age of Discovery in the Middle Ages.

In the center of global attention, every step of the world's manned space exploration was accompanied by unpredictable, huge risks.

It was on a clear morning that the history of manned space flight started.

A small village along the Volga River was bathed in serenity of sun rays. Farmers were ploughing in the fields, cattle and sheep grazing, and children running around. Suddenly, with a strange sound, a huge colorful parachute descended from the sky. Curious eyes were fixed on the fall.

At 10:55, the parachute landed in a field. What was that? Villagers hurdled curiously. Then Yuri Gagarin in an orange spacesuit came into sight.

Soviet astronaut Yuri Gagarin took off on the Oriental spacecraft from the Baikonur launch site in the Kazakh steppes over an hour before. Having experienced space and unimaginable zero gravity, he returned to Earth safely 108 minutes later.

Gagarin was fortunate to be the first man to fly into space in history and became a space hero in the hearts of people all over the world.

April 12, 1961 was no more than an ordinary day for the villagers, but it was groundbreaking for the world. Gagarin's 108-minute journey in space stunned the planet. For the first time, a human had broken away from the cradle and the epoch of manned space flight began.

The day has been designated as the International Astronomy Day to celebrate this epoch-making achievement of mankind in the exploration of the cosmos.

The Soviet Union took the lead. The United States caught up right away.

On May 5, 23 days after astronaut Gagarin's first space flight, Americans used the spacecraft Mercury-Redstone 3 to send astronaut Alan Bartlett Shepard Jr. into space. At an altitude of 186 kilometers, the flight lasted for 15 minutes and 23 seconds.

John Fitzgerald Kennedy, then President of the United States, asserted "Whoever controls space controls the earth." Hereafter, the United States and the Soviet Union started a fierce competition in manned space flight.

On February 20, 1962, American astronaut John Glenn became the first American to enter Earth orbit by flying the Mercury-Atlas 6 three times around the Earth.

On June 16, 1963, Soviet astronaut Valentina Tereshkova made the trip and became the first female astronaut in the world.

On October 12, 1964, Soviet cosmonauts Vladimir Komarov, Boris Yegorov, and Konstantin Feoktistov circled the earth 16 times in Voskhod 1, the second-generation manned spacecraft and returned 24 hours and 17 minutes later. A new record was set, a three-astronaut manned flight was achieved for the first time.

On March 18, 1965, Soviet astronaut Alexei Leonov, who boarded the spacecraft Voskhod-3KD No.4, conducted the first spacewalk.

On March 16, 1966, the American Gemini 8 spacecraft carrying astronauts Neil Armstrong and David Scott docked with a target vehicle named Gemini 8 in flight, which was the first space docking.

On January 16, 1969, the Soyuz 4 spacecraft docked with Soyuz 5 successfully. For the first time, two spacecraft docked in space.

However, in 1969, the greatest surprise to the world was the American mission Apollo 11, in which the lunar module traversed 380,000 kilometers in 102 hours and 39 minutes and landed safely on the Moon at 16:11:40 a.m. Eastern Time on July 20.

About 600 million people around the world watched Armstrong walk on the Moon on television that night.

The gangway ladder leading to the Moon had only nine stairs. Armstrong walked carefully for three minutes. At 22:56:20, his feet touched the surface of the Moon. And the first human footprint was left on the Moon.

From the distant Moon, Armstrong uttered the words that will always be remembered: "That's one small step for man, one giant leap for mankind."

Immediately, Armstrong and astronaut Buzz Aldrin set up a stainless-steel memorial plaque on the Moon. It bears the signatures of President Richard Nixon and the two astronauts and has an inscription that reads:

Here men from the planet Earth first set foot upon the Moon July 1969, A. D. We came in peace for all mankind.

Nixon said in the phone call made from the Oval Office to the Moon "Because of what you have done, the heavens have become a part of man's world."

America's successful landing on the Moon pushed the competition for manned space in the world to a climax. From 1969 to 1972, the United States sent 12 astronauts to the Moon in five flights in Apollo spacecrafts.

Hereafter, the world's manned space flight has been developing at faster pace.

In 1976, the Soviet Union withdrew from the construction plan of space station Mir. In February 1986, it was sent up by carrier rocket Proton K.

In 1983, the United States envisioned an international space station. In November 1998, Zarya, the first module of the International space station, was launched into space.

Statistics reveal that up to the launch of China's Shenzhou 5, 240 manned space flights had been carried out in the world, more than 200 astronauts have entered space successively, and Russian astronauts also set a record of 483 consecutive days in space.

The Late Coming Manned Spacecraft: Shuguang

Compared with manned space flight in the world, China had a late start, years behind.

Manned space flight is the embodiment of a country's comprehensive national strength. It is a game of great powers. In plain language, without the technology, there is no going to space.

For new China, with a large population but a weak economy, manned space flight was beyond reach for a long time despite a long craving.

Fan Jianfeng, the first generation of space science and technology workers in China, participated in the demonstration and design of the overall plan of China's first manned spacecraft.

Fan was at the rocket test site the day he learned the news of Soviet astronaut Gagarin. He recalled he had trouble sleeping that night. He stood in front of the window and stared at the stars in the sky. He fantasized making a manned spaceship overnight that would send Chinese astronauts into the sky early in the next morning.

However, China was stuck in three years of natural disasters. Hunger was the priority issue to address.

From that day on, Fan began to track and study manned spacecraft of the Soviet Union and the United States. He later found Qian Xuesen and suggested that manned spacecraft be developed as soon as possible. Qian turned out very supportive. "Anyway, we should first start the preparations for manned space flight."

In May 1967, the Chinese Academy of Man-made Satellite Spacecraft was formally established. In January 1968, the Symposium on the Conception and Demonstration of the Master Plan of China's First Manned Spacecraft was held. In April, the General Design Chamber of Manned Spacecraft was officially founded by the Chinese Academy of Space Technology, with Fan as director.

At the same time, taking Qian's advice, the Institute of Aerospace Medicine and Engineering was formally established. The secret unit codenamed "507" is the predecessor of the Beijing Institute of Aerospace Medical Engineering, China's current astronaut training institute.

After the establishment, Fan summoned experts to embark on the designs as soon as possible. Later, more than 200 experts from all over the country participated in the demonstration of the overall design scheme of manned spacecraft.

Soon, the manned spacecraft program was officially in operation and the spacecraft was named Shuguang (meaning dawn).

Opinions differ concerning the name. Some believe it was Qian's sole idea while others argue it was the result of a discussion amidst a group of experts. Fan recalled that the name Shuguang was a decision of the Office of the Central Committee, which Qian disclosed to them.

When the Apollo spacecraft landed on the Moon in 1969, China was plunged into the political storm of the Cultural Revolution The development of the Shuguang manned spacecraft were in stagnation.

In April 1970, as China's first artificial satellite, Dongfanghong 1, was about to launch, aerospace experts gathered at the Beijing Jingxi Hotel for a meeting to discuss the Shuguang program again. Shuguang manned spacecraft was conceived as a two-compartment spacecraft whose appearance was modeled on the Gemini 8 of the United States. Scientists and technicians also made a full-scale model.

During the meeting came the news of the success of Dongfanghong 1. It greatly encouraged the attendants. Morale was high. They proposed the vision of sending the first manned spacecraft to space in 1973.

In May, the, Fifth Academy of the Ministry of Defense and the Air Force jointly drafted a request report on the development of Shuguang and submitted it to the State Commission of Science for National Defense and the Central Military

Commission. On July 14, Chairman Mao Zedong drew the last circle of approval on the report.

Consequently, China's first manned spacecraft program was officially activated, with the codename Project 714.

One year later, unexpectedly after the "9.13" defection of Lin Biao, Project 714 was halted. The reason was claimed to be political, but at a deeper level, the weak economy was the culprit.

At that time, Chairman Mao decided to get the tasks on the planet done before embarking on an extraterrestrial adventure. In August 1978, Deng Xiaoping gave important instructions on the development policy of China's missile and space technology. now that China should not join in any space contest, a Moon landing was not urgent; a developing country should concentrate its efforts on urgent and practical application satellites.

The cancelation of Project 714 turned out a wise call provided that China's weak economy back then would not be able to afford the enormous cost of manned space flight.

Deng Xiaoping's Close Contact with Space Flight in 1979

China and the United States established diplomatic relations on New Year's Day, 1979.

On January 28, 1979, the first day of China's lunar year, Deng Xiaoping visited the United States. A five-star red flag was raised for the first time on the south lawn of the White House.

Deng's visit to the United States was memorable. There was a classic photo in which he waved to the Americans in a cowboy hat while watching a rodeo performance in Simonton, Texas.

However, Chinese aerospace workers were more impressed with Deng's close contact with space travel during the trip.

On the first afternoon, Deng and his delegation toured the Air and Space Museum in Washington, DC.

The next day, they flew to Houston to see the Houston TT&C center, which accomplished the Apollo lunar landing mission, and then to the Lyndon Johnson

Space Center, also known as the US manned space base. John Glenn, the first American astronaut to orbit the earth, guided Deng to see Apollo 17's command module, lunar rover, and a replica of a three-story lunar lander.

During the visit, Deng was highly intrigued. He held the hand of Astronaut Glenn and quipped, "I just met a god in the flesh."

Aged 74, he later vigorously strode onto the flight simulator of the shuttle cockpit. Under the instruction of an American astronaut, he put on headphones, simulated flight at twice the speed of sound, and experienced the landing of a space shuttle from 100,000 meters above earth.

He also made inquiries regarding the details about the military, economic, and scientific significance of space flight at a lunch with NASA officials and experts.

It was James Schlesinger, then Secretary of Energy of the United States, who accompanied Deng Xiaoping throughout the visit. A reporter tossed a question at Schlesinger whether he considered the Chinese were left too far behind in space technology to catch up with the Americans and the Soviets? He shook his head, "I don't think so."

A few days later, at a welcome banquet in Texas, Deng concluded in his speech, "I have had an unforgettable day in Houston. I met with some American scientists, scholars, and astronauts at the Space Center and temporarily became an "astronaut" with vice Premier Fang Yi. We had a trip in space. I'm glad. There is so much to learn from America ..."

There is no doubt that Deng's close contact with the American space industry has deepened his understanding of human space flight. I assume that Deng's support for China's manned spaceflight program may have originated from this trip.

Allegedly, Deng regretted he did not live long enough to witness two tasks completed: one was the Three Gorges Project, the other manned space flight.

An aerospace expert confided in me that when the news of Deng experiencing space flight simulation in the States was brought to his knowledge, he felt embarrassed and a sense of mission and urgency arose spontaneously.

To narrow the gap, one must spare no effort.

Project 863 Restarted Manned Space Flight Program

In the spring of 1983, President Reagan proposed the Strategic Defense Initiative (SDI), also known as Star Wars. It was a great surprise to the world.

Though defined as a strategic defense initiative against the Soviet Union's military threat, it in fact was another high-tech development program of the United States after the Apollo project. The cross-century project would take 25 years and cost $1 trillion in order to achieve the strategic commanding heights of high-tech in the 21st century for the United States.

A stone stirs up a thousand waves. Following America's action of Star Wars, countries have come up with their ambitious high-tech strategic plans.

Japan in November 1984 put forward their *Outline for Revitalizing Science and Technology*, confirming the basic national policy of building a country by science and technology.

Led by France, 17 Western European countries jointly formulated the Eureka plan in April 1985. Eureka, in Greek, means: "I found it." Over 2,000 years ago, when Archimedes discovered a great scientific theorem in a bathtub, he ran out naked, shouting, "Eureka, Eureka ..." Since then, the word has become synonymous with a great discovery.

Closely thereafter, the Soviet Union released the Outline of High-Tech Development and India the Statement of New Technologies Policy ...

In the global high-tech tide, how could China stay put?

Four scientists forward came who had made great contributions to the creation of atom bombs, hydrogen bombs and man-made satellites. Wang Dajun, Wang Lianchang, Chen Fangyun, and Yang Jiacung, jointly signed the Letter of Suggestion on Tracking the Development of Strategic High Technology in the World and submitted it to Deng Xiaoping and the Party Central Committee.

Every word of the Letter was valuable, as it was the fruit of the combined wisdom of four scientists.

High technology was in the center of international competition for national power, and China was no exclusion; the priority was to possess high technology as the presence would make a big difference; and cutting-edge technology was beyond money's reach...

Caught in a fierce global competition, if we slack off a little, we would be finished for good. Investment in high-tech programs was better sooner than later. When the whole world was accelerating the development of new technologies, sluggishness would lead to unimaginable consequences ...

On March 3, 1986, the proposal was presented to Deng.

Deng left a directive on March 5 confirming the significance that Comrade Ziyang should be invited to chair the discussion with experts and relevant personnel, and to put forward opinions for wiser decision-making; the matter should be addressed without delay.

In April, more than 200 military and civilian experts were organized by the leading team of the Science and Technology of the State Council, the State Science and Technology Commission and the National Defense Science and Industry Commission, together with relevant ministries and commissions. Preparations of the plan for high-tech research and development in China were commenced. And four months of efforts gave birth to the Outline of National High-tech Research and Development Scheme.

The Political Bureau of the Central Committee convened an expanded meeting to formally approve the Outline of National High-tech Research and Development Scheme in October. And 10 billion yuan was granted for the implementation of the Outline.

On November 18, the State Council officially issued the notice on the Outline of National High-tech Research and Development Scheme, and China's strategic high-tech development plan for the 21st century was officially made public. And the scheme was named Project 863, as the date of the submission of the Letter by four scientists and Deng's approval was March of 1986.

Project 863 selected 15 main projects in seven fields, such as biotechnology, aerospace technology, and information technology, as the focus of future development. The goal was to narrow the gap with foreign countries in the next 15 years and strive to make breakthroughs in advantageous areas.

Referred to as "863-2," aerospace technology was the second of the seven fields. It aimed at developing a manned space station system. Experts believe that the space station system is defined as a base for long-term manned space science experiments in near-Earth orbit and will become one of the significant symbols of great power in the 21st century.

Project 863 has ignited the hope of developing high-tech in China. And the long-set-aside Chinese manned space flight program was restarted in 1986.

The Rise of Shenzhou

Manned space flight was a road of thorns for China.

A few steps on the road would lead to a fork, at which choices must be made. Every choice was difficult, as it was either a short cut or a big detour; and every choice determine if it would either win or lose.

The first intersection appeared soon after the project activation. Which of a spaceship or a space shuttle would make a better launch vehicle?

What would be the choice? Chinese experts have never had a dispute as heated.

Spacecraft voters believed that it was the simplest and cheapest tool with a shortest development cycle for exploring space; it could transport astronauts and materials for future space stations; China's full harness of the recovery of returning satellite with a high success rate was conducive to ensuring the safe return of astronauts; however, the technology involved in a space shuttle was complex; the investment was large; the development cycle was long; the risk was high; and the success rate was lower.

Space shuttle advocates argued the starting point of manned space flight in China should be set higher; in terms of technological development, the trend was space shuttle and a spacecraft was something from the 1960s; it would combine the advantages of rockets, satellites and spacecrafts, and could launch vertically like a rocket, orbit the earth like a satellite, and return to the ground like an airplane; in the long run, a space shuttle with higher technology involved could be reused, which is more economical and cost-effective than one-time use.

Neither could persuade the other. Eventually, the dispute lasted for over three years and spread to Zhongnanhai, the residential compound in Beijing housing top party leaders.

Liu Jiyuan, Vice-Minister of the Ministry of Aeronautics and Space, delivered Recommendations for the Development of Manned Space Technology in China to Deng on January 30, 1991, at the China Aerospace High-tech Reporting Conference. It read:

> *We believe that whether to embark on manned space flight is a political decision, not a purely science and technology; scientists and technicians are not to make the call ... The development of China's aerospace industry was made on the shoulder of the older generation of proletarian revolutionaries. The hard-won international*

status of aerospace is at risk of being lost. We urge the Central Committee to make a decision as soon as possible.

China's actual national conditions, funding, development cycle, security risks, and many other factors have led to the final decision: a spacecraft. Of course, it was not to follow the old road of the Soviet Union, but to break out a new path with Chinese-features.

Ren Xinmin, as a senior adviser to the Ministry of Aeronautics and Space, went through the entire process of debate and decision-making. He made two trips to Zhongnanhai to report to the central leadership.

Ren had clear memories of two significant days:

One was March 15, 1991, when Ren and Qian Zhenye, the group leader of the Manned Space Project Expert Team, went to Zhongnanhai to give a briefing. Having heard out their elaborations, the central leadership responded that manned space flight should start with a spacecraft and we should strive to have it done before the 50th anniversary of the founding of the People's Republic of China.

The Central Special Committee convened a meeting in Zhongnanhai to gather the opinions of the Expert Committee on Space Technology of Project 863 on the development of manned space flight in China on January 8, 1992. Ren Xinmin concisely explained to the central leadership using the manned spacecraft model. It was the Memorial Day of Premier Zhou Enlai. When he stepped into Zhongnanhai, Ren had a flashback of Dongfanghong 1, when he had reported to Premier Zhou, and now it was China's manned space flight that was about to shine ...

The rise of Shenzhou began in Zhongnanhai.

On August 1, 1992, the Central Committee convened another meeting to draw the three-step plan for China's manned space project.

The first step was to launch unmanned and manned spacecrafts, complete a preliminary supporting pilot-manned spacecraft project and carry out space application experiments. The second was to achieve a smooth docking between a manned spacecraft and space vehicle after the successful launch of the first manned spacecraft, and apply manned spacecraft technology to refit and launch a space laboratory to address the issue of space practice limited to a certain scale and short-term care. The third step was to create a space station to address the issue of space practice extended to a large scale and long-term care.

On September 21, 1992, the meeting was extended by Standing Committee of the Political Bureau of the Central Committee of the Communist Party of China. Official approval of the feasibility report of manned space engineering project was issued. Therefore, the project was officially a go. As the approval date was September 21, 1992, it was named Project 921.

Independent Chinese Spacecraft

Manned space flight was a national event, whose scale was never so enormous.

The project was divided into seven systems: astronauts manned spacecraft, space application, launch vehicle, launch site, TT&C communication, and landing site.

The plan urged to have the first test unmanned spacecraft completed in 1998 and its first flight in 1999.

Faced with arduous tasks to be done urgently, who would qualify as the chief engineer to command an enormous science team?

Rumor has it that the leaders specially solicited Qian Xuesen's opinions. After meticulous consideration, Qian recommended Wang Yongzhi.

The recommendation resulted from the deep impression Wang left on Qian.

In the summer of 1964, when the Dongfeng 2 missile was launched, the desert was scorching. High temperature gasified and expanded the fuel, so it could not be replenished in storage. Common sense it is a missile cannot reach the range if there is not enough fuel. They racked their brains about how to inject enough fuel.

As a lieutenant, Wang came up with an idea: the release of 600 kilograms of fuel would enable the range of the rocket to be reached. "That makes sense. We found the way." Qian affirmed his converse thinking. Since the weight of the projectile affects the range, the release of fuel would reduce it and therefore have the range reached. As a result, Wang's method worked. Range was reached and the target was accurately hit.

It was also well known that Wang dared to shoulder the responsibility. To enter the international space launch market, he publicly took a military vow at the Oath Committee of Long March 2E's Development that he would gladly accept dismissal of his position should the mission fail.

On November 15, 1992, Wang was officially appointed Chief Designer of Manned Space in China under the recommendation of Qian after the investigation of various organizations. That year, Wang Yongzhi just turned 60.

Wang was born in a poor village in Changtu County, Liaoning Province, in 1932. His father had a 100% peasant background. In a childhood of poverty, the eldest brother secretly took him to the village school for enrollment. Given that he did not even have a proper name, a schoolteacher, pushing his glasses and staring, gave him the personal name Yongzhi after quoting an old saying that where there is a will there is a way.

And the name was widely accepted ever since. In 1952, he was admitted to the Department of Aeronautics, Tsinghua University. In 1955, he went to the Moscow Aeronautical College in the Soviet Union for further study. Academician Vasily Pavlovich Mishin, Wang's tutor when he was studying abroad, was a well-known rocket and missile expert in the Soviet Union and chief designer of the Soyuz spacecraft. Wang was a straight-A student during six years of studies. And his graduation thesis "Intercontinental Missile Design" got the highest score. In his thesis defense, Mishin profoundly said to his Chinese disciple, "This is your first time as the chief designer of intercontinental missiles. I hope this is not the last."

The excellence of Wang went beyond Mishin's expectation. In April 2001, Wang Yongzhi went to Russia to celebrate the 40th anniversary of astronaut Yuri Gagarin's space journey, and returned to the Moscow Aviation Institute, his alma mater, to receive an honorary doctorate and the gold medal of outstanding graduates. At the ceremony, Wang Yongzhi was invited to introduce the current landscape of manned space engineering in China. Mishin proudly said, "You have heard that Chinese spacecraft was independent."

Manned space flight cost an unprecedented amount of money. Take the United States for example. The U.S. Mercury cost $390 million; Gemini $1.28 billion, the Apollo lunar landing project $25 billion and the space shuttle project $15 billion. For China, money was the problem. The total grants by the central government were only RMB10 billion.

Manned space flight had never faced with technical barriers so high. Take the Long March 2F rocket as an example. The rocket has 10 subsystems and more than 200,000 components. To enhance the reliability of a component from 90% to 97%, more than 200 repeated tests needed to be carried out without a single failure. Therefore, China could not be more rigorous with the product quality. How rigorous was it? For example, it only took 600 seconds from the rocket to go from launch to orbit, but the design life of the rocket reached 546 hours.

Manned space flight had a scale as large and management as intricate as never. Take the Apollo lunar landing program as an example. Statistics reveals that 43,000 scientists and engineers participated in the implementation together with 420,000 direct staff members. There were 20,000 companies and 120 Universities involved in the development of Apollo spacecraft.

The scale of China's manned space project was vast beyond the imagination of the general public. Over 100 military and civilian departments and units directly undertook engineering tasks, and more than 3000 units undertook engineering research and development cooperation. There are millions of experts, engineers and technicians, ordinary workers and soldiers directly or indirectly involved in manned space engineering. The seven major engineering systems are composed of about 10 subsystems, and the subsystems contain more subsystems.

How to achieve more while spending less? How to achieve technical breakthroughs when the industrial base is weak? How to avoid detours and take leaps?

"If we follow the old path of the Soviet Union and the United States, we will always lag behind others." Chinese aerospace experts headed by Wang believed that China's manned space project must embody Chinese characteristics in general and take a leapfrog development path with a high starting point, high efficiency, high quality, but low cost.

The Spacecraft After 40 Years' Development

The first major decision that China's manned space flight had to make was between developing spacecraft or space shuttles. Three years of debates led to the final choice of spacecraft.

The second major decision to make was the design of the spacecraft between a two-module scheme and a three-module scheme.

And the debate over it went more than a year.

"I'm under a lot of pressure, and I've been thinking about how we can make a spaceship that still excites people decades after the first creation." Wang Yongzhi, chief designer of China's manned space project recalled every bit of the tension and uneasiness at the beginning.

By October 1993, five expert groups, including Ren Xinmin, had made the final choice, thanks to a 3:2 voting result. The three-module scheme won, including a propulsion module, a return module, and an orbital module.

Qi Faren was once again appointed as the chief designer of the spacecraft system. The first time I met Qi was at the Jiuquan Launch Site. Aged 59, his silver hair made him look like a tugboat captain, who has always been pulling the big ship of China aerospace forward.

Qi was born in Fuxian County, Liaoning Province, and graduated from the Department of Aircraft, Beijing Aviation College in 1956. In the Dongfanghong 1 project, he was enlisted as one of the 18 warriors. From Dongfanghong 1 to III, he has been on the satellite research frontier for more than 20 years.

Now that the development of a spacecraft started from scratch, countless difficulties awaited, large and small. And the Shenzhou manned spacecraft has 13 subsystems with 120,000 components.

China did not take the path of from small to large, from single to multi-crew, from single cabin to multi-cabin as did the Soviet Union and the United States, but directly set the goal as the third generation of internationally advanced manned spacecraft.

A leap requires double the efforts. The incredible tenacity enabled Qi and his colleagues to, under the circumstance of relatively less sophisticated design and manufacturing level, have made a spacecraft with an advanced performance that outshone 40 years of foreign wisdom and technology.

Compared with foreign spacecraft, China's Shenzhou has the same significant originality: after the astronauts' return in the capsule, the orbital module with power supply, propulsion system, and control system can also be used as a space experimental satellite, and stay in orbit for half a year; it also lays the foundation for future docking.

Qi worked tirelessly for 11 years since the activation of the manned space program in 1992 to the first manned space flight of Shenzhou 5.

At the moment of success, Qi humbly said: "It seems as if, we created history. Yet, in fact, we were merely standing on the shoulders of giant predecessors." He made a profound and vivid analogy. When a person is hungry, one slice of bread is slightly helpful, the second slice eliminates more hunger, and the third drives it away. Seemingly, the third takes all the credit, but without the first two, the third would not work as well. The first generation of aerospace workers returning from the United States, represented by Qian Xuesen, can be regarded as the first slice of bread, while the second generation returning from the Soviet Union as the second, and himself the third.

While the spacecraft was still under development, the picking of a proper name had begun amid the Chinese aerospace workers, who treated it as if naming their own child. In 1993, the China Manned Space Engineering Office issued a notice to all the units participating in the development to solicit more ideas. Soon, many names with Chinese characteristics, such as "Huaxia," "Jiuzhou," "Tenglong,"

"Shenzhou," were recommended. And repeated discussions led the China Manned Space Engineering Office to use the name "Shenzhou."

The literal meaning is the magical boat of the galaxy, and it is also the homonym of another Chinese word meaning the divine land of China. It represents the reception of national support and the product of great cooperation from all walks of life. At the same time, it has the meaning of spirited, indicating that the spirit of the Chinese nation would cheer up.

Later, Comrade Jiang Zemin inscribed the word for the manned spacecraft. It was a passionate dream of a thousand years. Hereafter, Shenzhou, a glorious name carrying the Chinese nation's dream of flying, has spread throughout the country.

A Space Highway

"If an athlete starts late in a race, the only thing we can do is run faster." Recalling the full-scale launch of the manned space project, chief designer Wang Yongzhi spoke with passion.

On July 3, 1994, China's first manned space launch site broke ground at the Jiuquan Satellite Launch Center in the Badain Jilin Desert.

A vertical assembly test plant is the core building of a launching site and the symbol of the vertical launching mode. There has never been a 96-meter giant architecture built in the desert without any pull-through floors in the middle in Chinese nor Asian architectural history. A new scheme of the "reinforced concrete mega-frame-multi-cylinder space structure system" was proposed by the officers and soldiers of the special engineering installation brigade of the General Equipment Department of the Chinese People's Liberation Army. Its successful implementation not only saved over RMB40 million, but also had better fire resistance, heat preservation, and sound insulation than a steel structure. And it earned the site the name of a Chinese miracle amongst foreign experts.

The gate of the assembly vertical test workshop was the biggest in Asia. It is 74 meters high, 8 meters wide on the upper side, and 14 meters wide on the lower side. The overall weight exceeds 350 tons. In order to build it with high quality, science and technical personnel worked together to solve a series of technical problems such as transportation, installation of giant components, airtight and wind resistance in windy and sandy environments, and created a precedent for installing super-large-scale door equipment with simple construction tools.

Eight months later, a nearly 100-meter-high, world-class launching tower stood on the desert. The overall technical scheme with Chinese characteristics: vertical assembly, vertical testing, and vertical integrated transportation, is unique in the world space industry.

To ensure the first flight to be complete in 1999, the seven systems of China's manned space engineering went hand in hand, and each step embodied the magical Chinese efficiency.

On October 28, 1994, the construction of Beijing Space City began. And mere 300 days witnessed the completion of a modern space flight control center that integrated command and communication, information processing, control calculation, and flight control.

The Chinese Academy of Launch Vehicle Technology has organized over 300 technical research projects and conducted repeated tests to improve the reliability of the Long March rockets from 0.91 to 0.97. China therefore became the third country to master manned launch vehicle technology.

China began to multitask, selecting and training astronauts, designing and producing space environment control and life support equipment, carrying out medical supervision and support research, and developing spacesuits and space food. Breakthroughs up to international standards were made on short notice. Consequently, China had built the third astronaut center following Russia and the United States.

Meanwhile, China adopted highly intensive system optimization design to build the world's third astronaut search and rescue system, the Shenzhou landing site system.

Scientists and technicians of space application systems have developed 135 types of 188 shipborne science instruments and equipment successively. As a result, it opened up six science fields: space materials science, life science, astronomical observation, earth environment monitoring, space environment detection and prediction, and fluid physics.

A new Yuanwang ship entering the water and new ground TT&C equipment being installed, China's new generation of manned space TT&C network also came into shape quickly ...

On the thorny road to space, barriers were taken down one after another as Chinese aerospace workers bore hardships, persevered, and were dedicated. In the shortest time, a space highway was laid.

Shenzhou 1

1999: For a while, *Victory without War*, a book by Richard Nixon, former President of the United States, used to top the bestseller lists.

By 1999, however, the world's political landscape had not changed as described in the book. Socialist China, thriving in the east, was still growing stronger at an incredible rate.

1999 witnessed three globally focused events: China's 50th National Day, the return of Macao to the PRC, and the first flight of Shenzhou spacecraft.

Regarding the first two, vivid scenes of details still linger in the memory of the Chinese people.

As for Shenzhou's first successful flight, domestic media did not cover the story as much as it deserved. On November 21, 1999, brief words occupied a golden spot on the front page of domestic media:

> *At 6:30 a.m. yesterday, China's first space test spacecraft Shenzhou, named by Comrade Jiang Zemin, General Secretary of the Central Committee of the Communist Party of China, Chairman of the State and Chairman of the Central Military Commission, lifted off with a new launch vehicle at the Jiuquan Satellite Launch Center. After over 20 hours of on-orbit operation, it landed safely in central Inner Mongolia at 3:41 a.m. today. The Central Committee of the Communist Party of China, the State Council, and the Central Military Commission telephoned with congratulations.*

What was made known to the Chinese people was the astonishing take-off. Little was known about the stories behind of the scenes.

I had the privilege to witness all the excitement throughout the launch of Shenzhou 1.

On July 23, 1999, a special train carrying Shenzhou 1 departed from Beijing and arrived safely at the Jiuquan Launch Center three days later. And almost a month later, the Long March 2F rocket left the factory and arrived there too.

The launch concerned the future development of manned space flight and the honor and dignity of the country. Therefore, the General Command Office of Manned Space Engineering has ordered that should be no failures.

To eliminate the hidden danger of technology safety, ground training was organized for each system. Unexpectedly, the test of the core spacecraft system detected a serious problem: a directional gyroscope was stuck inside for unknown reasons. Since the directional gyroscope was the "eye" of the spacecraft, a malfunction would result in the spacecraft's failure to fly along the predetermined orbit and return.

The spacecraft condensed seven years of hard work by tens of thousands of scientists and technicians. And the news hit them off guard. What next? To solve the problem, the only way was to open the heat-proof bottom of the spacecraft and fully test the equipment inside.

Before the heat-proof sole of the spacecraft was closed and sealed, every wire, screw, and solder joint inside had gone through a series of tests. It was risky to open it for re-examination as a solder joint broken or a wire misconnected due to the re-examination would disable the whole spacecraft.

To open or not? The General Engineering Command Office convened an emergency meeting. Heated debates led to a consensus: the problem must be addressed before the launch; regardless of the high risk, it must be opened.

When the scheduled day of the opening finally came, the site was deadly silent with high tension. Wang Yongzhi, chief designer of manned engineering, came to the front line to supervise the action. Qi Faren, chief designer of spacecraft, inspected key spots over and over again.

The process didn't last long, but everyone on the site felt it was much longer than the seven years in which the spacecraft was built.

The lifting was conducted millimeter by millimeter. When the spacecraft was lifted 360 millimeters off the ground and operators put a cushion on, there was a long sigh of relief in the room.

The heat-proof sole of the spacecraft was then dismantled. Fortunately, the malfunction was not perplexing: only a signal line was broken when the sole was closed. It was a huge relief, and many technicians on the scene shed tears.

However, there was a new problem. The launch window was postponed from November 8–12 to November 18–22 due to the delay of opening the heat-proof sole. Upon the settlement of date, two pieces of bad news ensued. First, the Yuanwang 1 surveying vessel encountered strong storms at sea, thus was unable to reach the mission area as planned. Then, the Space Environment Forecast Center of the Beijing Academy of Chinese Sciences telegrammed that there would be a Leo meteor shower on November 18.

A meteor shower is the natural enemy of space flight. If Shenzhou 1 were to be launched on the 18th, it would most likely encounter the meteor shower. Therefore,

Shenzhou 1 was vertically shipped from the assembly plant to the launch tower on November 16, 1999 (Photographed by Qin Xian'an)

General Engineering Command Office held another emergency meeting and decided to postpone the launch date to November 20.

On November 18, the meteor shower arrived as expected. Consequently, the 2-day delay was a wise decision.

In the midst of a cold wind, the Long March 2F carrying Shenzhou 1 was erected on the launch tower as if a master sword was pointing at the sky at 6:00 a.m. on November 20. At 6:30 a.m., the desert trembled. Rockets and test spacecraft broke free from the earth and began the maiden voyage of China's manned space flight.

After 14 circles around the earth, Shenzhou 1 made it back to the ground at 3:41 a.m. on November 21. The actual landing point of the return module was 11 kilometers away from the theoretical one, and all the equipment inside was well.

"Shenzhou 1 was a merely prototype." Qi Faren recalled that Shenzhou 1 had designed the minimum configuration scheme, where only eight of the 13 subsystems were utilized.

"It's risky to make the first prototype go." Qi used the idiom "Days wear on like years" to describe his deep anxiety during 21 hours of the operation. "My heart did not land until the re-entry capsule landed."

After the recovery of the re-entry capsule, Qi rushed to the hospital to show his wife Jiang Fuling in the hospital the newspaper. The news of the successful launch of Shenzhou 1 was on the front page. Jiang, suffering from lung cancer, took Qi's hand and shed tears of joy. By then, it had been four months since Qi left for the mission. And the success was the best gift for his 66th birthday.

Hong Kong media spoke highly of the spacecraft. The *Dagong Daily* used over two full pages to cover the many aspects of the successful launch. The editorial titled "Shenzhou Exhilaration" commented that Hong Kong people, as every Mainlander, were washed over with joy for the successful mission. China's possession of independent spacecraft ended the domination of space by the United States and Russia. China earned itself a spot. It was indeed the most precious welcoming gift for the Chinese at home and abroad stepping into a new millennium.

The launch involved two flags on the Shenzhou 1: the national flag of the People's Republic of China and the regional flag of the Macao Special Administrative Region.

On December 20, 1999, the day of Macao's return, the flag of the Macao Special Administrative Region, which has special significance, was handed over to the officials of the Special Administrative Region by the Aerospace Industry Corporation.

As the new year sun rose on January 1, 2000, the first space flag of China made by aerospace workers, was raised on Tiananmen Square.

Shenzhou 2

December 31, 2000, the last day of the 20th century, was the darkest day for Huang Chunping, commander-in-chief of the rocket system of the Shenzhou 2 launch mission.

In the afternoon, the telephone rang annoyingly. "Commander Huang, the rocket was hit." The words from the other end of the phone triggered a massive sweat and soaring blood pressure for Huang.

It struck so abruptly as lightning that no one was prepared.

Prior to that, all preparations for Shenzhou 2 went well.

On November 8, 2000, Shenzhou 2 was sent to the Jiuquan Satellite Launch Center by a special plane. Following that, the Long March 2F rocket arrived safe and sound by rail.

On December 31, the rocket had successfully docked with the spacecraft, and the fuel of the spacecraft had been added. As planned, vertical transshipment would be conducted the next day and the launch would be in four days.

What now? Huang's team has made numerous contingency plans. But none expected that the rocket would be damaged after the most challenging transportation from Beijing.

Nothing similar had ever happened throughout Chinese space flight history. Huang hurried to the site with his team and discovered that it was an operator's misconduct that caused the hit.

At this point, the rocket was wrapped in an eleven-story platform. Starting from the top, technicians made the investigation story by story. Full attention was concentrated to scrutinize every corner. The rocket was so thin-skinned that anything stiff would pose a threat. "Counting shows the rocket suffered 18 hits, leading to a 20-centimeter position deviation. I was sad to see the wounds as if it were on myself," claimed Huang.

Immediately, the General Command Office convened an emergency meeting.

The safest option was a cancellation of the launch mission, which would take time to replace all the parts that had been hit, and to reassemble and re-test them. However, the clock was ticking. Another two months to have all the actions above done would miss the deadline.

Instead, they decided to run a "physical examination" of the rocket. As long as the damages were not too severe to repair, the launch could still go on.

That night, heavy snow fell on the desert. The "examination" must be completed in the shortest time. Missing the best launch window would ruin the Shenzhou 2 launch. Furthermore, all subsequent launches would suffer delay and disruption.

All-nighters were pulled with no second spared. On January 3, the expert group finished the "physical examination" and concluded the rocket body was normal. The rocket's "windpipe" was lucky to be a few millimeters away from the hit. Should the windpipe be hit, the rocket would be totally discarded as useless.

What a silver lining. Next, technicians repaired the fairing where the rocket was obviously impacted and replaced the parts that were not 100% certain of remaining intact.

Three days later, the General Command office summoned a meeting to issue a "work permit" for the restored rockets and set the launch window at 1 a.m. on January 10, 2001.

"Ignition!" On that cold night, I witnessed the splendid scene of the lift-off.

The launching tower, in the illumination of nearly 100 spotlights, shone as a delicate crystal palace. The milky white Long March 2F burst out tangerine flames

Shenzhou 2's successful launch from the Jiuquan Satellite Launch Center at 01:00 January 10, 2001. (Photographed by Qin Xian'an)

and carried the spacecraft across the night sky, roaring like the thunder of spring while sending the first greetings of the Chinese nation to the vastness of space in the new century.

Ten minutes later, the precise parameters of the spacecraft's entry into orbit were transmitted from the offshore survey ship. At once, the hall and the observation deck roared with applause.

The duration of the flight increased from one day to seven in this mission. On January 16, Shenzhou 2 returned to the ground smoothly after 108 laps in space.

Shenzhou 2 was the first flight model in China of an unmanned spacecraft, whose technical status was basically the same as that of manned spacecraft. "That was the first comprehensive assessment of the project and its systems." Yuan Jiajun, then commander-in-chief of the spacecraft system, said that after the re-entry capsule had been recovered, the orbital module of Shenzhou 2 had been in orbit for about half a year, and much valuable data had been obtained.

Shenzhou 3

Compared with the first two spacecraft, Shenzhou 3 had two strengths.

Firstly, although Shenzhou 3 was still an unmanned spacecraft, a human figure was place on the spacecraft: – a "Simulated Man."

"It is equipped with human metabolic simulator, anthropomorphic physiological signals, and a body dummy. Therefore, it can simulate important physiological parameters such as breathing and heartbeat, blood pressure, oxygen consumption, and the heat generation of astronauts, and explore the way for astronauts to enter space." Su Shuangning, chief commander and designer of the first astronaut system, cxplained that unlike the United States and the Soviet Union, which first put animals into space, the "Simulator" was a Chinese creation. Facts have proved that the "Simulator" successfully fulfilled the mission.

Shenzhou 3 was being hoisted at the vertical assembly plant on March 2002. (Photographed by Qin Xian'an)

The service tower was closing after being the vertical transportation of Shenzhou 3 to the launch tower. (Photographed by Qin Xian'an)

Secondly, the central leadership demanded to watch the launch on site. This meant that the launch of Shenzhou 3 would make a public appearance (the first two launches had not been made public).

Obviously, the mission of Shenzhou 3 burdened the aerospace workers under greater pressure.

On September 30, 2001, Shenzhou 3 arrived at the Jiuquan launch site by air. As planned, the mission would be launched before the end of the year, thus there was not a minute to spare. To ensure perfect preparation, no one enjoyed any days off on the National Day holiday, but immediately launched into the technical testing of various systems of the spacecraft.

An unexpected problem appeared.

On the third day of the technical test, it was found that there was a different socket connection on the pressure sensor.

It was not a trivial thing. There were more than 70 sockets of the same kind on the spacecraft. Over one thousand technical points were involved. Was it that one socket which malfunctioned or all? Field technicians found it difficult to make the judgement.

An anxious atmosphere clouded the head of all the personnel involved. They understood that there was the possibility that all sockets had to be replaced. Whereas, there was unavailability of this kind in the factory. It would take at least three months to produce them.

Authoritative expert groups soon rushed from Beijing for inspection. On October 22, the conclusion was drawn that unfortunately it was all. As a result, eight months of preparation were in vain; the launch of Shenzhou 3 had to be suspended, and the launch team would return to where they came from.

No other withdrawal like this had ever happened. A small socket let tens of thousands of scientists and technicians' eight months of efforts go to waste and the launch team consisting of over 100 specialists return empty handed. The entire launch plan was delayed. It was a painful lesson to learn.

"Quality bears no carelessness. And no matter is too little when it comes to a spacecraft." Drawing lessons from past failures, the Institute of Space Technology decided that the salary of all staff of the Institute should be reduced by 10%, and that of all participants of the Shenzhou 3 mission by 15%.

Three months of hard work contributed to the full repair of Shenzhou 3. At the same time, the rocket arrived at the launch site through a special train ride. On March 18, 2002, the spacecraft docked with the rocket and all set.

On March 25, 2002, at 22:15 hours, the rocket lifted up Shenzhou 3 and went straight into space. The launch hall burst with joy.

At 16:51 hours on April 1, the safe landing was completed at the landing site of Dorbod Banner, Inner Mongolia, seven days after orbiting the earth.

Shenzhou 4

The Shenzhou 4 mission was the last rehearsal of China's launch of manned spacecraft.

Limited national power impeded China from launching nearly 10 or even more unmanned test spacecraft before manned flights than the developed countries.

The headquarters stressed that Shenzhou 4 must focus on "manned." Everything related to that must be 100% guaranteed.

Therefore, Shenzhou 4 was in the same technical status as that of manned flight. All systems were activated including the total activation of emergency life-saving areas and even the clothes for changing. The problem of excessive harmful gases found in the first three unmanned flight tests was thoroughly addressed on Shenzhou 4.

On October 30, 2002, it entered the site by air. Everything went well. Neither the spacecraft nor the rocket had the same difficult technical problems as before.

But God arranged a test. An unprecedented cold wave swept over the launch site, covering the whole site with heavy snow On December 21, it was only eight days to go before the scheduled launch date. The Jiuquan Base Meteorological Department made a forecast: the cold front, which was rarely seen in 30 years on the desert, would lead to continuous temperature drop in the next few days. Possibly, the lowest would reach minus 30 degrees Celsius.

The department further accurately estimated that if the launch were scheduled for December 29, the lowest temperature would be minus 24 degrees Celsius. This would pose a serious challenge to a safe and effective launch.

A launch below minus 20 degrees Celsius had never happened, not only in China but the world. Technical documents stipulate that the minimum temperature should not be lower than minus 20 degrees Celsius when launching a rocket. A low temperature would have many adverse effects on a rocket, such as the failure of a sealing ring, bad contact of a cable plug and so on. The impact of low temperatures on rocket engine is even more fatal.

The most typical example is the 1986 Challenger accident. An engineer had suggested that launching at low temperatures could easily cause rubber seals on the space shuttle to fail. The opinion was ignored by senior executives, resulting in the tragedy of the explosion and the deaths of seven astronauts.

Shenzhou 4 spacecraft was successfully launched at Jiuquan Satellite Launch Center at 00:40 hours on December 30, 2002. (Photographed by Qin Xian'an)

To keep the rocket warm, a temporary "Rocket & Spacecraft Cold Resistance Team" was set up at the launch site. They utilized more than 20 high-power air conditioners, heating the rocket and the spacecraft 24 hours a day and night. Then foam plastic was wrapped around them and electric light bulbs also provided some support too. Rumor has it that a total of more than 200 quilts were used to keep the rocket warm, even where the air leaked from the launching tower was blocked by quilts. "The approach was old-fashioned yet effective," summarized Huang Chunping, commander-in-chief of the rocket system.

The meteorological department made another forecast that the temperature on the 30th was higher than that on the 29th, which could meet the lowest launch conditions. Thus, the launch was postponed one more day.

At 00:40 on December 30, 2002, the rocket took off with Shenzhou 4 and soared to a sky full of stars. The flame on the tail glowed brightly in the night sky. Ten minutes later, a message came from Qingdao TT&C Station that the separation was complete, and the spacecraft had entered orbit.

On January 1, 2003, Shenzhou 4 sent New Year greetings to China via the Tiandi Voice System: "Happy New Year." This first greeting left the country in deep pride and excitement.

After 108 circles around the earth, Shenzhou 4 received return instructions and landed safely at the Dorbod Banner landing site at 19:30 on January 5, 2003.

In the re-entry capsule were seated two smiling simulator figures responsible for assessing the adaptability of the environment inside. Real-time data collected indicated no abnormality in physiological signals and metabolic indexes provided by the simulator figures, and the safety of the spacecraft met the medical requirements of a manned flight.

The success of Shenzhou 4 was a solid step taken towards China's realization of manned flight. Associated Press claimed that China's launch of two spacecrafts in nine months was a sign of increasing confidence in space development. Agence France-Presse predicted China's first manned flight was around the corner.

Shenzhou 5

Immediately after the return of Shenzhou 4, China embarked on the Shenzhou 5 mission.

"We must safely send China's first astronaut into space!"The closer to the dream, the more craving and sensitive the aerospace workers became.

Shocking news came from the other side of the Pacific on February 1, 2003, on the first day of the Lunar New Year. he US space shuttle Columbia exploded, killing all seven astronauts on board.

It astonished the Chinese aerospace team. Qi Faren, chief designer of the manned spacecraft system, was sleepless all night. Great pressure burdened his shoulders. "No mistake is allowed, otherwise I will be a shame of the nation."

The Columbia accident sounded another alarm bell.

On the road to space, danger lurks everywhere. In spite of successes and glories achieved, every launch is new and maximum caution must be given.

More importantly, Shenzhou 5 was the first manned flight of China.

And it was only 10 months away from the launch date. Mourning for the tragedy of the Columbia spacecraft, the world looked to the East: Can the Chinese make it happen? The Chinese also asked themselves: Can we do it?

Yes, or no? Actions would give the answer.

Yuri Gagarin, the world's first astronaut to travel in space, died in an airplane crash at the age of 34. Today, his former office has become a holy place for many astronauts to make a pilgrimage. Rumor has it that astronauts visiting the office would touch his wide table to pray for safety and success.

Qi Faren told the story to the researchers of the spacecraft team. He warned that we should not encourage Chinese astronauts to place their lives on empty prayers, but on solid work. Thereafter, they solved the landing impact overload of the re-entry capsule, which further improved the safety and comfort of Shenzhou 5.

In the spring of 2003, SARS (Severe Acute Respiratory Syndrome) swept China. In this ravaging plague, Chinese aerospace undertakers did not panic, but held to their respective positions. No preparation work was interfered with for a day.

Shenzhou 5 arrived at the Jiuquan Satellite Launch Center on August 5, 2003. On August 23, the Long March 2F rocket followed. And all the aerospace experts participating in the launch also joined the arrival at Jiuquan.

However, on this very day, another tragic space launch happened.

A Brazilian rocket burst into flames and exploded, killing 21 people on the spot. In tears, Brazil's president announced that this was the saddest day in Brazil's space history.

The news tightened sensitive nerves of the Chinese aerospace team. The Brazilian explosion must be taken as a warning. Maximum caution should be given with no mistakes allowed.

The next day, the Shenzhou 5 mobilization oath division meeting was held. At the meeting they read out a letter written to the astronaut about to go on the mission.

Dear comrade astronaut:

In this golden autumn, entrusted by the motherland and people, when you board the Shenzhou spacecraft designed and manufactured by us Chinese, please accept the highest tribute and best regards from more than 500 crew members of the China Aerospace Science and Technology Group on behalf of the 100,000 employees.

...

Dear astronaut, Despite our different posts and responsibilities, we share the same missions, the same goals of the great rejuvenation of the Chinese nation and the same dream of space flight.

Please believe that 500 hearts are connected with yours, and 500 hearts will accompany you in space. And please rest assured that we will practice our solemn commitment with practical actions to ensure the accurate orbiting of Shenzhou, the normal operation and your safe return.

We wish you well. When you return to the embrace of your motherland safely, the people will surely welcome you with the most solemn courtesy: the hero who returns with honor!

The letter was the solemn promise of more than 500 test crew members and the solemn guarantee of 100,000 employees. Without 100% certainty, no promise could be made.

The moment in great anticipation finally came. At 9:00 a.m., on October 15, 2003, the Long March 2F carrying the Shenzhou 5 spacecraft rose in a roar of thunder, as if a dragon ejecting tangerine flames across the sky of autumn desert.

At 9:10, separation pulled through, and the spacecraft entered a predetermined orbit. The vastness of space ushered in the first Chinese astronaut, Yang Liwei. Three

At 5:35, astronaut Yang Liwei was ready to go on the flight. (Photographed by Qin Xian'an)

minutes later, for the first time, the voice of a Chinese traversed in space. Yang Liwei reported: "The flight is in normal operation."

At 09:42 hours, the commander-in-chief of the manned space project announced the spacecraft's entry to the intended orbit, thus the success of the mission. The control room roared with excitement. The 241st space flight belonged to China.

During the next 21 hours of space travel, astronaut Yang had a number of Chinese firsts:

At 11:00, the first meal in space; at 12:00, the first rest in space; at 19:58, the first space-to-earth conversation; at 21:00, the first photograph of earth from a spaceship ...

It will not be forgotten when Shenzhou 5 started its seventh lap, astronaut Yang, holding a five-star red flag and a blue United Nations flag, sent greetings in Chinese and English for the first time in the space, 343 kilometers above the ground. "Greetings to the people of all countries in the world, to our peers working in space, to the people of our motherland, to our compatriots in Hong Kong and Macao, Taiwan and overseas, and thank the people of the whole country for your concern."

On October 15, 2003, China made its first manned space flight. Shenzhou 5 was launched at Jiuquan Satellite Launch Center. (Photographed by Qin Xian'an)

The 21-hour space trip was very short. Astronaut Yang had acted to update the record of China's space flight time after time. The elegance of China's space flight was in display. Twenty-one hours was also long for the people on the ground, who were anxiously waiting for the safe return of astronaut Yang.

Finally, it was time to go home. At 6:20 on October 16, 2003, on the grasslands of Dorbod Banner, Inner Mongolia, as if a colorful cloud, the Shenzhou 5 parachute as big as 1,200 square meters came into sight.

"It's back!" When the spacecraft landed, the sun had just risen. The grassland was covered in golden rays. Slowly it descended. Cheers echoed in the air.

The door opened and Yang appeared. A Mongolian woman presented flowers and a Khatag (a good luck scarf) to the astronaut.

Yang appeared in good health after stepping out the capsule. He spoke, "The spacecraft is good. I feel well. I couldn't be prouder of my motherland."

Symbolically, the landing time of the Shenzhou 5 re-entry capsule coincided with the time when the national flag was raised in the morning in Tian'anmen Square in Beijing.

On October 15, at 19:58, Yang's wife Zhang Yumei and son Yang Ningkang held a 4:50 second conversation with him at Beijing Aerospace Command Center and exchanged regards. (Photographed by Qin Xian'an)

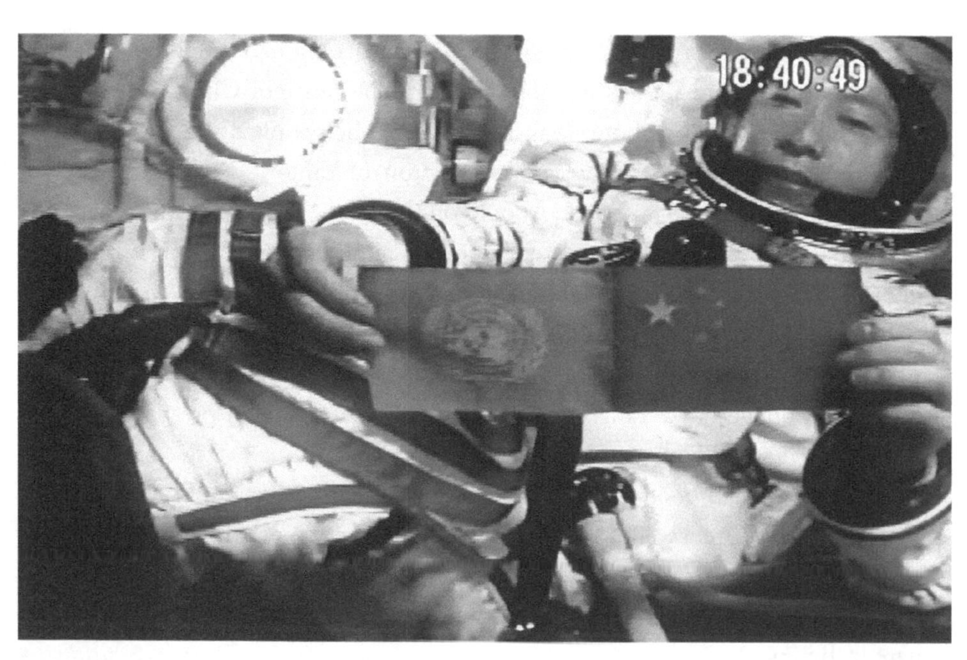

Yang displayed the United Nations and Chinese national flags in space. (Photographed by Qin Xian'an)

First Chinese Astronaut in the re-entry capsule. (Photographed by Qin Xian'an)

Having stepped out the capsule, Yang received a warmest welcome. (Photographed by Qin Xian'an)

Up to that moment, it had been 21 hours and 23 minutes from Yang's entry into space, three years and 329 days from the launch of China's first test spacecraft, and 11 years from the start of China's manned space project.

It took China 11 years to achieve what developed countries had reached in 40 years.

On October 16, 2003 is a great day to remember. On this day, the historic breakthrough in China's manned space made another brilliant milestone in China's high-tech field, following the atom bomb, the hydrogen bomb, and man-made satellites. China has thus become the third country in the world capable of sending astronauts into space, after Russia and the United States.

NASA Administrator Sean O'Keefe issued a statement on China's successful launch of the Shenzhou 5 manned spacecraft: "This launch is an important achievement in the history of human exploration. Following Russia and the United States, China is the third country to successfully launch human beings into space. The Chinese people have a long and outstanding exploration history. NASA wishes China a continued safe human space flight program."

Meanwhile, countries sent congratulations. Russian media claimed that China's successful launch of the Shenzhou 5 manned spacecraft showed that China's science and technology had made great achievements and its economy had developed rapidly. Jean-Jacques Doldan, Director-General of the European Space Agency, commented that China had become the third country to send humans into space alone, which shows the reliability of space technology; China's success in manned space flight would usher in a new era of international space cooperation.

UN Secretary-General Kofi Annan congratulated Shenzhou 5 on the successful launch. He expressed that space travel has no borders; the launch of Shenzhou 5 was a step towards space for all mankind.

Shenzhou 6

At the Great Hall of the People in Beijing, a party was held to celebrate the success of Shenzhou 5's first manned space mission on November 7, 2003. Simultaneously, a seminar with a theme of the development of Shenzhou 6 was being held in the office building of the Chinese Academy of Space Technology, dozens of kilometers away from the Great Hall of the People.

The celebrations did not delay embarking on the next project. Aerospace workers did not have the time to savor the joy of success; instead, they raced against time to embark on a new journey.

It may have been noticed that the chief designer of the spacecraft system had been replaced. Zhang Bainan, 42, took over the heavy burden from Qi Faren, 71, as chief designer of the Shenzhou 6 and chief director of China's future space station.

Zhang Bainan, a native of Qiqihar, Heilongjiang Province, graduated from the Department of Solid Mechanics, the National Defense University of Science and Technology in 1984. At the age of 35, he became deputy designer of the manned spacecraft system. He was 1.8 meters tall, simple, and plain. At first glance, he seemed be a laborer. His colleagues' impression was "Zhang has hardly been overwhelmed." Qi was content with his successor. His impression of Zhang was, "He is like me, but knows more."

In fact, the spacecraft system not only experienced a "change of command." The whole research team had changed from old to new. Technicians under 40 years old accounted for 80% of the whole team. The average age of key technical personnel positioned above deputy director of the subsystem was 32.

On November 7, 2003, the Great Hall of the People held a meeting called the Report on Advanced Events of Manned Space Engineering. *(Photographed by Qin Xian'an)*

Apart from the spacecraft system, the seven systems of the Shenzhou 6 had become "young;" the average age of the commanders and chief designers was 48.7, 5 years younger than the average age of those involved in Shenzhou 5. Huang Chunping, a rocket expert, handed the baton to Liu Yu, a man 24 years younger. At the media conference I attended, Liu Yu, 43, joked "Although there are veteran journalists of astronautics, I believe none has seen me." Scientists and technicians under 35 years old in the rocket system test team have accounted for 80%. The rise of a new generation of talent had made new breakthroughs in China's aerospace missions and accumulated a strong momentum for development.

From Shenzhou 5 to Shenzhou 6, only a small "1" has been added to the digital sequence, but the great changes it implies show the soul of Chinese aerospace workers: independent innovation.

After the success of Shenzhou 5, the Chinese determined the mission objectives of Shenzhou 6: a multi-astronaut, multi-day flight during which the astronauts first live and work in the orbital module, conduct the first space science and technology experiments with human participation, and run the first assessment of the spacecraft under a heavy load for a long duration.

It was a bold move that came with great challenges. From "one astronaut a day" to "two astronauts five days," the load of power supply, oxygen supply, gas supply, and dehumidification of Shenzhou 6 increased significantly. In a microgravity environment, to enable astronauts to live and work normally and conveniently was almost a new territory for us, many of which were difficult or even impossible to verify directly in ground tests. In addition, the flight time of Shenzhou 6 was 5.5 times that of Shenzhou 5 while the probability of failure was proportional to the operation time.

"Regardless of the success of both Shenzhou 1 and Shenzhou 5, the technology was not mature enough," said Wang Yongzhi, chief designer of China's manned space project. "Every flight is improved on the basis of the previous one. This is how the overall project and its seven major systems gradually achieve performance optimization."

Song Zhengyu, Deputy Designer of the Long March 2F, noted that in 12 seconds, the rocket would turn; in 139 seconds, the booster would detach itself; in 159 seconds, the stage 1 rocket and stage 2 rocket would separate; in 588 seconds, the rocket and the satellite would separate ... For a breathtaking 10 minutes, many ground tests were conducted. Mathematically, one hour of testing was done for each second of the flight. The ratio corresponded to the old saying that a day in heaven is ten years in the human world. Long March 2F has undergone 75 technical improvements on the basis of previous ones. The reliability and safety design have met advanced international standards.

To make astronauts a comfortable "home" in space, spacecraft engineers should not only participate in the design of spacecraft, but also operate it. An operational motion often involved dozens of trials. In one trial, it was found that the seat-cushion-lifting button could cause the astronaut's arm to be bruised, so the button was adjusted to the astronaut's hand, completely eliminating the potential hazard. Compared with Shenzhou 5, Shenzhou 6 made up to 110 technical improvements.

Perfection without leaving any defects and regrets was the consistent pursuit of Chinese aerospace workers. And spacecraft recovery was the last stick of the safe relay. A 1,200 square meter Shenzhou parachute was sewn up by 1,900 pieces of parachute cloth. More than 3 million stitches were on the parachute, and their length exceeded 10 kilometers. To ensure the parachute's absolute intactness, they checked every stitch and thread dozens of times. When the stitches were insufficient, they re-did it; when the parachute cloth was found wrinkled, they replaced it without hesitation.

From Shenzhou 5 to 6, the seemingly minor changes were actually qualitative. Back and forth, 187 technological improvements and innovations were made to the spacecraft and rocket in the Shenzhou 6 project and various systems. The launch site, landing site, and TT&C communication network were optimized and perfected in more than ten respects. And more than 160 plans in case of failure were formulated for the flight.

At 5:37 a.m. on October 12, 2005, the Chinese made another great voyage into space, beginning with the Astronaut Apartment of the Jiuquan Satellite Launch Center.

Two years ago, I witnessed astronaut Yang Liwei go into space for the first time at the same place, fulfilling the thousand-year dream of manned space flight.

Two years later, I stood here again seeing off astronauts Fei Junlong and Nie Haisheng.

God laid a test in front of the expeditors. An unexpected snowfall brought Jiuquan Satellite Launch Center from a golden autumn to a cold winter. Temperatures suddenly dropped by more than ten degrees Celsius.

"Comrade-in-chief, we have been ordered to carry out the Shenzhou 6 manned space flight mission. We are ready, please give further instructions. This is Fei Junlong, astronaut of the Chinese People's Liberation Army Astronaut Brigade."

"This is Astronaut Nie Haisheng."

"Commence!"

"Yes, sir!" A firm reply ensued. Standard military etiquette and Chinese astronauts' demeanor were once again fixed in the history of human conquest of space.

Snowflakes fell with blessings while drizzle moistened the journey. Accompanied by the "Welcoming Song," astronauts Fei Junlong and Nie Haisheng, shouldering the mission of the Chinese nation to explore the mystery of the universe and peacefully develop and utilize space resources, embarked on the road to space in a snowstorm.

Fei Junlong, of Han ethnicity, was a citizen of Kunshan city, Jiangsu Province. He was born in May 1965, attended college, and enrolled in the army in 1982. He ranked as a third-class astronaut and colonel in the Chinese Astronaut Corps. He was also the highest-ranking pilot among the first batch of pilots in China, a super-class pilot, who had been flying safely for 1,599 hours and 22 minutes continuously and won the second-class merit.

Nie Haisheng, also of Han ethnicity, came from Zaoyang city, Hubei Province. He born in September 1964, attended college, and enrolled in the army in 1983. He was ranked as a third-class astronaut and colonel of the Chinese Astronaut Corps. He has flown three types of aircraft with 1480 hours of safe flight and been rated as a first-class pilot. He has won the third-class merit twice. He is one of the first members of the first flight echelon of manned space flight in China.

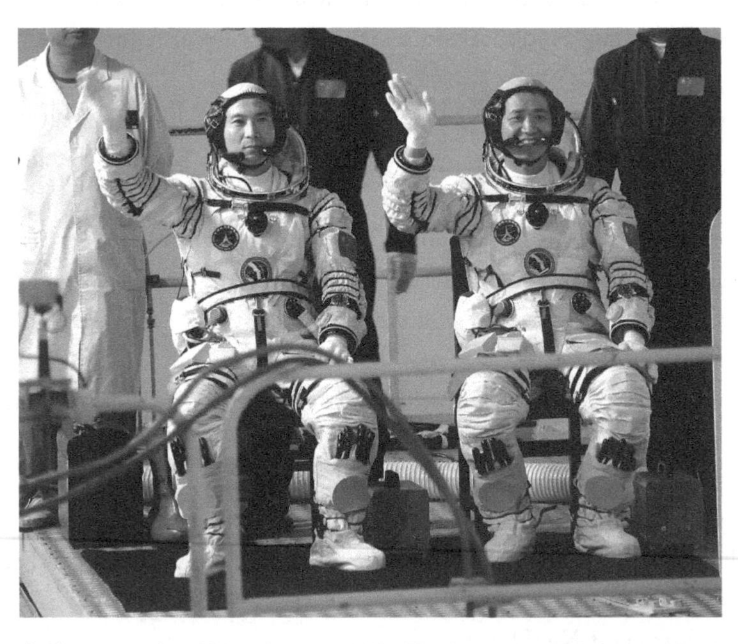

Fei Junlong (left), commander of Shenzhou 6, and Nie Haisheng, pilot of the spacecraft, waved before they left for the expedition. (Photographed by Qin Xian'an)

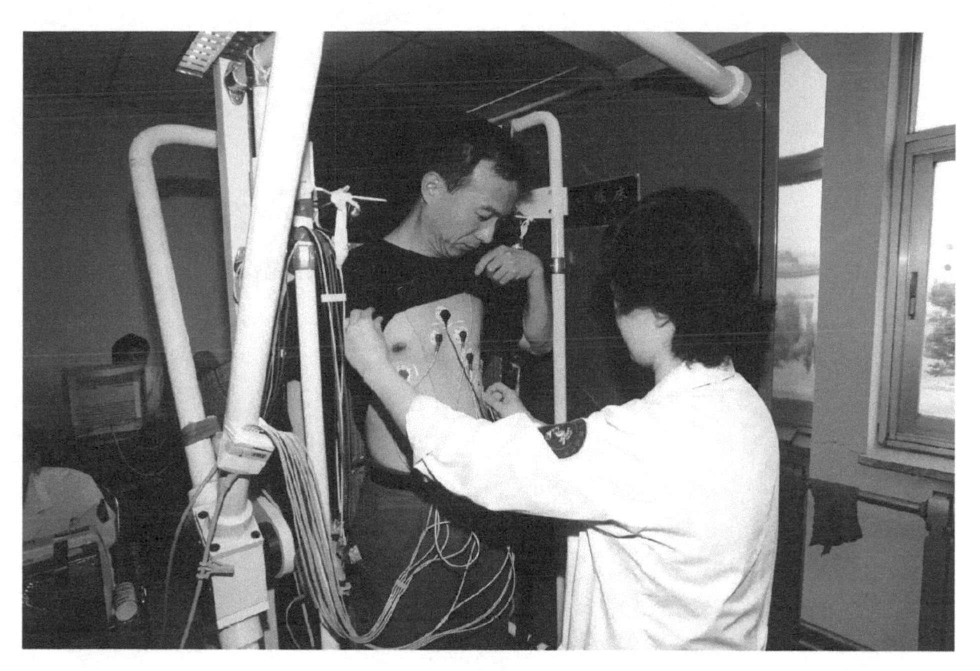

Fei Junlong received blood redistribution training. (Photographed by Qin Xian'an)

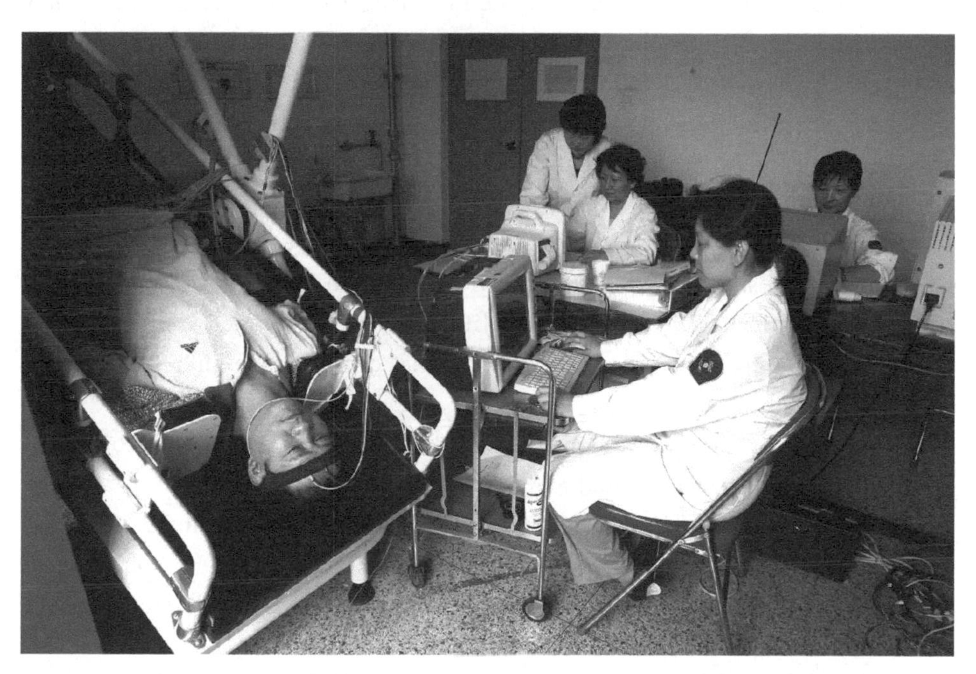

Nie Haisheng received adaptive training of blood redistribution. (Photographed by Qin Xian'an)

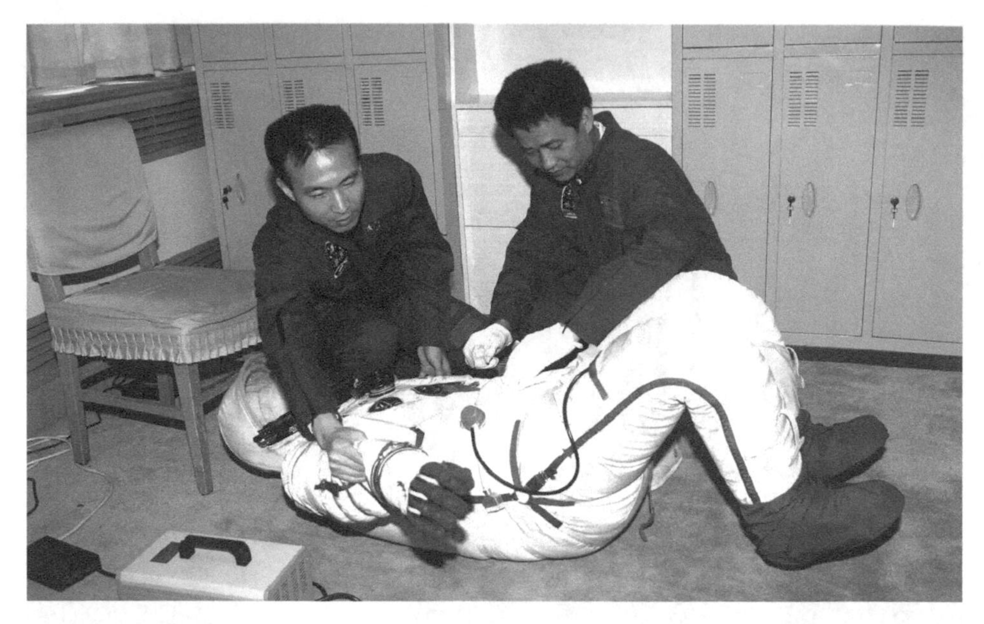

Fei Junlong and Nie Haisheng checked the airtightness of spacesuits before training. (Photographed by Qin Xian'an)

Two hours before, Fei Junlong and Nie Haisheng, who were designated to carry out the Shenzhou 6 manned space mission, got up on time. Faced with a risky space journey, they were as calm as merely going for an ordinary business trip, with a constant heart rate of around 70 bpm. After a rigorous examination and selection and seven years of unimaginably hard training, they stood out from 14 astronauts to go on the motherland's second space adventure.

A national flag was waving in chilly wind. In military uniforms, dresses, and colorful ethnic costumes, people were waiting quietly in the night on the square in front of the astronaut's apartment of the Jiuquan Satellite Launch Center. Central leadership came, comrades-in-arms came, coaches came, and children too. They all came to see the heroes off.

At 9:00 a.m., the rocket was ignited, and Shenzhou VI rose from the ground. Fei Junlong and Nie Haisheng soared together to the vastness of space.

They spent 115 hours in space, almost five days and nights; covering 3.25 million kilometers, equivalent to four roundtrips to the Moon. During the 115-hour journey, the cooperation between Fei and Nie was perfect, creating one unforgettable classic moment after another.

A birthday was celebrated in space. On the second day of the flight, astronaut Nie celebrated his 41st birthday with the blessing of hundreds of millions of Chinese.

He said to his daughter, "It feels wonderful up here!" This was also the first Chinese birthday in space.

They shaved their beards in space. On the third day, astronaut Fei shaved his beard with special shaving cream applied in front of a mirror on the bulkhead. His movements were very careful and in a relaxing manner.

A somersault was performed in space. At 16:30 on October 14, astronaut Fei suddenly made a forward flip in the spacecraft. Still in the mood, he made two more forward flips in succession. Nie took pictures. "Space somersault" therefore became the signature move of astronaut Fei. A reporter calculated that Fei spent about three minutes in the four flips there; at the speed of 7.8 kilometers per second of Shenzhou VI, 351 kilometers was covered in one flip.

Having traversed five days and nights in space, the two astronauts created a series of new records in Chinese space flight.

Shenzhou 6 was successfully launched at the Jiuquan Satellite Launch Center at 9 a.m. on October 12, 2005. (Photographed by Qin Xian'an)

On October 12, 2005, Nie Tianxiang, daughter of Nie Haisheng, wished her father happy birthday in space. (Photographed by Qin Xian'an)

For the first time, Chinese astronauts took off their spacesuits and entered the orbital module; for the first time, space medical experiments were performed; for the first time, pressure suit putting on / taking off experiments were performed... The series of firsts symbolize that China's space flight has entered a new stage of human involvement in space flight testing.

It merely took two years for China's manned space flight to advance from one astronaut to two, from one day to five, and from the return module to the orbital module in terms of the scope of astronaut activity. The world has really witnessed China's speed. "It's a perfect flight." concluded Tang Xianming, then director of the office of China's manned space project.

At 5:38 a.m. on October 17, 2005, the Shenzhou VI capsule landed safely somewhere less than nine kilometers away from astronaut Yang Liwei's landing site. In the illumination of helicopter searchlights, astronauts Fei and Nie stepped out of the capsule and had their feet on the soil of their motherland.

With the return, global attention once again centered on the Dorbod Banner grassland.

Surrounded by a crowd of welcomers, astronaut Fei depicted: "Our space flight was very smooth, the working and living environment in the cabin was very good,

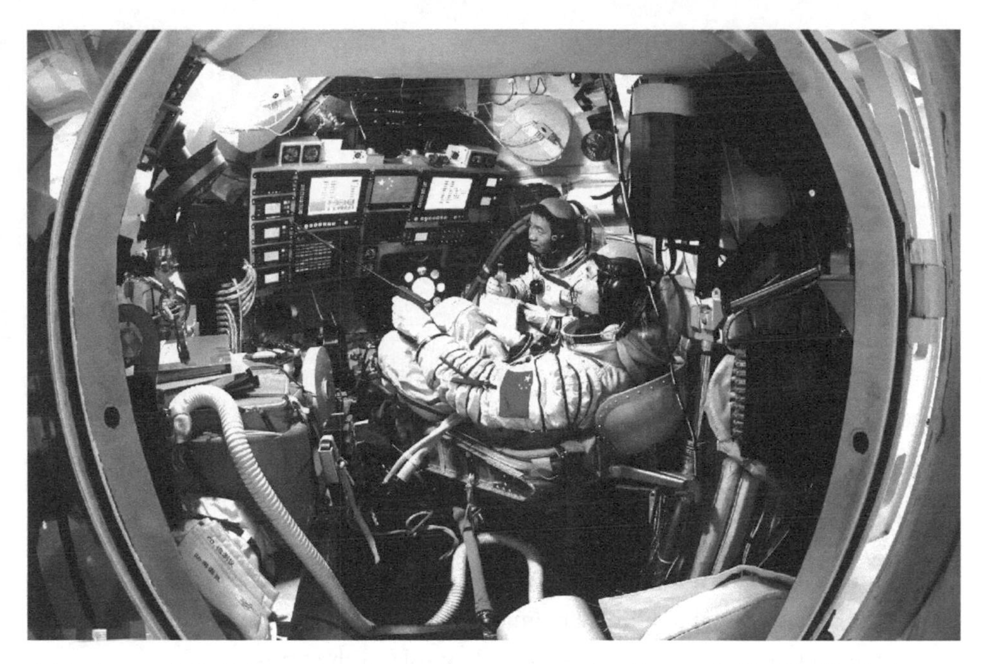

Fei Junlong and Nie Haisheng operating in the re-entry capsule. (Photographed by Qin Xian'an)

and we feel well." Astronaut Nie added: "When we were up there, we can feel the concerns of countless people, and we thank you and the motherland for your care and love."

The completion of the Shenzhou VI mission marks that China has stepped into the second phase of the vision. Specifically, after becoming the third country in the world to independently master manned space technology, China has also become the third to independently carry out science experiments in space.

Domestic and international media has commented on the achievements of China's Shenzhou flights. Space Web, America's biggest platform of space flight news, commented that the success of Shenzhou VI made space flight a part of China's national pride; in addition, during the trip of Shenzhou VI, astronauts carried out science experiments and tested new space equipment to prepare for the future construction of their own space station and exploration of the Moon.

However, Chinese were not lost in victory. Instead, they embarked on the next mission. At interviews, Qi Faren, a veteran of China's aerospace industry, humbled, "To cover 90 percent of one's destined distance brings the traveler no farther than the midway point. This is a good start indeed. But more important work lies ahead."

Shenzhou 7

In 2008, it was already the season of golden Euphrates poplar.

In the hinterland of Badain Jilin Desert, white rockets stood tall piercing the sky on the launching tower near the riverside.

Shenzhou 7 was good to go.

At a press conference held at the Jiuquan Satellite Launch Center, a reporter asked, "What difference does it make other than three astronauts instead of two in the Shenzhou 7 mission?" Zhang Jianqi, Deputy Director-General of China's manned space engineering, responded proudly: "It is more than a simple increase in quantity, but a qualitative leap."

Manned spacecraft, a space laboratory, and a perpetual space station were the three-step strategic goal of China's manned spaceflight. The Shenzhou 7 mission, as the first action of the second step, played a key role in the transition.

Zhou Jianping, who succeeded Wang Yongzhi as chief designer of China's manned space engineering, explained that the Shenzhou 7 shouldered three missions:

Shenzhou 7 was hoisted in an assembly workshop. (Photographed by Qin Xian'an)

carrying out China's first extravehicular activity through making breakthrough in the technology of extravehicular activities; carrying three astronauts for the first time in a multiple-day flight; completing a series of space science and technology experiments such as relay satellite communications and releasing small satellites.

Zhou Jianping, who succeeded Wang Yongzhi as chief designer of China's manned space engineering, explained that the Shenzhou 7 shouldered three missions: carrying out China's first extravehicular activity through making breakthrough in the technology of extravehicular activities; carrying three astronauts for the first time in a multiple-day flight; completing a series of space science and technology experiments such as relay satellite communications and releasing small satellites.

Compared with Shenzhou 6, Shenzhou 7 was a new challenge. In the past three years, aerospace workers were swamped with busy and delicate preparations.

To ensure absolute safety, researchers worked hard for three years. More than 300 key technical problems were tackled, more than 350 experiments were performed, and 36 new technologies were adopted to improve the safety and reliability of the rocket to 98%.

Shenzhou 7 transformed the orbital-retaining module into an air lock with access to space. The interior of the spacecraft was well designed and decorated. More than 1,200 hours of testing before launching were performed, the longest of the seven spacecrafts, but the problems in the testing were the least.

In terms of TT&C technology, relay satellites came out, and various training facilities for extravehicular activities have been put into use one after another. Every subject was new.

To realize a spacewalk, the most critical and difficult new technology needed was the development of a domestic extravehicular spacesuit.

In July 2004, the Extravehicular Spacesuit Development Project was officially established.

It was more than a simple spacesuit, but a "small space vehicle" with not only the function of a spacecraft, but also higher requirement of weight, volume, and power consumption. It was a collection of advanced space science and technology.

In accordance with the common practice in the international space industry, the development cycle of a new mature spacecraft product should range on average at least from seven to ten years. Yet, there were only four years left.

"When we received the task, we felt as if there was a heavy rock pressing on our heart." Recalling the situation, Liu Xiangyang, director of the Extravehicular Spacesuit Research Department of the China Astronaut Training Center kept fresh memories.

Astronaut Zhai Zhigang checking the cabin hook. (Photographed by Qin Xian'an)

However, with the order out, there was no turning back. To finish the new project, it obviously did not make sense to follow the usual practice. What to do? Calm contemplation led the project team to an idea: make the best use of relevant mature technologies in China and integrate relevant achievements for innovation.

At an incredible speed in China, they completed the research and development of the extravehicular spacesuit. Liu Xiangyang told me: "The numbers show we had only worked for less than four years, but in fact, the working hours in the four years were more than a person's average working time in eight."

There was a slogan during the four years of the development of extravehicular spacesuits: "We have no right to be sick!" The simple words described aerospace workers could bear hardships and tackle key problems.

I have closely appreciated the splendor of the extravehicular spacesuit at the Chinese Astronaut Research and Training Center, with a cost of 30 million RMB and a weight of 120 kilograms.

All white, from top to bottom, there was the helmet, upper limbs, trunk, lower limbs, pressure gloves, boots. From the inside to the outside, there were six layers: comfortable layer made of cotton fabric with special anti-static treatment, backup air-tight layer with rubber texture, main air-tight layer composed of composite joint structure, restriction layer made of polyester fabric, heat insulation layer through thermal reflection and outermost outer protective layer. The four limbs of the suit were equipped with adjusting bands. Anyone from 1.60 to 1.80 m. could fit by adjusting the length of upper arms and lower limbs.

The aluminum alloy trunk shell with a wall thickness of only 1.5 mm was densely packed with various instruments: electronic console, gas-liquid console, gas-liquid combination socket, emergency oxygen supply pipe, and electric umbilical cord. There were nine switches in the slightly-more-than-ten-centimeter square electronic console, such as lighting, digital control, and mechanical pressure gauge, and more than 20 valves in the gas-liquid console.

China's extravehicular spacesuit was characterized by flexibility, as it was not bulky regardless of the heavy weight. Designers used air-tight bearings at shoulders, elbows, wrists, knees, and ankles, which enabled the astronauts' hands and feet to rotate freely while the airtightness was strictly guaranteed. Extravehicular gloves appeared to be very thick. It would look like a boxing glove if four fingers apart from the thumb were sewn together. The back was white, the palm and fingers grey: dense gray rubber convex particles generate an anti-skid and heat insulation

At 21:10 on September 25, 2008, Shenzhou 7 was successfully launched by Long March 2F at the Jiuquan Satellite Launch Center. (Photographed by Qin Xian'an)

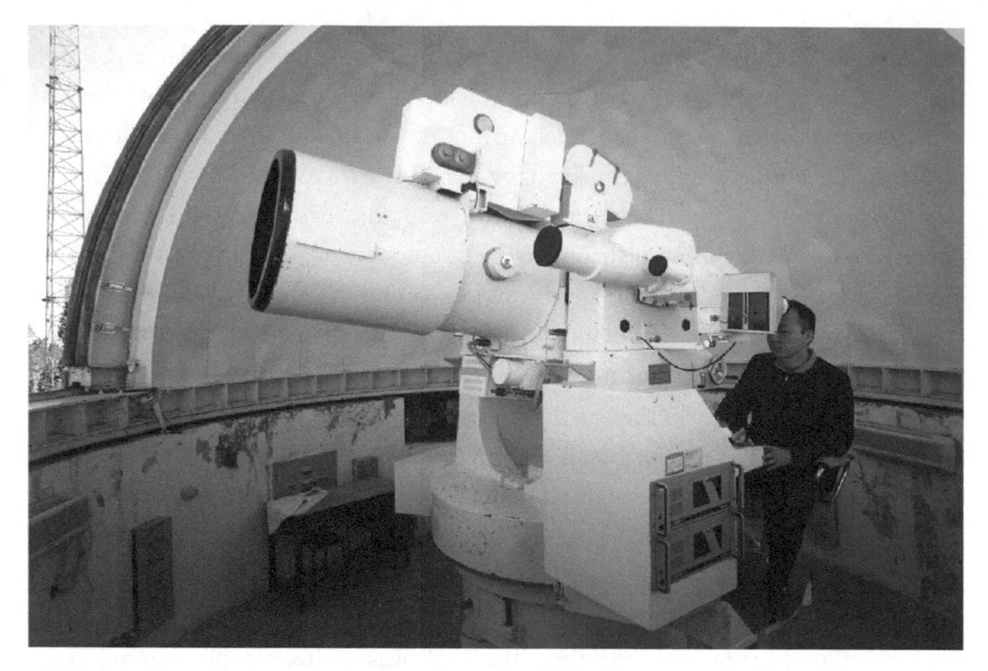

A technician at the optical radar station tracked the spacecraft. (Photographed by Qin Xian'an)

effect. Although space suits were produced in batches, gloves were tailored for each astronaut using the internationally advanced 3D digital scanning. The gloves of China's extravehicular spacesuit were world-class in flexibility, so astronauts could easily hold objects with a diameter of 25 mm.

On the eve of Shenzhou 7's launch, journalists from more than 10 overseas news agencies, including Reuters, the Russian News Agency, and the Associated Press, gathered at the Jiuquan Satellite Launch Center. For the foreign press, this was the first open access to the Shenzhou missions.

Behind the high transparency stood the high confidence of Chinese aerospace workers. The confidence came from strength. Faced with the Shenzhou 7 mission, which had a wide technological range and high risks, the constantly low-key veteran chief director of various systems has repeatedly promised to the media: "Success is a sure thing."

At 21:10 on September 25, 2005, with the order of Commander No. 0 Guo Zhonglai, console operator Xu Wenxi calmly pressed the ignition button.

In an instant, the desert quaked. The "dragon" soared into the air. Shenzhou 7, carrying astronauts Zhai Zhigang, Liu Boming, and Jing Haipeng, pierced across the vast night sky of and gradually shrunk into a small bright spot and disappeared in the clouds.

At 21:33 on September 25, the spacecraft entered its scheduled orbit. And three astronauts began their work and life in space.

Astronaut Jing Haipeng took knife and fork to eat his first meal in space at 08:18 on September 26. Wearing heat-proof gloves, he took food out of a beige space food heater, which was similar to a domestic electric oven, and floated back into the space living room, where he sat down steadily. Although it was merely a night snack, it was not un-serious at all. A plate, a spoon, and a fork were provided.

At 10:20 on September 26, astronauts Zhai Zhigang and Liu Boming began to assemble and test extravehicular spacesuits in the orbital module. Suddenly, a soft white light filled every corner of the "space home."

At 13:30 on September 26, astronaut Jing Haipeng started to play with the SLR digital camera on the spacecraft. Facing the porthole, with the camera pointed at the vast and charming universe, he kept pressing the shutter.

There was a relay in space. Up there, every star was a witness; down on the ground, millions of eyes were looking.

On-orbit inspection and procedure training for astronauts before extravehicular activities. (Photographed by Qin Xian'an)

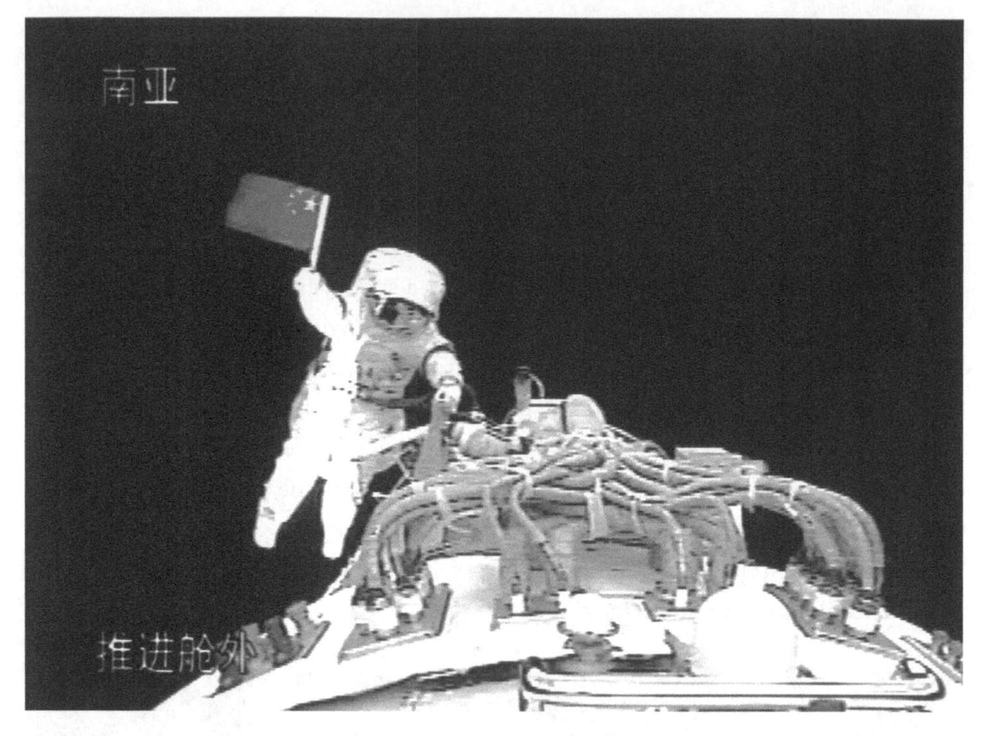

At 17:38 on September 27, 2008, astronaut Zhai Zhigang, assisted by astronauts Liu Boming and Jing Haipeng, successfully completed China's first space extravehicular activities. Pictured is Zhai's spacewalk. (Photographed by Qin Xian'an)

Finally, the moment came when the door of Shenzhou 7 was opened.

This is a historic moment worthy of the pride of the Chinese nation: 16:34 on September 27, 2008.

Grasping the handrail, astronaut Zhai Zhigang, lightly floated out of the cabin, taking the first step of the first Chinese spacewalk. The bright five-star red flag on the extravehicular spacesuit attracted worldwide attention.

"Shenzhou 7 reporting, I have been out of the cabin, feeling well."

"Shenzhou 7 greets the people of the country and of the whole world. Motherland, please rest assured that we will resolutely accomplish the task."

Space, the Chinese are coming. Zhai Zhigang, the 354th astronaut in the world, was 43 years later than Leonov, the Soviet astronaut who first carried out the spacewalk.

We can go as far as the dreams take us.

Over a month ago, at the opening ceremony of the Beijing Olympic Games, we used the huge "footprints" formed in the night sky with bright fireworks to announce to the world that the century-old Olympic dream of the Chinese nation has come true. Today, astronaut Zhai told the world with his footprint in space that China's millennium dream of walking in space has come true.

In space, Zhai Zhigang took toddler steps. Cooperating with astronauts Liu and Jing, Zhai walked outside for 19 minutes and 35 seconds. The distance he covered by the capsule door was insignificant for the vast space, but a new milestone for China's space adventure.

When Armstrong landed on the Moon, he said "One small step for man, one giant leap for mankind." Today, it can be said that Zhai's small step is a big step for China's manned spaceflight, and China has become the third country in the world to independently master the technology of extravehicular activity.

At 17:38 on September 28, 2008, the Shenzhou 7 re-entry capsule, carrying astronauts Zhai Zhigang, Liu Boming, and Jing Haipeng, landed steadily at the main landing site of Dorbod Banner in Inner Mongolia.

Zhai was the first to step out of the cabin door, almost without help. Smiling, he went straight to his seat, head titled back, like sitting in a big chair at home. Although Liu Boming's legs were a little soft when he left the cabin, as soon as he was seated, he crossed legs with the right leg on the left, fiddling with the blue belt tied to his ankle and chatting with Zhai Zhigang. Jing Haipeng came out of the cabin and waved to people as soon as his feet touched the ground. Even when sitting down, he didn't stop the action.

"Fantastic!" People roared in cheers as they witnessed the splendid performance of space heroes. Zhai Zhigang laughed, Liu Boming laughed, and Jing Haipeng laughed. At this moment, the sunset gilded the horizon a beautiful hue.

"The vast space welcomed visitors, three "horses" galloped in space." This poem was composed by Zhang Jianqi, deputy director of manned space engineering in China. It delivers a hidden message that all three astronauts, aged 42, belong to the Chinese zodiac sign of the horse.

A French news agency commented that the launch of Shenzhou 7 was historic and flawless, which indicated that China had taken another big step towards space.

Shenzhou 7 astronauts Zhai Zhigang (middle), Liu Boming (right), and Jing Haipeng (left) trained in the re-entry capsule. (Photographed by Qin Xian'an)

A victorious return to Space City. (Photographed by Qin Xian'an)

Tiangong-1

Following the breakthroughs in manned spacecraft's round trip between earth and space and extravehicular technology, Chinese aerospace workers quickly centered their attention on another core technology necessary for the establishment of a space station: space docking.

Tiangong-1 is a space target vehicle developed by China for the construction of a space station. The main tasks are to complete space docking flight test as a docking target, to ensure safe work and living of the astronauts during their short-term on-orbit stay, to perform space applications and space medical experiments, space science experiments, and space station technology experiments, and to preliminarily establish a space test platform capable of short-term manned operation and long-term independent and reliable operation.

And it took three years for China to crack this hard nut.

After the Shenzhou 7 mission, Zhou Jianping took charge from Wang Yongzhi as chief designer of China's manned space engineering. Zhou showed that as the beginning of the second stage of the "three-step" strategy of China's manned space project, Tiangong-1 was the first Chinese space laboratory; it would dock with Shenzhou 8, Shenzhou 9, and Shenzhou 10 that subsequently would be launched.

Zhou Jianping said: "Three docking missions only require four launches and is cost-saving and economical. It is the first time that China specialized in the development of a target vehicle and the N+1 docking test model."

As the foreign scholar Brian Harvey commented on China's manned space flight in his book, this was a very typical Chinese-style space program; every time they take a big step forward, and repeated efforts happen rarely.

The name Tiangong originated from an allusion in the classic novel *The Journey to the West*, one of the four great books of ancient China, which embodies the aspirations of the Chinese nation for space exploration since ancient times. Tiangong-1 was initiated in 2005. The whole project lasted six years from design to initial sample to correct sample.

On August 18, 2010, the Tiangong-1 target vehicle was finished being assembled. It was composed of an experimental module and a resource module, with a total length of 10.4 meters, a maximum diameter of 3.35 meters and a takeoff weight of 8.5 tons. It was designed to work in orbit for two years. The experimental module, able to accommodate 2 to 3 astronauts, was China's LEO vehicle with the longest lifespan so far.

Tiangong-1 was the first "apartment home" in space. On the spot of the Tiangong-1 Test Hall of the Jiuquan Satellite Launch Center, researchers and designers showed me around the vehicle.

A huge cylinder, standing in the empty test hall, had a surface wrapped in heat-controlled materials of different colors. Had the researchers not told me that this was Tiangong-1, I would have thought it was just a part of the rocket. The ordinary appearance of the vehicle was unexpected, and so was the difficulty of the construction. To adapt to the harsh environment of space, Tiangong-1 had very high requirements for sealing. And researchers have racked their brains out in material and welding technology to meet the requirements.

Hou Xiangyang, the chief designer, said to me, "When the outer wall of a room is built, various hydroelectric pipelines need to be installed. These devices are arranged around the cylinder. Adding structural plates divided it into many layers, each of which as installed with different cables and instruments."

In his eyes, building the vehicle was no fun, because from every screw to the invisible air inside, all required new technological innovation. It took a large number of researchers, who frequently worked overtime, six years to create a brand new piece; the whole vehicle had 13 subsystems and over 500 devices.

Compared with the spacecraft, the vehicle was very spacious with an overall pattern of "two bedrooms and one living room." The decoration had an Oriental style. The rooms were equipped with air conditioners. The scientific research and experimental area, operation area, instrument display area, storage area, in addition to fitness area, were set in the "living room." Academician Qi Faren said that the upgrade was another major breakthrough in Chinese spacecraft construction.

In terms of technical difficulty, Tiangong-1 can be regarded as the most difficult "room" to build, because almost every component involved a technological innovation. From the perspective of construction cost, it can be regarded as the most expensive "apartment," because almost every corner was an unprecedented design.

On June 29, 2011, Tiangong-1 was shipped to the destination. On July 23, Long March 2F arrived. Before the launch, 234 contingency plans were formulated for Tiangong-1; ground simulation verification and failure walkthrough were strengthened to ensure a success.

At 21:16 on September 29, 2011, at the Jiuquan Satellite Launch Center, tangerine flames erupted from the bottom of the rocket. Long March 2F lifted Tiangong-1 off the ground, roaring towards the sky like thunder.

At 21:16 hours on September 29, 2011, Tiangong-1 Target Vehicle was successfully launched by Long March 2FT1 at the Jiuquan Satellite Launch Center. (Photographed by Qin Xian'an)

From unmanned experimental spacecraft to manned flight, from multi-person multi-day flight to a spacewalk, China's first Tiangong vehicle has embarked on a new journey of adventure. In just 12 years, China's space industry had made one historic leap after another.

At 21:38, Tiangong-1 went smoothly into orbit. China's first space docking test pulled through; thus China's space industry had entered the era of the space station.

"When is the Moon full and bright? Raising a wineglass, I ask the blue sky. I know not what year it is tonight; the heavenly palace is so far away." The Chinese poet Su Dongpo's famous poem vividly depicts the ancient yearning and imagination for heaven where the gods fare. Now, the "Heavenly Palace" described in the verse that has been through nearly a thousand years is getting closer to grasp. With the Long March rocket spitting dazzling flames, Tiangong-1 is hung amidst the stars.

Shenzhou 8

Thirty-two days later, at 5:58:07 on November 1, 2011, Shenzhou 8 set off on a long journey to pursue Tiangong-1.

This was a pre-advertised space "date," to which the road was highly risky. During the next two days of space flight, Shenzhou 8 had to undergo five orbital changes and four berths to dock with Tiangong-1 as planned.

Shenzhou 8 was the first "space bus" in China. It was 9 meters long, weighed 8,082 kilograms, and had a maximum diameter of 2.8 meters and a maximum wingspan of 16.9 meters. This superior manned spacecraft excited chief designer Zhang Bainan: "This is the finalized version of China's future Shenzhou series spacecrafts."

"It was a great leap." Zhang Bainan said, "Compared with the US and Russia, who adopted the standard transport system after two models, Shenzhou has been aiming at this goal from the beginning of the development. The efficiency was greatly improved with less time spent."

"Shenzhou 8 has made the biggest improvement in its history. Among 600 sets of devices, 15% were newly developed, 40% improved, and 51% faced with technical changes. It was a brand-new space-to-Earth shuttle vehicle." Speaking of his own "child," Zhang Bainan was full of vitality.

At 5:58 a.m. on November 1, 2011, Shenzhou 8 was successfully launched at the Jiuquan Satellite Launch Center by Long March 2F. (Photographed by Qin Xian'an)

Yang Haifeng, general design engineer of Shenzhou 8, showed me the change of the spacecraft: to complete the docking task, the orbital module was equipped with a new type of measuring and communication device, which realized the full coverage of the relative measurements in a larger distance. The front end was equipped with docking mechanism, whose internal channel had a diameter of 800 mm.

The new generation of CNC computer made Shenzhou 8 more acute, more capable, and more intelligent. From Yang Haifeng's point of view, the most pride was in the fact that all the single-machine equipment of Shenzhou 8 were domestic.

Shenzhou 8, the space bus, took the route that was familiar to many. It started at the Jiuquan Satellite Launch Center and ended at the landing site of Dorbod Banner in Inner Mongolia. From Shenzhou 1 to 7, seven spacecraft have a 100% safety record on the route between space and earth. There was no doubt about the safety of road conditions.

An ordinary highway is made of reinforced concrete meter by meter. Instead, the "highway" between space and earth was paved inch by inch with the most advanced information technology by Chinese aerospace scientists and technicians. And the construction has been going on for over two decades.

"Docking put forward a series of new requirements, including high-precision orbit determination, high-frequency guidance control, dual-target measurement and control management." Xie Jingwen, deputy designer of the TT&C communication system, disclosed that scientists and technicians had carried out a new round of upgrading and transformation of the "highway" to create a safer and smoother "road condition" for Shenzhou 8.

It is worth mentioning that China's self-designed, developed, and constructed trinity of land, sea, and space-based manned space TT&C communication network formally and jointly escorted the docking mission for the first time. With the addition of two relay satellites, the coverage of TT&C was increased to 69%.

Tick tock. Shenzhou 8 chased Tiangong-1. The two "Chinese stars" were moving closer to each other in space.

At this time, many media commented that Shenzhou 8 and Tiangong-1 would have their "space kiss," which would be one of the world's most spectacular scenes this year.

At 1:23 on November 3, 2011, China's first space docking reached the climax, with Tiangong-1 and Shenzhou 8 approaching and leaving only 30 meters to go.

Thirty meters is a short distance on the ground that can be covered with merely a dozen steps. Yet, the thirty-meter distance in space took years to come. Since the launch of Shenzhou 8 on November 1, it has been flying day and night for almost two days. And since the launch of Shenzhou 1 in 1999, Chinese astronauts have been through thick and thin for 12 years.

Nowadays, this distance became the center of global attention. The dream was within reach that China would have its first space laboratory, a "Chinese home" up there.

Since then, after five thousand years, the country has a home up there and down here.

The final "sprint" began. Shenzhou 8 and Tiangong-1 crossed the night sky above the Himalayas and entered the sky over the Chinese territory. At this moment, Zhang Mingli, a soldier of the Aliplan Frontier Guard Company guarding the highest mountain in the world, stood at the sentry post looking up at the sky, hoping to see two "Chinese stars" meet in the night sky.

Together heaven and earth celebrated tonight, and China was sleepless. At this moment, just like him, millions of families were watching all this on the TV screen, in great anticipation of that moment.

28 meters, 26 ... The gap was narrowing.

In space, Shenzhou 8 and Tiangong-1 were flying at an absolute speed of 7,800 meters per second, eight times faster than rifle bullets and 22 times faster than sound. The docking error must be controlled within 18 centimeters, as if needle

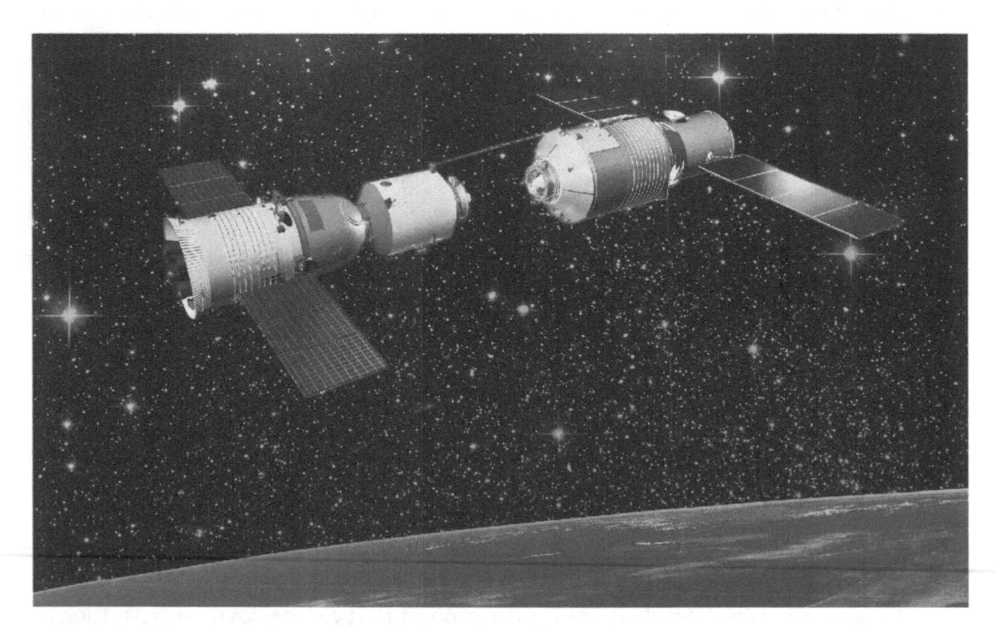

At 9:58 on May 27, 2011, Shenzhou 8 was 30 meters away from Tiangong-1. (Photographed by Qin Xian'an)

piercing in ten thousand miles. With a slight mistake, Shenzhou 8 and Tiangong-1 would miss one another.

Tick tock, the sound became clear; boom-boom, hearts beat faster.

At 1:28 a.m., Shenzhou 8 accurately embraced Tiangong-1. "This happened right above the sky of Dongfeng Space City," said Lu Lichang, chief designer of TT&C communication system for manned space engineering.

Tonight, excitement was displayed on everybody's face. The lights shone upon the quiet Ruohe River and the rustling poplar forest. This is where China's dream of space flight was born and where Shenzhou 8 and Tiangong-1 took off. This place witnessed the hardship, sweat, wisdom, and heroism of Chinese aerospace workers in making the dream come true.

A home up there and a home down here.

Sound of cheers spread from a farmyard in Tianshui, Gansu Province, hometown of Fuxi, one of the earliest legendary rulers. The old man, Yan Zongyi, and his wife were watching live TV. Yan Jun, the old man's son, as a TT&C engineer, was busy working for the space project on the Yuanwang No. 5 in the Pacific. Staring at the night sky, he told the journalists on board that his father must be smiling with pride.

In spite of the small size, the home up there brought a huge happiness to the home on the ground. The old man believed that the docking was a great cause that his son was devoted to. A few days earlier, Yan Jun made an overseas phone call to his family. Father told son with great joy that the winter wheat planted had budded. The old man did not know that not far from home was the national Aerospace Breeding Engineering Center of Gansu Province. The wheat in the field was a new variety cultivated by aerospace technology.

Closer and closer. In the vastness of space, Shenzhou 8, weighing 8 tons, approached Tiangong-1, which weighed 8.5 tons. It buffered and calibrated. With the help of sensors, two giants adjusted their postures lightly. Shenzhou 8 and Tiangong-1 began to embrace each other in space at a relative speed of 0.2 m/s.

At this moment, they were flying over the Great Wall across northern China.

Tonight, an embrace happened up there. Down here, new lives were brought to the world. Li Shiwen, a baby girl, was born in the PLA General Hospital. Mother Li Fenghong was embracing her daughter and soon to enter a sweet dream.

This new-born did not know that the population of the planet she came into had reached 7 billion. Meanwhile, Shenzhou 8 was heading to space on the mission of 17 joint experiments of space life sciences carried out by Chinese and German scientists. These experimental results would make the future world's babies healthier.

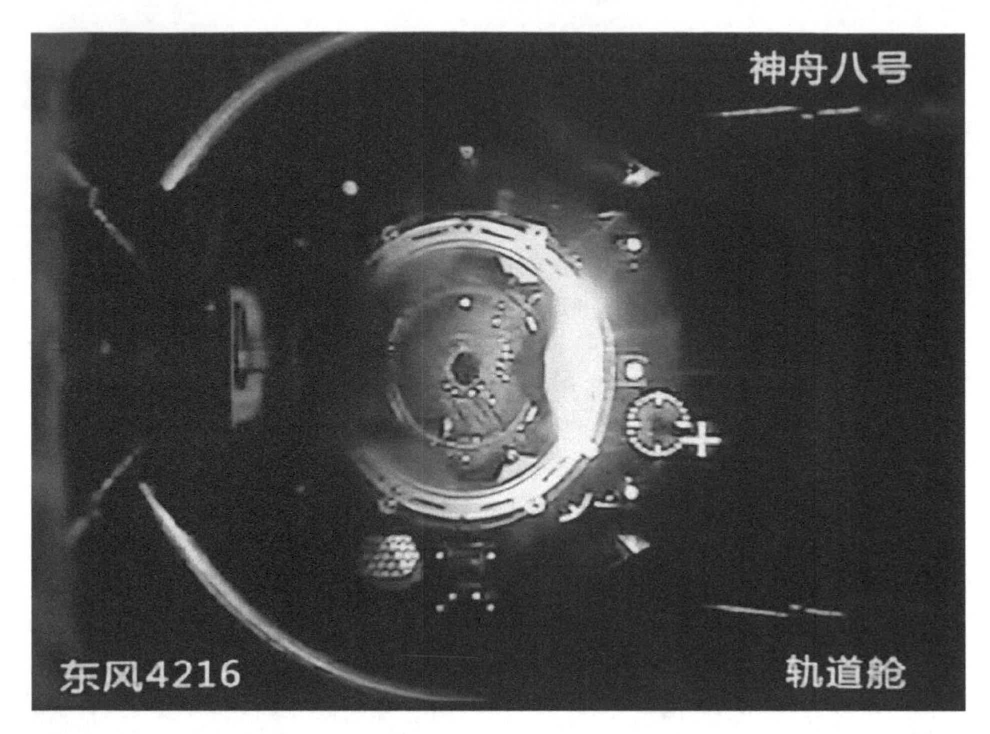

Tiangong-1 and Shenzhou 8 docked. (Photographed by Qin Xian'an)

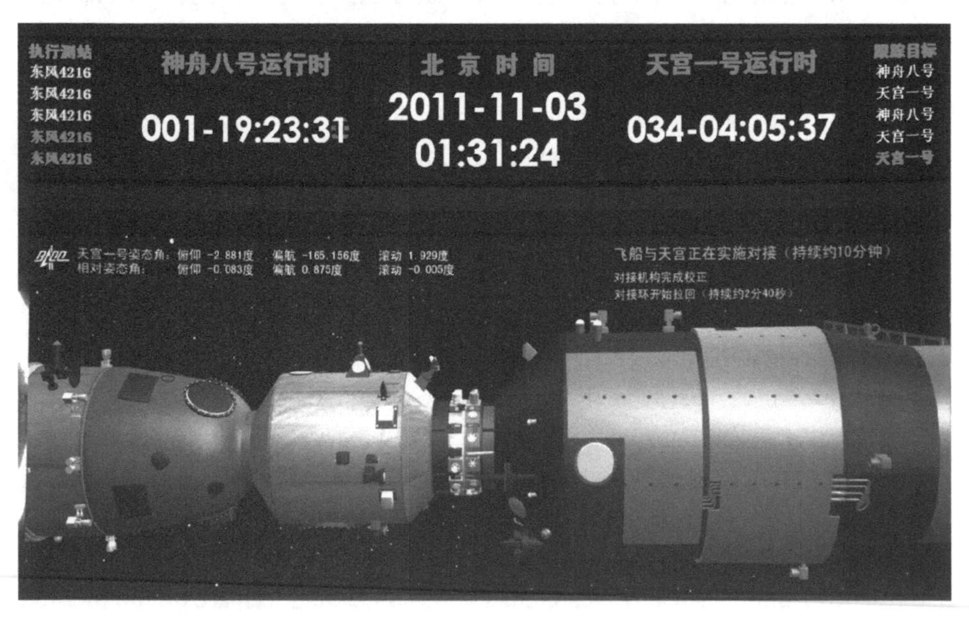

Tiangong-1 and Shenzhou 8 docked. (Photographed by Qin Xian'an)

And the 7 billion people on earth would benefit more from scientific and technological achievements from space.

The mother and daughter asleep did not know that Wang Yue, China's first "Martian," was about to complete a long "journey to Mars" and return home. In the future, more and more Chinese would enter the China Space Laboratory, from which they would guard the future of the world.

At 1:36 a.m. on November 3, 2011, Shenzhou 8 and Tiangong-1 completed docking.

In the Flight Control Hall of Beijing Space City, there was thunderous applause. Among the crowd, software engineer Yu Tianyi and his spouse Huan Pei embraced in laughter. To complete the docking, the couple spent countless hours.

In Dongfeng Space City and on the Pacific Yuanwang survey ship, it was not only couples who embraced happily in laughter. The younger generation and the older generation also exchanged hugs.

The Tiangong-1 control complex began to orbit. Radio waves traveled through space, sending in a picture – Tiangong-1 and Shenzhou 8 bulkheads were hung with Chinese knots with the Chinese character *fu*, meaning "happiness," in the center.

Chinese knots have thousands of families connected together. It is a meaningful picture – at this moment, China had happiness hung in space.

Compared with the land of 9.6 million square kilometers, the homeland in space was very small, especially when Shenzhou 8 and Tiangong-1 added up to less than 30 cubic meters. Compared with the 1.3 billion people on the ground, the homeland could only accommodate two or three. However, our space industry is closely linked to the happy life and bright future of millions of folks.

On the same day when China had the first docking success, China Space Post Office was opened in Beijing Space City, making it possible for Chinese to send mail into space. Yang Liwei, a space hero, served as director. The post office adopted the combination of virtual and physical business mode. The physical post office was located in the Beijing Space City while the virtual was in the manned spacecraft. And 901001 was the postal code.

On November 14, 2011, Tiangong-1 and Shenzhou 8 successfully conducted the second docking test. At about 19:00 on November 17, the Shenzhou 8 re-entry capsule returned safely. On November 18, Tiangong-1 was transformed into a long-term operation mode at an altitude of about 370 kilometers, waiting for docking with Shenzhou 9 and 10 the next year.

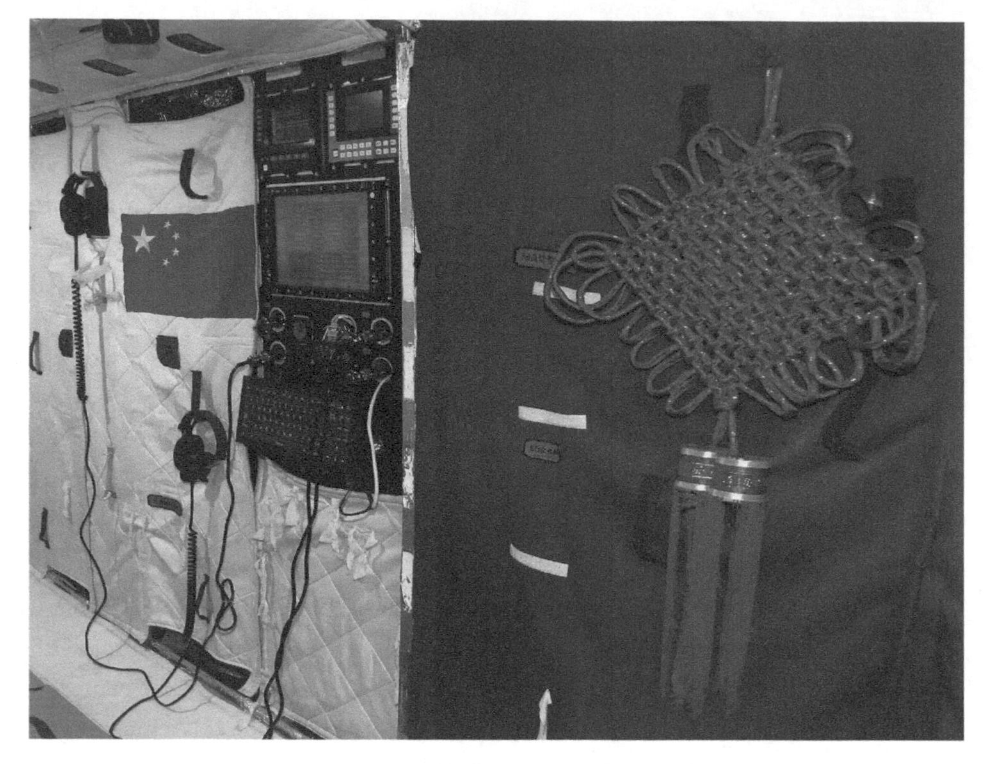

Measuring and controlling instruments, national flag and Chinese knot inside Tiangong-1. (Photographed by Qin Xian'an)

Shenzhou 9

"*Sink to the ocean to catch the turtles, soar to the sky to hug the Moon.*"

On June 19, 2012, the same day 40 years before when Chairman Mao composed the poem, the vision came true.

On this day, Jiaolong, the first manned diving vehicle independently designed and integrated in China, carried out the fourth submarine test at 7,000 meters, creating a new "Chinese depth." It was also the greatest diving depth of the same type of manned submersible in the world. "Jiaolong's success of diving 7,000 meters deep proves that it can achieve long-term seabed navigation, photography and filming, and sediment and mineral sampling in 99.8% of the world's seabed," said Liu Feng, commander-in-chief of the sea test site. "A major breakthrough in deep-sea technology."

At 7:41 a.m., Shenzhou 9 astronauts Jing Haipeng, Liu Wang, and Liu Yang successfully completed manual docking, re-stationed on Tiangong-1, and once again refreshed the "height of China." Wu Ping, spokesman for China's manned space project, commented at a press conference held by the Information Office of the State Council that cracking and mastering docking technology determined the realization of the second-stage goal of China's manned space development and laid a solid foundation for the following construction of the space station.

At the nodes of history, the weight of time can be felt.

Shenzhou 8 having cracked the technology of automatic docking, Chinese immediately threw themselves on the next target of the manual docking.

"Why is there the necessity in mastering manual docking technology while automatization has been achieved?" Zhou Jianping, chief designer of China's manned space engineering, explained that docking technology, together with manned space-to-Earth shuttle technology and astronauts' space extravehicular activities technology, constituted three basic technologies for manned space development. The comprehensive mastery indicates the ability to build a space station. In the design of Shenzhou spacecraft, manual control was a redundant measure of automatic control. Manual docking can improve the reliability of the mission. It was of great significance to master both. Only when manual docking technology is cracked and mastered, will docking technology be fully mastered.

The Shenzhou 9 astronauts carrying out the manual docking mission attracted great attention: astronaut Jing Haipeng, the first Chinese astronaut to go twice; astronaut Liu Wang, the first to perform manual docking in space, and astronaut Liu Yang, the first female astronaut in China.

"Salute!" At 15:51 on June 16, 2012, with a solemn military salute of Shenzhou 9 astronauts Jing Haipeng, Liu Wang, and Liu Yang, another great Chinese expedition into space began at the Jiuquan Satellite Launch Center.

By that day, Tiangong I had been in space for 260 days. Over these 260 days, the long-term management team of Tiangong-1 in the Beijing Space City has been with it every second, carefully monitoring the first Chinese "space home."

"There are more than 7,000 parameters in Tiangong-1's routine inspection, and 900 parameters should be confirmed every day." Xu Hongbing, deputy director of the aircraft management office, clicked open the Tiangong-1 remote sensing page on a computer. He pointed to the dense data that was constantly being updated, saying "These parameters reflect the real-time state of Tiangong-1, which is how we monitor it." As the "home field" of the space mission, it has been fully prepared for spacecraft's docking and astronaut's living.

On June 16, the crew of the Shenzhou 9 astronauts waved to the people. (Photographed by Qin Xian'an)

"10, 9, 8 ... Ignition! Take off! "At 18:37 hours, Long March 2F ejected a huge tangerine flame and rose from the ground, lifting Shenzhou 9 straight to the sky.

At the moment, the three astronauts lying inside the spacecraft saluted at the camera. This was their solemn commitment to the motherland and the people.

With the spacecraft flying into space, the attention of the world's major media was on the space trip of Chinese astronauts. This time, it received different treatment as it was given more news coverage than previous docking with the International Space Station and the launch of Russian manned spacecrafts.

Historic was the frequently used word on the media. More attention was centered on China, the successor of manned space flight in the world.

Why?

Because this time, if all went well, Shenzhou 9 would dock with China's space laboratory, making it the third country to complete manned space docking, after the United States and Russia.

Because this time, the space flight would last 13 days, the longest for China's manned space flight.

Ignition after ignition, we witnessed these great journeys: one astronaut, one day of Shenzhou 5; two astronauts, five days of Shenzhou 6; and three astronauts, three days of Shenzhou 7. And the duration of Shenzhou 9's trip was planned to reach 13 days.

At 18:37 on June 16, 2012, Long March 2F was successfully launched at the Jiuquan Satellite Launch Center. (Photographed by Qin Xian'an)

What does 13 days mean in space? The spacecraft flew 16 laps a day. In terms of day and night, the astronauts would experience 208 cycles. It was a big step.

"In addition to cracking the technology of manned docking, this mission also had a very important feature, that is, the transition from short-term to medium-term and long-term." Chen Shanguang, director of the China Astronaut Center, said, "It is a great challenge and test for astronauts and their life support, as well as the lifespan and quality of products in the aircraft."

The journey was long and challenging. Astronauts would be challenged in technology, living, weightlessness, psychology, and experiments.

Among them, the most difficult and risky was the implementation of manual docking.

The manual docking would commence when the distance between Shenzhou 9 and Tiangong-1 narrowed to 140 meters.

This was the first time for Chinese astronauts to navigate a spaceship in space. And they were required to accurately dock with Tiangong-1 that was 140 meters ahead. How difficult was it for them to dock a spacecraft at a speed of 7,800 meters per second with Tiangong-1 at the same speed?

"As if space relay race, where the front high-speed runner was holding an embroidery needle and the back high-speed runner was to pass a silk thread through the eye of the needle. Figure it out." Huang Weifen, deputy designer of manned space engineering astronaut system, described vividly: "It directly tests a astronaut's sense of space, ability of space orientation and coordination, quick judgment of aircraft orientation and attitude, operation accuracy, emotional stability, and emergency handling. As if driving a car, where the throttle, brake and steering wheel are all controlled by one person, the spaceship was also navigated by an astronaut. His "steering wheel" consisted of two handles, one of which controlled horizontal movement while the other controlled posture. Therefore, the dynamics of six degrees of freedom and twelve directions were in control. The speed must not be too fast or too slow.

The exciting moment arrived.

At 12:38 on June 24, 2012, Liu Wang, a 24-year-old astronaut, piloted the spacecraft at an absolute speed of 7,800 m/s, aiming at Tiangong-1. It was moving eight times faster than a rifle bullet and 22 times faster than sound.

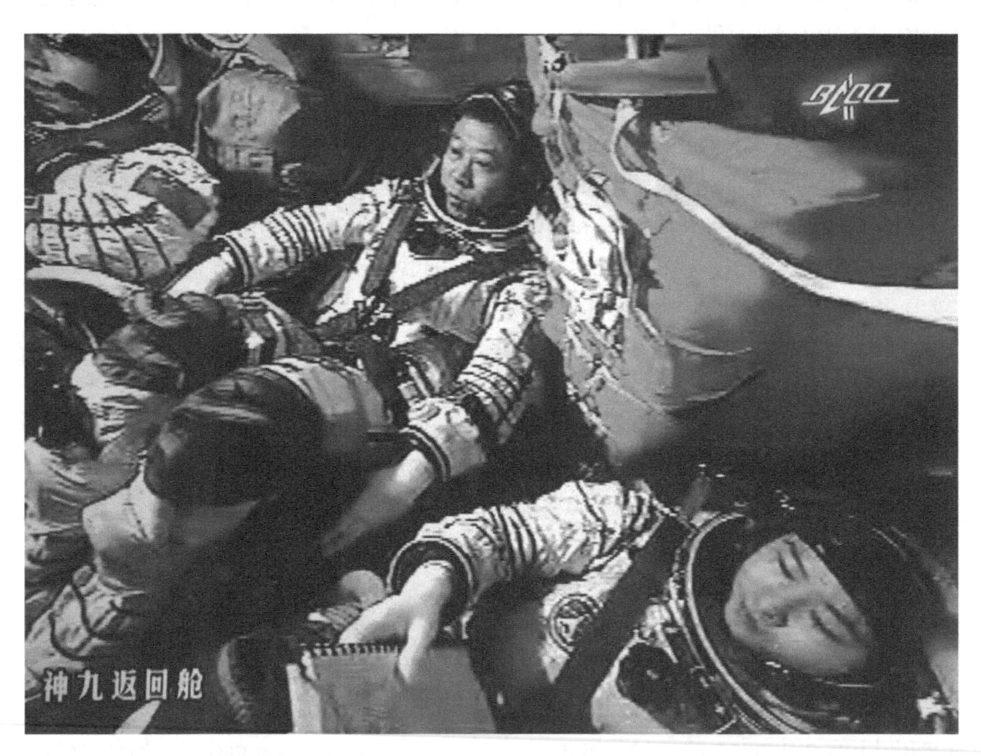

At 12:38 on June 24, 2012, the first manual docking between Shenzhou 9 and Tiangong-1 began. This is Liu Wang's manual operation. (Photographed by Qin Xian'an)

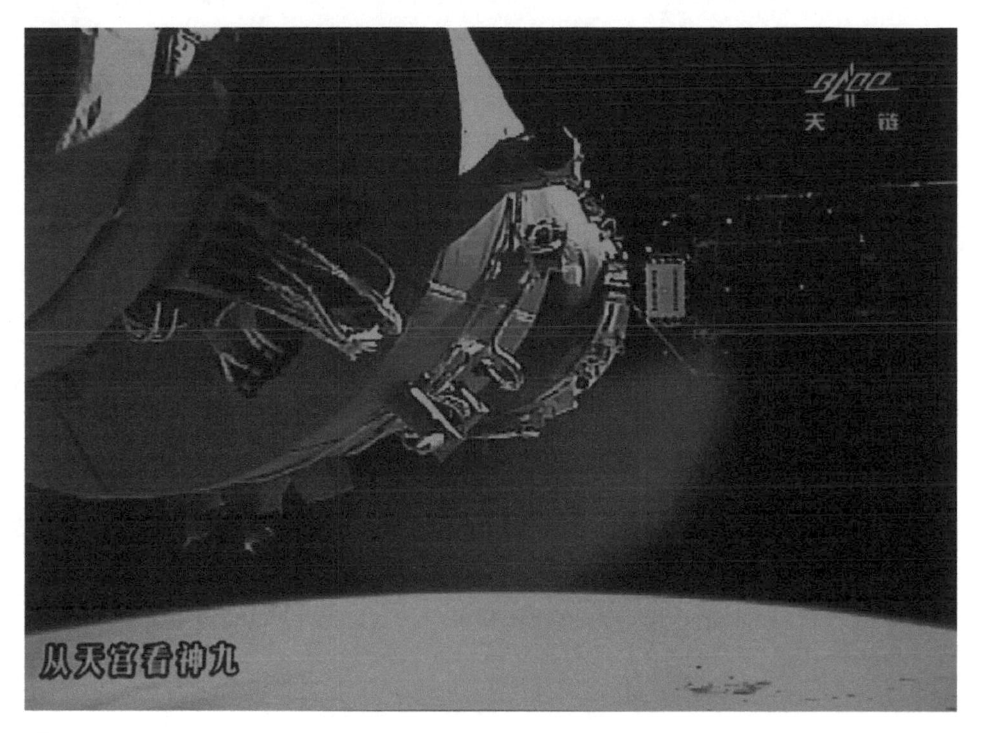

The sight of Shenzhou 9 seen from Tiangong-1. (Photographed by Qin Xian'an)

This was a special "space target shooting," 10 meters away, 5, 3 … on a manually controlled TV, it was approaching the center of the target cross. And in a soft halo, Shenzhou 9 flew to Tiangong-1.

The "target shooting" was significant. Liu Wang had to strictly control the angle between the spacecraft and the Tiangong-1 within one degree. And the lateral deviation should not exceed 0.2–0.3 meters. Otherwise, two aircrafts weighing more than eight tons might collide. Should the docking mechanism be damaged, the astronauts would not be able to enter Tiangong-1 again. This newly built Chinese "space home" would be forced to be discarded as useless …

The pressure was enormous. At that moment, billions of eyes on the earth were staring at Liu's every move. He has become a "special shooter" with the largest audience in history. Liu controlled the handle while observing the instrument: more than 1,500 simulated training practices on the ground enabled him to remain confident and calm.

At 12:48, Shenzhou 9's docking ring accurately captured Tiangong-1. The two "crosses" swayed slightly and joined tightly together. Immediately, 12 docking locks on both aircrafts were activated.

"Shenzhou 9 reports, instrument displays that docking has been completed."

"Attention, this is Beijing. According to Beijing telemetry data, Tiangong-1 astronauts are in good condition and the manual docking is complete."

At 12:55, Shenzhou 9 and Tiangong-1 embraced each other tightly again. China's first manual docking turned out a success.

Immediately, Beijing Aerospace Flight Control Center was roaring with applause. At the visiting table, Chen Shanguang, director of the China Astronaut Center, could not hide his excitement: "Today, we shot a bullseye, perfect, just like a normal training."

On the big screen, Jing Haipeng, Liu Wang, and Liu Yang held their hands high above heads. Before the expedition, Liu Wang said to reporters: "I am 100% sure of the success." Today, he finally fulfilled his solemn promise, which was made before the adventure, to his motherland.

"Following breakthroughs in space-to-Earth flight and extravehicular activities, China has fully mastered the last of the three basic technologies of manned space flight: space docking."

12:50 At 12:50 on June 24, 2012, the first manual docking between Shenzhou 9 and Tiangong-1 was completed. (Photographed by Qin Xian'an)

At 11:23 on June 29, 2012, the crew of astronauts that performed the docking mission between Tiangong-1 and Shenzhou 9 landed safely on the Amgurang grassland of Dorbod Banner, Inner Mongolia, and exited the cabin smoothly. (Photographed by Qin Xian'an)

Zhou Jianping, chief designer of manned space engineering in China, told me: "This means that we have the basic ability to build a space station, which is a solid step towards the three-step strategic goal of manned space flight."

At 10:02:12 on June 29, 2012, the Shenzhou 9 re-entry capsule, bathed in brilliant sunshine, landed in the soft embrace of the motherland on the green grassland of Dorbod Banner.

Jing Haipeng, Liu Wang, and Liu Yang, dressed in spacesuits, exited the cabin in turn with the assistance of the staff. They, in high spirits, waved greetings. Loud cheers broke out at the scene.

With flowers in hand, the three astronauts wore a brilliant smile on their faces. Jing Haipeng said: "We successfully completed the first manned docking mission and came back safely. We thank the people of the whole country for your concern and support." Liu Wang, as always, is not a fan of speeches: "It feels so good to be on the ground." Liu Yang was still full of positivity: "Tiangong-1 is like our home in space, very cozy."

"Shenzhou 9 and Tiangong-1 are '9 + 1', a perfect number," Chinese folks quipped on the Internet.

Foreign media commented that China, with a history of 5,000 years of civilization, could add the successful realization of manned space docking to the list of China's achievements.

Shenzhou 10

On June 29, 2012, Shenzhou 9 successfully returned to Beijing. When the world was cheering for the success, Beijing Space City was surprisingly quiet.

No gongs, no drums, no colorful flags, and no festooned vehicles. When the convoy arrived at the astronaut's apartment building, the welcoming ceremony lasted only five minutes. Then everything calmed down.

Not far away in the Long-term Management Hall of Tiangong-1, Beijing Aerospace Flight Control Center, scientists were busy as usual. Liu Junze, director of the management office, said, "We don't have time to indulge in the joy of success when Tiangong-1 is still flying in space. It will be docked with Shenzhou 10 in the future. We can't afford any carelessness."

In other spots of Beijing Space City, I saw that researchers were still working as usual to tackle key problems.

This quietness implied that in the eyes of Chinese aerospace workers, the success of Shenzhou 9 was only a new starting point for the construction of the space station. And they saw no pause in the development between Shenzhou 9 and 10.

In the news heatwave regarding Shenzhou 9, many have neglected two important pieces concerning the future of China's space industry.

The National Defense Science and Technology Bureau announced that a new generation of high-thrust rocket engines had been successfully developed. Yang Liwei, deputy director of the China Manned Space Office, told me: "The Chinese space station is designed to weigh 60 tons, which is more spacious than Tiangong."

For Chinese aerospace workers, there is a striking time coordinate 2020: building China's own space station.

With only eight years left, they had to pick up the pace.

Of course, the primary goal is to implement Shenzhou 10 well and fast.

On November 10, 2012, Niu Hongguang, deputy commander of manned space engineering and representative of the People's Liberation Army attending the 18th

National Congress of the Communist Party of China, revealed to the media that the Shenzhou 10 was scheduled to launch in early June 2013 and that three astronauts would visit Tiangong-1 again.

"A success is not maturity. Shenzhou 10 is by no means a repetition of Shenzhou 9." Niu Hongguang made a comparison with the United States, Russia and other countries that have implemented hundreds of space docking: China has only conducted four, and the assessment of technical breakthroughs was incomplete, thus requiring further verification.

On February 28, 2013, when the National People's Congress and the Chinese Political Consultative Conference were about to commence, the spokesman for China's manned space project officially announced that Shenzhou 10 would be launched from June to August this year at Jiuquan Satellite Launch Center. Three astronauts would take the spacecraft to conduct manned docking with Tiangong-1 in orbit.

Unlike the experimental flight of Shenzhou 8 and Shenzhou 9, Shenzhou 10 would pioneer the application flight of manned space in China. The so-called application flight is to perform normal transportation tasks, similar to the American space shuttle and Russian Soyuz spacecraft. The same as high-speed rail, it has to be operated on a trial basis at the beginning and adjusted according to the conditions of the trial before being put into normal operation.

"Starting from the Shenzhou 10 mission, a new stage will the space-to-Earth transportation system composed of the Shenzhou spacecraft and the Long March launch vehicle enter. We can now provide a fully functional manned space-to-Earth transportation system to transport personnel and materials to and from manned space facilities in orbit. This is a very important milestone," said Zhou Jianping, chief designer of manned space engineering in China.

Flight duration was a highly concerned factor for China's manned space flight as well as an index to measure its pace. The flight of Shenzhou 10 was planned to last 15 days, two days longer than that of Shenzhou 9. From 13 days to 15, behind these two-day increase was a qualitative leap. Jiang Guohua, chief engineer of the China Astronaut Center, said, "International practice implies 15 days is the start of the medium-term space flight, thus the Shenzhou 10 mission is the first medium-term space flight in China."

The US space website gave a more straightforward evaluation: "Shenzhou 10 will be China's final test of mastering space docking technology, which will enable China to build and operate large space stations around 2020."

Shenzhou 10 was being transported vertically to the launch area. (Photographed by Qin Xian'an)

On March 31, 2013, Shenzhou 10 arrived at the Jiuquan Satellite Launch Center. On May 2, Long March 2F arrived. On June 10, the Yuanwang survey fleet composed of Yuanwang 3, 5, and 6 were deployed somewhere in the Pacific Ocean and would jointly carry out the maritime TT&C mission on Shenzhou 10.

It took only one year for China to complete the preparations from Shenzhou 9 to Shenzhou 10. The pace was always faster than what the world's peers expected, much faster.

At 17:38 hours on June 11, 2013, the Long March 2F rocket rose off the ground amid the earth-shaking roar and lifted the Shenzhou 10 manned spacecraft into space.

Shenzhou 10 astronauts Nie Haisheng, Zhang Xiaoguang, and Wang Yaping underwent the adventure together.

In this mission, Major General Nie Haisheng, 48, served as commander responsible for manual docking operations; Zhang Xiaoguang, 47, acted as commander's assistant, cooperating with him to complete the tasks of spacecraft piloting, manual docking, spacecraft evacuation, as well as a cameraman for space lecturing; Wang Yaping, 33, a female astronaut, was in charge of aircraft status monitoring, space experiments, equipment control, crew life care, and space lecturing.

This was the 10th launch of Shenzhou spacecrafts, just 10 years away from Yang Liwei's first flight in space aboard Shenzhou 5. During this period, a total of 10 Chinese astronauts set foot to space.

On this mission, Nie Haisheng was the first Chinese astronaut to enter space as Major General. In response to a question from a reporter in the *PLA Daily*, Nie Haisheng said, "I am not only an astronaut, but a soldier. It is a soldier's duty to obey command in all actions. In the army, whether general or soldier, their glory comes from the mission. As long as the motherland and the people need them, they will bravely march forward without hesitation at a command."

On June 12, it was Dragon Boat Festival when Chinese commemorate Qu Yuan. Over 2,000 years ago, Qu Yuan asked a question in his poem Question to the Universe: "The universe is above cloud nine, how shall it be measured?" Today, Chinese astronauts' action was the perfect answer to the ancestor's inquiry.

While weightlessness was being explained, Nie Haisheng performed "meditation in suspension." (Photographed by Qin Xian'an)

At 13:00 that day, three astronauts of Shenzhou 10 held up a board with the words "Happy Dragon Boat Festival," sending sincere greetings to the people of the whole country and the Chinese people all over the world through live feed.

At 16:17 on June 13, the door of Tiangong No. 1 was opened. After 360 days of Shenzhou 9 astronauts' visit, China's "space home" once again ushered in its master. It felt good to be back "home" – commander Nie Haisheng waved gently and "swam" into Tiangong. Then, Zhang Xiaoguang and Wang Yaping joined the "swim."

On June 13, Barbara Morgan, the world's first female space teacher and former NASA astronaut, wrote to Wang Yaping, the female astronaut who would teach in Tiangong, through the media. The letter is quoted as follows:

Dear Wang Yaping:

On behalf of teachers and students around the world, I send you greetings of honor and love as your orbit our Earth and prepare to teach your lessons from space. We are proud of you. We wish you and your crewmates safety and success.

You will be very busy up there, but please remember to take time to look out the window. China and all of this world are beautiful.

Sincerely yours,
Barbara Morgan

On June 20, the biggest highlight of the Shenzhou 10 mission was presented in high expectation – China's first lecture in space.

At 10:11 a.m., came a sweet Chinese voice from space: "This is Wang Yaping. I will be giving the lecture. Weightlessness is a unique phenomenon in space…"

With the assistance of commander Nie Haisheng and photographer Zhang Xiaoguang, space teacher Wang Yaping, demonstrated the physical phenomena of object motion and liquid surface tension in weightlessness through five experiments, including mass measurement, pendulum motion, gyroscope motion, water film and water balloon. She also answered the students' inquiries regarding water use for spacecraft, space garbage disposal, weightlessness confrontation and space scenery.

Over 330 primary and secondary school students, including those of minority ethnics, children of migrant workers and student representatives from Hong Kong, Macao and Taiwan, participated in the physics class in the ground classroom of the secondary school affiliated to People's University of China. And more than 60

Wang Yaping taught space weightlessness via gyroscope demonstration. (Photographed by Qin Xian'an)

When Wang Yaping taught space weightlessness, a ball of water reflected her image. (Photographed by Qin Xian'an)

"Shenzhou 10 Science Popularization Education Activity" was set up in the high school affiliated to People's University of China in Beijing. (Photographed by Qin Xian'an)

million teachers and students from about 80,000 middle schools across the country watched it simultaneously on live television.

Three hundred kilometers away, astronauts sent word to the students.

Nie Haisheng said, "May you study hard, increase your knowledge and contribute to the Chinese Dream!" Zhang Xiaoguang said: "Deep space is full of infinite mysteries and endless exploration. Let's work on them together." Wang Yaping said, "The dream of space flight never fades; the dream of science is boundless."

This class was given on the highest platform: located on Tiangong-1 that was 300 kilometers above from the Earth.

This class had the largest audience: 60 million primary and secondary school students in 80,000 schools across the country.

Big time, big dream. The largest class in the world happened in the space age of mankind. Liu Cixin, a science fiction novelist, said: "The greatest thing about this class is not the knowledge points that were demonstrated, but that it acted as a brush to draw a space world different from the gravitational world for children."

At 10:17 on June 23, the accurate operation by astronauts Nie Haisheng and Zhang Xiaoguang together with the cooperation with Wang Yaping resulted in the successful manual docking between Tiangong-1 and Shenzhou 10.

The second manual docking was by Chinese astronauts. And the expected success of the docking had hardly aroused any excitement nor attracted attention from the public.

One success is only one success. Australian space expert Maurice Jones said: "For any space program, docking is a dangerous and complex operation regardless of previous experience." In this sense, the second docking success is as significant as last year's first, which again means more maturity and stability in the technology.

On the morning of June 25, few noticed that Chinese astronauts discreetly staged a wonderful "space dance." Shenzhou 10 circled above Tiangong-1 to its rear and completed the close docking. China's first spacecraft orbiting docking test was achieved.

Again, the test confirmed the profound interpretation of the development of China's space industry by foreign experts: "They hardly repeat the efforts as each flight involves a new step up." The orbiting test, seemingly a small step, was actually a big one. Experts explained that the main purpose of the test was to verify the technology of spacecraft's orbiting and multi-direction docking, which was one of the essential capabilities for China's subsequent space station construction.

This was another new historical moment in China's manned space project.

Shenzhou 10 crew Nie Haisheng (middle), Zhang Xiaoguang (left) and Wang Yaping (right) in the re-entry capsule. (Photographed by Qin Xian'an)

Shenzhou 10 astronauts saluted people with flowers after exiting the cabin. (Photographed by Qin Xian'an)

At 8:07 on June 26, 2013, facing a golden sunrise, the Shenzhou 10 re-entry capsule, carrying three astronauts, slowly landed in the embrace of the soft grassland of Amgurona, Inner Mongolia.

"The whole mission was smooth perfection." Wang Zhaoyao, director of the China Manned Space Engineering Office, commented that it had perfectly fulfilled the second step of China's manned space engineering, thus paving the way for the construction of a manned space station.

For the always low-key, cautious aerospace workers, "very" is rarely used. However, this time, many believe it was well deserved: Shenzhou 10, very perfect. Around 2020, China will build a space station.

Tiangong-2 and Shenzhou 11

On June 26, 2013, the day when Shenzhou 10 mission succeeded, China dropped two "bombs" on the public.

China's new-generation carrier rocket was expected to make its maiden flight during the 12th five-year plan period. According to Yuan Jie, deputy general manager of the China Aerospace Science and Technology Corporation, Long March 5 was the largest carrier rocket under development in China. With a carrying capacity of 20 tons for near-earth orbit, it would mainly be used to launch manned space stations. With a carrying capacity of 13 tons in near-earth orbit, Long March 7 would carry cargo spacecraft during the manned space station project. At present, the development of the both carrier rockets were progressing smoothly.

The launch of China's Tiangong-2 space laboratory was scheduled to be around 2015. "The overall development scheme of China's manned space program plan is that the country will next carry out the development and construction of the space laboratory and plan to launch the Tiangong-2 space laboratory around 2015." Wang Zhaoyao, director of the China manned space office, disclosed that the manned space station project was also in simultaneously smooth progress as planned, with the plan to launch an experimental core module around 2018 and complete the construction of China's manned space station around 2020. During this period, a series of cargo and manned spacecraft need to be launched to supply the space laboratory and the space station while transporting astronauts.

It can be seen that Chinese aerospace workers did not slow down the pace, although the achievement they have made already left the world in awe.

"Urgency leaves no time for pride," said a young technician at the Beijing Space City "With less than seven years to realize the goal of building a space station by 2020, bigger steps must be taken."

Moving forward in silence with maximum efforts is how Chinese aerospace workers are always seen.

2016 was the 60th anniversary of the founding of China's space industry as well as a very important year in the development of China's manned space program. After three years' accumulation of strength, another great show was presented in 2016, a year with special historical significance.

On February 28, 2016, the spokesperson of the China Manned Space Program announced that from the middle of this year to the first half of the next, China would organize the implementation of the manned space laboratory mission.

The spokesman introduced the space laboratory mission as an important part of China's "three-step" development strategy for manned space flight. It was the mark of the entry of China's manned flight into a new stage. It served as the important transition. During the flight, it would test the key technologies for the construction and operation of the space station, such as cargo transportation, propellant replenishment in orbit, and astronauts' mid-term residence, and carry out large-scale space science and application tests. To achieve these goals, China's manned space program has developed the Tiangong-2 space laboratory, the Long March 7 carrier rocket and cargo spacecraft, and built the Wenchang Space Launch Site in Hainan province. And four launch missions were required to be organized.

At present, various preparations for space laboratory missions were being performed as planned. The Shenzhou 11 mission flight crew was composed of two astronauts, who received training for the mission; the Tiangong-2 space laboratory, Shenzhou 11 manned spacecraft, and two Long March 2F carrier rockets, which underwent final assembly tests; the newly developed Long March 7 carrier rocket went through final assembly; the development of the cargo spacecraft was basically completed, the first cargo spacecraft, Tianzhou-1, underwent a final assembly test; all kinds of space test loads, product production, and related preparations were completed; the Hainan Wenchang launch site, Jiuquan Launch Site, TT&C communication system and landing site system were on schedule to prepare for the mission.

That China would launch the Tiangong-2 and the Shenzhou 11 this year was made public by Wang Zhongyang, spokesman of the 5th Institute of China Aerospace Science and Technology Corporation, on the first China Space Day,

April 21. In 2017, Tianzhou-1 cargo spacecraft would be launched to dock with Tiangong-2. The development and launch of the Tianhe 1 space station core module would be completed around 2018, which was an important start for the construction of China's space station.

Zhou Jianping, chief designer of China's Manned Space Program, revealed that the Long March 7 was scheduled to be launched in June to test the correctness, function, and performance of the new generation of medium carrier rockets; the Tiangong-2 space lab was scheduled to be launched in September to receive visits from manned and cargo spacecrafts and to conduct space science and technology experiments. Shenzhou 11 was scheduled to be launched in October, carrying two astronauts to dock with Tiangong-2 for a 30-day stay.

Chinese aerospace workers, who have always kept a low profile, suddenly desired public attention. They had taken the initiative to speak out again and again, demonstrating a strong confidence. Over the years, from nervous to confident then to calm, China's aerospace workers made strides towards maturity.

Launch times became clearer and more precise, reflecting the faster preparation by the Chinese space crew.

The Yuanwang space survey fleet sailed the seas. (Photographed by Qin Xian'an)

At 22:04 hours on September 15, 2016, Tiangong-2 was successfully launched at the Jiuquan Satellite Launch Center. (Photographed by Meng Haochong)

On July 7, 2016, the Tiangong-2 space laboratory set off from Beijing. It arrived safely at China's Jiuquan Satellite Launch Center two days later. And the general assembly and test at the launch site successively took place.

On July 24, Yuanwang 6 and 7 ships left the China Satellite Maritime TT&C Dock with a long whistle. Yuanwang 5 had already set sail in advance. And three survey ships went to the Pacific and Indian Oceans to carry out maritime TT&C tasks.

The Long March 2FT2 rocket that would launch Tiangong-2 and Long March 2F that would launch Shenzhou 11 were safely shipped to the Jiuquan Satellite Launch Center on August 6. Therefore, China's new manned space mission officially entered the preparation phase.

On August 13, Shenzhou 11 manned spacecraft arrived at the launch site.

All the preparations were made according to plan.

Compared with the manned missions of Tiangong-1 and Shenzhou 10, those of Tiangong-2 and Shenzhou 11 would "fly higher, test more, and take longer."

A higher flight: was the manned mission closest to the future orbital requirements of China's space station. Previous manned flight and docking missions were carried out at an orbital altitude of 343 km above the ground. This mission was performed at an orbital altitude of 393 km above the ground, which was basically the same as that of future space stations. The orbit control strategy and mode were closer to the requirements of future space stations.

More tests: this has been China's manned mission with most space application projects by far. Tiangong-2 is the first real Chinese space laboratory. It has 14 space application loads, such as the space cold atomic clock. The mission would carry out space science experiments and applications in many fields, conduct a number of space medical experiments, and perform on-orbit maintenance technology verification for the construction and operation of the space station. The application and technology trial projects arranged in the task totaled more than 40.

Longer time: this was the longest Chinese manned flight. Shenzhou 11 is the sixth manned flight mission in China. After Shenzhou 11 docked with Tiangong-2, astronauts planned a 30-day stay in Tiangong-2. In addition to three days of independent flight, the total flight time increased from Shenzhou 10's 15 days to 33 days.

The Moon invited Tiangong, Shenzhou to share a dance. On the night of the Mid-Autumn Festival in 2016, hundreds of millions of Chinese descendants looked up to the sky, witnessing the birth of another new "home" in space. On September 15, at 22:04, Tiangong-2 was effectively launched, thus China's manned space flight entered a new stage of space application development.

"Tiangong-2 is a real space laboratory, which shoulders the task of verifying the technology of the construction of China's space station." Zhou Jianping, chief designer of China's manned space program, said, "In accordance with the three-step strategy of manned space project, China will build a space station around 2020. The dream that once stood far away is now within reach."

From Tiangong-1 to Tiangong-2, China's "home" in space has experienced a full technical upgrade.

The new "home" has secret equipment. Tiangong-2 is capable of a bigger load, but also equipped with more advanced devices. Apart from carrying a companion satellite, one of the most remarkable secret equipment is the robotic arm. "Generally speaking, the robotic arm is actually a kind of space robot." Researchers of the Fifth Academy of Aerospace Science and Technology Group introduced that it involved cracking a number of key technologies, which can be used for the collaboration between man and the robotic arm for future space stations.

The capability keeps expanding. Speaking about Tiangong-2, Sun Jun, deputy designer of the Beijing Flight Control Center, used three "Mores: more powerful, as the propellant replenishment technology is verified and on-orbit refueling is realized by docking with cargo spacecraft in the future; more agile, as the installment of visual navigation sensors enabled real-time rapid attitude adjustment when docking with Shenzhou spacecraft in the future; more optimized for system design, as the first use of modular technology enabled it to be quickly replaced and maintained in-orbit should any problems arise.

This new "home" has provided more comfort to live in. Zhu Zhongpeng, chief designer of Tiangong-2 from the Fifth Institute of Aerospace Science and Technology Group, said that it first systematically designed a manned livable environment for medium-term residence. From diet to sleep, from health to recreation, great efforts have been made in humanization to ensure that astronauts live more comfortably in space.

One month later, on the morning of October 17, while being watched by hundreds of millions of Chinese, Shenzhou 11 astronauts Jing Haipeng and Chen Donghao embarked on the expedition.

The dream was about to come true. On the expedition, all procedures were operated as planned: the melody of a song for the motherland echoed in their ears; the people who saw them off were in high spirits.

However, as a witness, I was careful enough to notice a certain difference: this time, the pace of the expedition had become faster, as it was the shortest time spent of all the expeditions.

There was less excitement than when Shenzhou 5 took its first flight, less anxiety than when Shenzhou 6 embarked on its journey, less pressure than when Shenzhou 7 took off, less tears than when Shenzhou 9 succeeded, and less concerning expectations than when Shenzhou 10 shot up.

However, this expedition came with more calm and ease; the pace was light, because the astronauts were calm. They were self-assured of a success.

On Oct. 17, 7:30:28, the rising sun was the audience of Long March 2F lifting Shenzhou 11 straight up into the sky. Astronauts Jing Haipeng and Chen Dong made their sixth space expedition on behalf of their motherland.

On October 19, at 3:31 a.m., Shenzhou 11 and Tiangong-2 successfully rendezvoused and docked automatically. At 6:32, astronauts Jing Haipeng and Chen Dong successively entered the Tiangong-2 Space Laboratory.

A space flight of 33 days was a new journey with unprecedented challenges. Astronauts should not only overcome the challenge of long-term weightlessness, but also complete various space experiments." Huang Weifen, Deputy designer of manned space engineering astronaut system, said, "Never has a manned flight

At 7:30 on October 17, 2016, Shenzhou 11 was successfully launched at the Jiuquan Satellite Launch Center. (Photographed by Wang Sijiang)

required such a high level of knowledge reserves for astronauts as this one." In-orbit testing involves many disciplines, such as space technology, space application, aerospace medicine, and so on, with an unprecedented span of disciplines. In order to master the relevant theory and operational skills of experiments in various disciplines, each astronaut has accumulated more than 3,000 hours of study and training.

On October 23, the Tiangong-2 accompanying satellite was successfully released. The satellite weighed about 47 kilograms and is the size of a printer. It has the ability of high-efficiency orbit control, flexible attitude pointing, intelligent task sequence processing and high-speed data transmission in TT&C communication. Accompanied by a satellite with a high-resolution full-frame visible light camera, it can be called the "self-timer artifact" of Tiangong-2 and Shenzhou 11. At 7:31 on October 23, the camera took its first photograph of the Tiangong-2 and Shenzhou 11 complex.

On the afternoon of November 9, Xi Jinping, General Secretary of the Central Committee of the Communist Party of China, President of the State, and Chairman of the Central Military Commission, came to the China Manned

Space Engineering Command Center, and made cordial calls with Shenzhou 11 astronauts Jing Haipeng and Chen Dong, who are on mission at Tiangong-2. On behalf of the Central Committee of the Party, the State Council and the Central Military Commission, and on behalf of the people of all nationalities, he expressed sincere greetings to them.

At 16:20 hours, Xi Jinping entered the command center and took a seat at the command post. The electronic screen clearly showed the real-time picture in Tiangong-2. Jing Haipeng and Chen Dong are carrying out on-orbit maintenance technology tests of man-machine cooperation for the robotic arm. Xi Jinping gazed at the big screen and watched the test operations of the two astronauts. Manipulated manipulator to predetermined position, manipulator and manipulator action, manipulator reset and data glove state recovery ... The astronauts completed a series of experiments accurately.

This was the first time in the world for man-machine collaborative in-orbit maintenance technology tests. Simulated disassembly of equipment was carried out in-orbit by using electric tools to screw and remove thermal insulation materials. Experience was accumulated for a space robot in-orbit service by exploring man-machine collaborative operation mode.

At 16:25 hours, Jing Haipeng and Chen Dong stood side by side at the video call location, saluting Xi Jinping. Xi Jinping smiled and nodded to the two astronauts. He picked up the telephone and talked to the astronauts. Kind voices were transmitted by radio waves to the combination of Tiangong-2 and Shenzhou 11, about 380 kilometers above the earth.

Xi Jinping: "Comrade Haipeng and Comrade Chen Dong, you have worked hard. On behalf of the Central Committee of the Party, the State Council and the Central Military Commission, and on behalf of the people of all ethnic groups in the country, I would like to extend my sincere greetings to you."

Jing Haipeng: "Thank you, General Secretary, and the people of the whole country."

Xi Jinping: "You have been working and living in space for more than half a month. Comrade Haipeng is the third time to carry out manned space mission. Comrade Chen Dong is the first time to go into space. People all over the country care about you. How is your health, life and work going well?"

Jing Haipeng: "Thank you for your concern! We are in good health and our work is progressing smoothly. We can also synchronously watch news broadcasts in space. The images are smooth and clear. Seeing the picture of the General Secretary at the Sixth Plenary Session of the Eighteenth Central Committee of the Party, we feel very warmhearted. China's manned space flight has entered a new height, and the working and living conditions of Chinese astronauts in space have been improved. We are proud of our great motherland."

Chen Dong: "General Secretary of the report, I have adapted to the weightlessness environment in space. I have a normal diet, living and working as planned. I will continue to make great efforts to complete the follow-up tasks satisfactorily."

Xi Jinping: "We are very happy to see you in good condition. You have united and cooperated to overcome difficulties, reflecting first-class and excellent quality. I hope you will continue to work hard, cooperate closely, and operate carefully to successfully complete the follow-up tasks. The motherland and people look forward to your successful return!"

CHAPTER 5

The Photo of Chinese Astronauts Together

It was a collective appearance that raised the spirits of millions of people.

On the Spring Festival Eve of 2017, on the stage of the CCTV Spring Festival Gala, 11 Chinese astronauts who went into space made a brilliant appearance. When thousands of families reunited, they came together to send their sincere wishes to the motherland.

They saluted the national flag, reported to the motherland, invited Chinese all over the world to build the Chinese space dream, and make the country strong.

Applause erupted, and the camera flickered. This historic scene had been captured in a group photo, which was trending on WeChat Moments for days. It took root in the Chinese memory.

Word on the street that the photo was so marvelous that a glimpse of it would fill the onlooker with great pride.

The world knows China through every vivid Chinese face. Chinese know China's space flight through every face of the fantastic astronauts.

From Yang Liwei to Fei Junlong and Nie Haisheng, from Zhai Zhigang, Liu Boming, Jing Haipeng to Liu Wang and Liu Yang, and from Zhang Xiaoguang, Wang Yaping to Chen Dong, the appearance of each new astronaut marked a new leap and a new height in China's manned space industry.

As the "cover" of China's manned space industry, the face of astronauts deserved to be called "the big face." Over the years, their brilliant work in space again and again wrote the history of China's space flight, won the world's attention and admiration, interpreted the changes of China's national image, and allowed confidence to climb to face of every Chinese.

Word on the street that the photo was of such historical significance that a glimpse of it would fill the viewer with great joy.

How many countries can achieve this? Eleven astronauts, all of whom were "homemade," have "gone to space with beautifully accomplished tasks" and "returned safely," were standing side by side on the stage of the Spring Festival Gala."

While bathed in the joy brought to us by the 11 Chinese astronauts, have you ever thought about the hardships and efforts Chinese aerospace workers have devoted?

At this moment, they had gone through thick and thin for 60 years. For an entire *Jiazi* (sixty-year cycle), generations of space scientists have sacrificed their youth and even lives, paving the way to space.

At this moment, since the initiation of China's manned spaceflight project in 1992, they had been working relentlessly for a quarter century. With unprecedented passion, they defined "China speed" to the world space history.

In a sense, 60 years of pain and glory in China's space flight are condensed in this special photo, which deserves the appreciation of the entire Chinese nation.

"I am proud of China!" This were the first words of Yang Liwei, the first Chinese space hero, upon landing. Yang believed that the real heroes were all the comrades who remain unknown yet selflessly were dedicated to the manned space front. They were the old experts, leaders, and scientists who have worked diligently for decades.

"On the way to space, every footprint of the astronauts is great." Behind Armstrong's small step on the Moon is a big step made by countless scientists with their lifelong wisdom.

China supported the Shenzhou series. Apart from Yang Liwei, every astronaut is well aware of the fact that the way to space was supported by millions of people; their wings to fly were forged by a vast number of astronautical scientists with their own wisdom and sweat.

Word on the street that it was a rare "family portrait" for Chinese astronauts, a glimpse of which would fill the onlookers with pride.

Family portrait was an incorrect statement. The media disclosed that the number of the China's first batch astronauts was 14, and that of second batch was seven (five men and two women), adding up to 21. Yet, there were only 11 astronauts on the stage.

So, what about the other 10 astronauts?

It was a long journey. Some of the first astronauts hard dedicated 20 years in order to go to space yet their names were little known. The vast majority of the second astronauts remained anonymous, as if it were a sharply polished arrow, ready to shoot into space any time …

Those who went were the Republic's space heroes. Those who did not were no less.

The risk of exploring space was notoriously huge. Since Yuri Gagarin went to heaven, around the world, 23 astronauts have devoted themselves to space. In the face of unknown risks and the call of the motherland, they never hesitated as if a soldier on an expedition closely tied personal values with the honor of the motherland.

It was a global tragedy when the space shuttle Challenger exploded in the sky, killing all seven astronauts. In mourning, President Reagan said that heroes were heroes not because of what we eulogize; it was their responsibility to practice in real life in the sacrifice of people lives with a sense of consistency and perseverance in exploring the miraculous and wonderful universe.

American astronaut Gus Grissom was killed in the fire on Apollo 1's test spacecraft. He once uttered some touching words "If we die we want people to accept it. We are in a risky business, and we hope that if anything happens to us, it will not delay the program. The conquest of space is worth the risk of life."

The voices of the Chinese astronauts were no different. "I love it more than my life. Going to space is to fulfill my vow to the motherland." Jing Haipeng, a three-time astronaut, said, "To go on behalf of our motherland is a career with not only a start but an end."

In the eyes of the ordinary, astronauts are heroes who face life and death with calm. But in the eyes of astronauts themselves, they are the ordinary. Liu Yang, China's first female astronaut, said, "We astronauts are a group of very real and flesh and blood, ordinary people."

Next, let's take a closer look at Chinese astronauts, who are, as if w national treasure, mysterious, great, yet also ordinary.

The Unfinished Space Dream

During the 2017 NPC and CPPCC sessions, the selection of Chinese astronauts had once again become a topic of public concern.

"With the completion of the overall plan, the selection of the third batch of astronauts in China is scheduled to be officially launched this year." Zhou Jianping, chief designer of China's manned space project disclosed in an interview that according to the progress of manned space missions, the team of Chinese astronauts would be expanded and diversified. "In the past, they were all pilots, and there will be flight engineers and load experts in the future."

Little was it known that it was in the 70s that China's first selection of astronauts took place.

It was a dusty history. In 1969, taking the Apollo spacecraft, American astronaut Neil Armstrong landed on the Moon. The success "provoked" the socialist countries. In 1970, China initiated the first manned spacecraft Shuguang project, code-named Project 714, immediately after the successful launch of Dongfanghong 1. Approving of the project, the Central Committee issued the following instructions, that is, to embark on the research and development of manned spacecraft, and to start the selection and training of astronauts.

Meanwhile, the central government decided that the Air Force should be responsible for forming a joint selection team from relevant departments to select astronauts from the air force pilots.

The first Soviet astronauts were selected from fighter plane pilots. The first American astronauts came from test pilots of the Air Force. They were required a flight experience that was no less than 1,500 hours.

As the selection team drew a lesson from the experience of the United States and the Soviet Union and combined with the actual situation of the Chinese Air Force, the flight time requirement was set at more than 500 hours. Other requirements included: age around 30, weight between 65–68 kg, height …

Selection began with profile screening, and proceeded with an investigation on political beliefs, flight skills, physical examinations, etc. The procedure was divided into primary selection and reselection. The former spanned from mid-October to mid-December, 1970 and the latter from February to May, 1971.

In the primary selection, 88 out of 1,800 were deemed qualified pilots.

The 88 selected candidates were gathered in a separate building of the Beijing Air Force General Hospital for re-selection. The process was conducted with great discretion throughout. As a witness, Fang Guojun recalled that when he arrived in Beijing, he had no prior knowledge what task he was going to perform; It was not until the day superiors showed him a documentary in which the Soviet astronaut Yuri Gagarin was on the Oriental spacecraft that the realization came.

The re-selection was more rigorous. After layers of sifting, 33 out of 88 remained. Then the selection team selected 20 of the 33 astronauts as the first batch of reserve astronauts. Fang Guojun, eventually made it to the list of the 20. Since then, he has dreamed of one day piloting a Chinese spacecraft into space.

Subsequently, the Air Force established the "Astronaut Training Preparatory Team," which scheduled two years to train astronauts and launch the Shuguang spacecraft at the end of 1973.

Yet, no one expected that on September 13, 1971, Lin Biao (Vice-Premier of the PRC, under Mao Zedong) defected, thus Project 714 was cancelled. The selected 20 reserve astronauts all returned to the original forces.

The plan of space flight was abruptly aborted. However, Fang Guojun never abandoned his dream. He evolved from an ordinary pilot to a general. He kept exercising until the end of his flight career in 1995. Fang said that he even pictured how he would say, "See you later, earth" before the flight. Unfortunately, Fang never made the dream come true.

On October 15, 2003, grey-haired General Fang watched on TV at home Shenzhou 5 astronaut Yang Liwei soar into space. Tears were shed regardless of efforts to hold them in.

After the safe return of Shenzhou 5, I interviewed Yang Liwei and told him about his predecessor Fang Guojun. Yang expressed his heartfelt admiration: "I admire him with great respect for his persistence in space flight."

99% Elimination Rate

Time flies. Twenty years have passed.

In September 1992, China's manned space project was in full operation. The selection and training of astronauts had once again been put on the agenda.

Manned space flight emphasized on "manned."

Among the seven systems of manned space project in China, the astronaut system was the "first system." The selection and training of astronauts was a big deal back then.

In the cold winter wind of 1995, the selection of Chinese astronauts commenced. The Selection Expert Team was composed of the Air Force and the National Defense Science and Technology Commission. The selected targets were active pilots of the Air Force. The selection criteria drew lessons from the practices of the United States and the Soviet Union.

For the selection, the investigation was very thorough: the vast number of checkpoints and the high standards were beyond imagination.

Basic requirements included: a strong will, dedication and good compatibility, height between 160–172 cm, weight between 55–70 kg, age between 25–35 years old, fighter plane and attack plane pilots, a total flight time of more than 600 hours, a college degree or above, excellent flight performance, and zero grade accidents. In addition, there were also preference over proper facial features, a clear tongue with no serious accent, and no smoking and alcohol addiction. Even a physical phenomenon as insignificant as hemorrhoids, armpit odor, or even snoring was to be investigated.

A detailed investigation of 1,506 Air Force pilots who met the basic requirements led to the decision of Selection Expert Team for 886 pilots to participate in the primary selection. After nearly half a year of inspection, 97 pilots were confirmed preliminarily qualified.

This was the real experience of surmounting all difficulties. All organs of the human body should be checked for function one by one, during which a slight defect would result in elimination. Next, the physiological functions for a space flight were examined, such as endurance in a centrifuge, low pressure chamber, rotating seat, etc. Every examination was a test of the extreme physical and psychological functions of an individual.

In August 1996, Yang Liwei passed the astronaut's primary examination in Qingdao Sanatorium. He was told to return to the army for further notice. Yang recalled: "Having passed the first round, I was waiting in great anxiety. During that time, I used to call Beijing to inquire about the situation in the fear of that it was a no for me."

Rigorous and scientific analysis and evaluations by the expert team had the 97 preliminary qualified candidates, including Yang, eliminated to 60. They were sent to Beijing Air Force General Hospital for centralized physical examination in a new round of a more rigorous selection.

Nie Haisheng recalled the month in the Air Force General Hospital, every day of which involved some check-ups. What was expected was checked, and what was not as well. For example, there were hundreds of checkups in blood test indicators, and there was blood drawn several days in a row; countless checkups were made for both eyes.

Physical condition being up to standard was the only the first round. Psychological quality was equally important since the astronauts, who were out of the earth facing a dark and lonely world, must have strong psychological endurance.

Psychological selection was an important part of the procedure. The conversation between the experts and the candidates may not necessarily revolve around the astronaut's expertise, around the flight, but everyday topics. And the potential ways of thinking and personal qualities of the candidates have been clearly perceived by the experts as it was not easy to deliberately conceal those. In addition to psychological interviews, psychological tests, situational tests, computer answers, the Minnesota Multiphasic Personality Inventory (MMPI) test questions were also conducted.

Sixty people were eliminated after the rounds, leaving only 20.

However, it was not the final result.

Subsequently, the relevant departments investigated family medical histories and ran physical examinations of their immediate family (spouses and children) as well as political inspection of the 20 candidates.

Wives of pilots, aware of the significance, were very supportive of their husbands. Zhai Zhigang's wife, Zhang Shujing, worried that her unqualified health may impede her husband from being selected, determined, "If I get in the way, I don't mind a divorce."

From April 14–18, 1997, under the direct leadership of the Manned Space Project Command, an appraisal meeting on the results of in-patient examinations and the first medical selection appraisal meeting for Chinese preparatory astronauts were convened successively. In particular, the latter brought together prestigious experts from major hospitals throughout the country. After repeated research and selection, experts recommended 12 candidates as reserve astronauts.

Finally, 12 outstanding flight talents made the cut. In addition to Wu Jie and Li Qinglong, who had been selected to study at the Russian Gagarin Astronaut Training Center, 14 individuals became the first batch of astronauts in China.

Internationally, the elimination rate of astronaut selection was 50%. In China, from the initial 1506 pilots to the final 14, the rate reached 99%.

The Mysterious Brigade of Astronauts

Beijing Space City, January 5, 1998.

Fourteen pilots from the Air Force took off the flight emblem on their chests and replaced it with the golden space emblem embedded with the Earth logo.

Facing the bright red national flag, they raised their right hands, clenched their fists, and pronounced their oath with a firm voice.

> *I volunteer to engage in the cause of manned space flight; becoming an astronaut is my supreme glory. In order to shoulder the sacred duty of an astronaut, I vow: to love the Communist Party of China, the socialist motherland, the Chinese People's Liberation Army, and the cause of manned space flight; and to obey orders and commands, study hard, train diligently, cherish weapons and equipment, abide by discipline and laws, and keep state secrets; to be brave, selflessly dedicated, unafraid of sacrifice, and willing to fight for the cause of manned space for life.*

"It was utmost excitement." One astronaut recalled that his hand was shaking when he left the signature on the national flag.

This little ceremony opened of a new and important page in the history of Chinese space flight – the formal establishment of China's first astronaut brigade. It was the third astronaut brigade in the world following that in the Soviet Union and the United States.

Henceforth, the 14 reserve astronauts embarked on an arduous adventure. They disappeared from public attention and lived a mysterious life isolated from the outside world.

On a sunny afternoon, I was honored to enter the camp of the Chinese astronauts' brigade.

The reddish orange new apartment for the astronauts radiated serenity in the transparent sunshine. At the gate, the banner that wrote "The motherland's interests above all" on the roof was particularly conspicuous. Fei Junlong, the captain of the brigade, told me that they went to bed with the concept literally hanging above heads every day. The sense of duty revealed in his words deserved deep admiration.

It was understood that the new apartment was only in operation for a year. The old apartment failed to meet the demand because of the new recruitment of the second batch of five male and two female astronauts. A five-day residence system was applied. The apartment thus was equal to the second home for the dwellers.

Stepping into the hall, I laid my eyes on the front wall where a three-dimensional space auspicious cloud pattern composed of a 2012 small blue earth hung. The elegant and delicate furnishings on both sides spoke of the space culture characteristics. At the moment, the astronauts had gone out for training. The corridor was empty. I had the urge to open one of the doors for closer inspection. However, I did not expect that fingerprint of each astronaut was required to open the respective door of the dormitory.

Opening the Interim Regulations on Astronaut Management, I noticed a series of forbiddens: forbidden to dine out, to sneak out secretly on holidays, to contact with unidentified individuals, to reveal their identity, to smoke. Each forbiddance spoke of the restraints they were imposed on their daily life.

A special gymnasium, which was covered with thick carpets, was built next to the apartment. Astronauts should exercise without any injuries. The fine division of sports equipment there enabled every muscle to be well exercised. And a turn after exiting the gym would lead to the swimming pool, which was available for astronauts. Swimming was the best option to keep the proper weight. Having toured around, I noticed the presence of an honor exhibition room, a multi-function hall, and a library. Every building hid the secrets of Chinese astronauts' space adventure; every path was engraved with the footprints of Chinese astronauts' persistence.

Following astronaut instructor Wang Quanpeng, I walked into the astronaut classroom. At first glance, it exhibited no difference from an ordinary classroom at school: small size, more than a dozen white desks in neat order, a projector hanging in front of the podium, and a blackboard on the side. A walk around allowed me to discover the hidden difference: at the upper right corner of every desk was neatly placed with green packaged activated carbon bags. Wang explained that it was used to absorb harmful gases inside, thus ensuring better health of the astronauts, because they often stayed up there studying almost every night.

An outstanding astronaut shall have both a strong body and a sharp brain. Huang Weifen, deputy designer of the astronaut system, told me: "Astronauts should not only learn to master the knowledge of space flight, but also space medicine, biology, and so on. In addition, as public figures, they should understand art and demonstrate eloquence, etc. It takes at least three and a half to four years to forge a mature astronaut."

Contrary to public speculation, these "national treasure" figures were not served by master chefs. Three meals were prepared by the kitchen squad in the military style. The chef of the astronaut brigade, Que Hongwu, was the oldest in the squad.

He had an experience of 11 years of cooking in the brigade. And he understood the personal taste of each astronaut. Que uttered to me some simple words: "Cooking was my sole task when the recipe was decided by nutritionists." I took a look at the recipe that evening. The name of the food was affixed with the information only comprehensible to professionals – the volume of calcium, the proportion of calcium and phosphorus, and the ratio of various vitamins ... To ensure adequate nutrition, three meals were cooked in strict accordance with the detailed recipe. Even though the tastes of astronauts varied greatly, the diet must be performed.

A banner hanging in the training venue left me a deep impression: "Your energy is beyond imagination!"

The Unique Astronaut Center

Training astronauts was what no Chinese had prior experience.

This burden fell on the shoulder of a mysterious department codenamed 507, the Beijing Institute of Aerospace Medical Engineering, established in 1968.

Since the establishment, the science researchers of the Institute have been exploring tirelessly to realize the dream of space flight despite changes in the external environment. By the time Project 921 was launched, over 100 veteran experts of the Institute had retired. Tears were shed. They dreamed of the job 30 years ago, but retirement came before it was complete.

On December 12, 1992, the Institute officially launched the development of Project 921. At a meeting, the Institute made a request to all personnel: no publicizing, no leaking any information to irrelevant personnel, staying unknown, concealing the specific code when publishing a paper, and performing with utmost confidentiality.

On October 12, 2005, when Shenzhou 6 astronauts Fei Junlong and Nie Haisheng were traveling in space, the Beijing Institute of Aerospace Medical Engineering was officially named the China Astronaut Research and Training Center (CASTC) and made a debut on the world stage.

To go to space, preparation must be conducted.

In order to train the 14 pilots to be qualified astronauts, the Institute had established two research rooms: the Astronaut Selection and Training Research Room, which was responsible for formulating the training outline and program, compiling training materials, and providing training guidance and flight operation guidance; and the Astronaut Medical Supervision and Insurance Research Room, which took care of medical supervision, examination, diagnosis, and treatment in the whole process from training to living to space flight.

The training of astronauts, as if a white paper to be painted, had to start from scratch.

It was basic work to formulate a training outline and program and compile training materials for the training session. It was of great difficulty. Despite references available in foreign materials, the instructions not only explained how to train astronauts but also the reasons behind the approach. The sole understanding of what to follow without why so would lead to a blind path. For manned space flight, blindness is a disaster. Therefore, the analysis of foreign materials must be conducted to find out the causal relation before compiling training manuals.

Efforts of the teachers led to the fruition of the understanding as well as innovation. They formulated a training program for astronauts, including 10 sub-syllabuses with detailed and substantial content; with Chinese characteristics, the training program was highly targeted, and even included the literary and artistic cultivation of the astronauts.

There was a great variety and load of training for astronauts, including basic theory training, physique training, psychological training, space environment endurance and adaptability training, professional technical training, flight procedure and mission simulation training, and rescue and survival training. Each major training included several or even dozens of specific trainings. As there was no ready-made experience for reference in the guidance of the work, they have maximized efforts to explore the key.

The development of ground simulation training equipment was a great difficulty. Take centrifuge for overweight endurance and adaptability training as an example. Imported from abroad, it was costly, yet the performance failed to meet the training requirements. "Self-resilience supports oneself the best." The science researchers of the institute, in cooperation with relevant domestic manufacturers, have been gnawing the "hard bone" for six years.

Hyperbaric oxygen bottle was a key product in the environment control and life support system. A minor malpractice in the development would have resulted in a massive explosion. Cao Hongnian, deputy quality director, ill at the test site, insisted on finishing the test before returning to the hospital in Beijing for an examination. The diagnosis turned out advanced cancer. A few months later, he passed away. Yet, before death, his mind was still on the progress of the development.

Science researchers prioritized to ensure the safety of astronauts. As chief designer of China's manned spaceflight project, Wang Yongzhi once told the chief designer of the seven major systems that the passenger of the first spacecraft should be the chief designers themselves. Only when they dare to pilot the flight should the astronauts board the spacecraft.

Having continuously tackled key problems, they gradually completed the construction of a dozen subsystems. From ground simulation training equipment to aviation food, from space flight medicine to pens, combs, paper towels, and vomit bags, the Institute basically realized the all-round guarantee of astronauts training by the end of 1998.

Sun Jinbo, an old expert, recalled: "The clock was ticking. We did the best picking up the pace, and at last we did not delay the whole process of manned space launch mission."

Admiral Sergei Krikalev, director of the Russian Astronaut Training Center and a Soviet aerospace hero, once visited Beijing Space City. Facing this huge city full of modern science and technology, he said in sincere praise that this was the third astronaut training center in the world.

The 27th Annual Meeting of the Association of Space Explorers was held in Beijing on September 13, 2014. Nearly 100 Chinese and foreign astronauts visited the China Astronaut Center. Alexei Leonov, the world's first spacewalker and Soviet astronaut, said that Russia's Gagarin Center only selected and trained astronauts, while the China Astronaut Center was more comprehensive as, apart from training, it also engaged in research in space medicine and ergonomics.

Gagarin's Unknown Mentor

The flight made the name of the astronauts.

Over the years, Chinese astronauts, from Shenzhou 5's Yang Liwei to Shenzhou 11's Chen Dong, set the fastest record for Chinese soldiers to rise to fame because of the space flights.

But who had trained them?

Except for the astronauts themselves, few knew the answer to the question. As Huang Weifen, the woman in charge of training Chinese astronauts, said, "Everyone knows Gagarin, but his mentor was unknown."

The training of astronauts was highly risky, as more than 10 astronauts have lost their lives in training. The first astronaut sacrificed in the history of human space flight was Soviet astronaut Valentin Bondarenko. On March 23, 1961, when training in a dark chamber filled with pure oxygen, he was burned alive because of a fire inside.

In the China Astronaut Center, apart from the astronauts' brigade as permanent staff, there was another team of simulated astronauts. As many training programs for astronauts are conducted for the first time in China, there were risks. Simulated astronauts were the guinea pigs. Only when they have proved it safe would the real astronaut get trained there.

"Astronauts are national treasures. Their safety, whether in space or at training, must be guaranteed." Wu Bin, director of the Astronaut Selection and Training Office, said, "We can train the astronauts correctly and well only when we have mastered it."

During the Shenzhou 7 mission, I had a conversation with the simulated astronaut Wang Zai.

Wang Zai, who graduated from Beijing Sports University in July 2001, was the chief designer of astronaut selection and training system. He was wearing a pair of black narrow-edged rectangular glasses with a friendly smile on his face. Speaking of his work, he said, "I have to do what astronauts are expected to do, and earlier."

In Shenzhou 7 mission, training was perplexing and dangerous. Procedure validation in low-ballast tanks was a key step in the implementation of the mission. Foreign experience had proved that some astronauts with excellent performance in atmospheric pressure environment would turn out otherwise once they enter the low-pressure chamber.

It was very dangerous to have a real person to run the test in a high-vacuum environment below 10 Pa. Should the air leak, an eardrum would be pierced immediately. There was no prior experience, but this was what must be done in order to conduct extravehicular activity. And Wang put his personal safety aside, and voluntarily asked for training in the low-pressure chamber as the first.

"He put on a heavy test suit, which was connected with a number of electrodes. Then he was suspended in a sealed low-pressure cabin, where instruments and electronic display screens were all around." Recalling the situation, director Liu Weibo admitted that he was so nervous that his hands were sweating insanely. In great peril, Wang was extremely calm. With an eye on the various posts and equipment, he carried out the operation accurately, according to the commander's instruction.

Since then, Wang has become a professional of low-pressure cabin training. And he was the first choice for every such training. Once, in a low-pressure test, the pressure gauge on Wang's suit suddenly stopped functioning, thus he could not grasp the pressure of the suit. Should it leak, the person in the cabin would be at risk of decompression sickness. Facing this terrible problem, he did not panic, but calmly communicated with his colleagues for confirmation, and conclusion was made that it was only the pressure gauge that had gone wrong. And his tap on the dial solved the problem.

In the face of danger, Wang never thought of giving up. Fearing that his family would be worried, he did not dare to mention his work to them, especially to Wang's mother, who was a doctor, who would certainly understand the peril ahead.

In the past two years, in cooperation with several colleagues in eight low-pressure chamber training sessions, each of which lasted four-to-five hours, he successfully completed various experimental tasks.

Having accumulated experiences, Wang Zai compiled the operation manual, emergency, and malfunction handling manual and technical instruction for extravehicular activities to guide astronauts on extravehicular training. Doubts were cast. "Is it worth risking it?"

He replied positively: "Absolutely. Seeing astronauts go to space, I feel like I am there too. I can totally relate to what they must be feeling."

Wang Zai, a simulated astronaut, was only one of the many mentors of Chinese astronauts. The China Astronaut Center was full of mentors who trained astronauts regardless of their personal safety as Wang.

Tian Liping is the head trainer of the simulator. Astronauts received a four-hour training one at a time. Fourteen astronauts meant 14 training sessions. After some time, he got an intervertebral disc disease, young as he was. He Yu, a simulated astronaut, entered the cabin for test, which should last continuously for three to four hours. Each stay inside left the head with a buzzing sound, thus very tormenting for the trainee. In a re-entry capsule tolerance test, Doctor Liu Jianzhong and a simulated astronaut bore the ocean wind and waves for 20 hours.

Since there was no zero-gravity aircraft in China, weightlessness flight training would be conducted at the Russian Astronaut Training Center. Chinese astronauts piloted a total of nine weightlessness flights. Xie Junshui, a doctor who tagged along for medical experiments with the team, persisted in his work after vomiting. Suzu Shuangning, an institute director, asked for an experience on the spacecraft. The Russian side gave him a thumbs up: "Your bravery set a good example for the boys."

After the success of the first flight, astronaut Yang Liwei expressed emotionally to the media: "Coaches spent no less effort than us astronauts. In a simulator training, we could rest after one's session was done, yet it was 14 sessions to repeat for them. The workload was much larger than that of us. In the past, every time I finished training, I had to report to the coach, and this time I would like to report too. It's nothing but normal."

The ladder to space was long, every section of which was assembled with the efforts of countless astronauts. Yang Liwei and every other Chinese astronaut in space shared such a feeling: I am not flying alone, but with the vast number of comrades-in-arms on behalf of the motherland.

Training on the Ground

A new leap forward was from soaring in the sky to a journey in space.

In order to realize the transformation from pilots to astronauts, the first generation of Chinese astronauts had been preparing for five years. The difficulties and obstacles they encountered were beyond imagination.

"The first lesson took place in the canteen." After the astronauts finished their first meal as their personal diet was intended, the leader of the astronauts' brigade invited nutrition experts over to instruct on how to cat properly.

Starting from scratch in dining, astronauts, began five years of hard learning and training.

The daily routine of Chinese astronauts was roughly arranged like this:

Every day, getting up at 6:30, breakfast at 7 a.m., training at 8 for 4 hours in a row, a break at noon, then training from 2 to 6 p.m., dinner at 6 p.m.; after dinner, self-study was usually arranged, or sometimes special training; next, return to the astronaut's apartment and bedtime. Only on weekends were they allowed to reunite with their spouse and children.

"I remember when I first arrived at the astronauts' brigade, the first batch of trainees advised me often to endure loneliness and persevere in order to realize the dreams." Liu Yang, the first female astronaut, confided in me that astronaut's life was often very prosaic. As if a high school senior was preparing for the college entrance exam, the schedule for the following day was made known one day earlier. And if it rained, it would strictly follow as scheduled.

The Soviet astronaut Alexei Leonov, who was selected with Yuri Gagarin in that year, once called the training stage the "ladder to space." His companion, Valerie Bekowski, added: "Not a short one."

The first stair on the "ladder" was fundamental theory of space science. Yang Liwei said: "We have a long list of courses to learn, including astronomy, astro-mechanics, aerodynamics, aerospace medicine, psychology, foreign languages, and pertinent knowledge of the seven major systems of manned spaceflight, involving over 30 branches of learning and over 10 categories."

Yang Liwei hardly slept before 12 p.m. for more than two years in order to overcome the difficulty. "To prevent dozing off, Yang bought a large cup to prepare strong tea, with the help of which he stayed fresh-minded. "I passed all tests with flying colors."

Immediately after mastering the basic theory began physical training that challenged physiological limits and space simulation environment adaptation training. All the overweight endurance training, hypobaric hypoxia training, vestibular function training, and weightlessness flight training was overwhelming.

Swivel chair training was crucial. An average individual would feel dizzy and throw up after a few spins. Every time the astronauts were trained for it, they chose the longest time and maximum intensity. Deliberately without pillows to sleep on at night, they raised the mattress to adapt to the redistribution of blood as if in space weightlessness. Astronaut Jing Haipeng said, "In the end, I was good enough to exempt from the training. As for centrifuge and treadmill training, I also insisted on the most difficult and intense. My load resistance and impact resistance exceed 8G, equivalent to 8 times my own weight at present."

Astronaut Wang Yaping disclosed that in each training for overweight endurance, facial muscles would be deformed, tears would flow out unconsciously, and the chest would feel extremely depressed and it would be difficult to breathe. Under the circumstance, they also had to perform all kinds of actions according to the regulations. Every second felt like a year.

In fact, a red button was made available within the reach of astronauts. Should they feel unable to proceed with the training, they could request a suspension at any time. But over the years, none has touched the button.

The absence of air in the vastness of space means that sound is unable to be transmitted. Spacecraft orbit in real silence. To nurture in astronauts the psychological quality to endure loneliness, they had to enter an isolation cabin to undergo a special training. They stayed alone inside the sealed and narrow isolation compartment

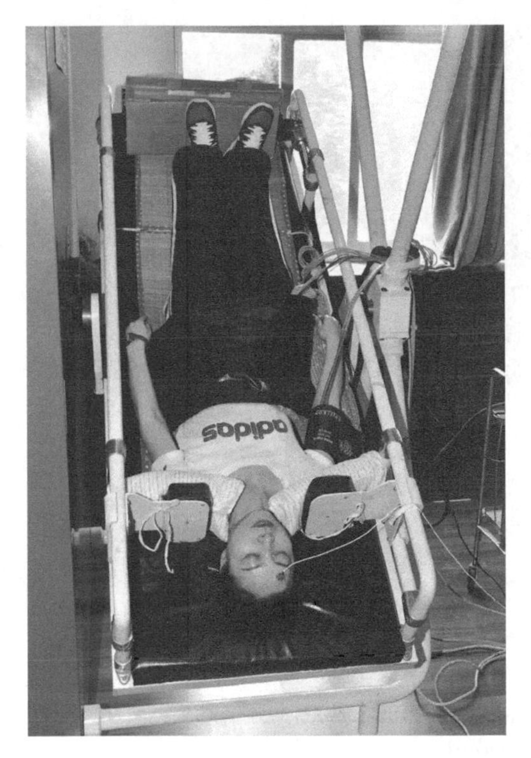

Wang Yaping conducted cerebral blood flow distribution test. (Photographed by Qin Xian'an)

for several days, consequently unable to distinguish between the alternation of day and night. Also, they had to complete all kinds of work up to a high standard. Yet, there was no one to discuss nor communicate with. At times, their eyelids, as heavy as lead, were about to close. But as soon as they close, a monitoring bell would ring. Sometimes, the stay in isolation capsules lasted 72 hours without rest for a second. And food was passed in through a small window. A physically and mentally exhausted astronaut, having stepped out of the isolation module, said, "On the first day, smearing a bit essential oil would help. On the last, it doesn't matter how much oil."

Time after time of physical and psychological overload trainings, Yang Liwei gradually discovered a rule: "When it was pushing the limit, success is around the corner."

There is a saying in the space industry: skills for space are trained on the ground.

Yet, the difficulty to train some of the space skills on the ground was beyond imagination.

In space, the biggest challenge was weightlessness. The most difficult to simulate on the ground was also weightlessness.

Astronauts in extravehicular spacesuits were trained in simulated weightlessness training tank for extravehicular activities. (Photographed by Qin Xian'an)

Coaches and medical supervisors in the control room of simulated weightlessness training were observing the astronauts' extravehicular operations. (Photographed by Qin Xian'an)

At the China Astronaut Center, my outlook was broadened in the witness of the largest neutral buoyancy simulated weightlessness training tank in Asia. The principle of tank simulating weightlessness environment is not complicated: when human body is immersed in water, a feeling similar of weightlessness is triggered by increasing or reducing the counterweight to make the gravity and buoyancy equal.

By the edge of the tank stood a steel hoisting device. Above was a tank suit for training, weighing over 240 kilograms. Astronauts could not move with it on. Only the device could put the astronauts into the water.

Seven divers, plus a ground crew of more than 50, were required to escort each training. Although astronauts have already been trained to qualify for deep-water diving, there was still a great risk in 10-meter underwater training in a tank suit. Once there was malfunction in a suit, equipment, or operation, astronauts' cardiopulmonary system would suffer permanent damage, and would be thus forced to withdraw from the team.

"To ensure the safety of astronauts and guide the training, there were four underwater cameras installed in the tank a mobile camera, a voice communication system, and a physiological parameter monitoring system. Every move underwater was under control." Field staff said that every training lasted at least two to three hours. Astronauts sweated heavily with a heart rate as high as up to 160 bpm. The load was high as well as the risk.

To walk properly in space, astronauts received hard training on the ground.

Statistics from the United States and Russia revealed that the time spent on ground training was generally 10:1 compared with the that spent on extravehicular activity. That is to say, a one-minute task in space would require 10 minutes invested in simulation training. When the US astronauts repaired the Hubble Telescope, they underwent 400 hours of training in the tank.

It agreed with the old Chinese saying practice makes perfect.

I remember in front of the manual docking simulator; I could not help widening my eyes: the two handles on both sides of the seat have been used so much that it shone.

Chen Shanguang, director of the China Astronaut Center, described the two handles as "steering wheels," one for translation control, and the other for attitude control. By operating them, astronauts should accurately control the six directions of the spacecraft in high-speed flight: front and rear, up and down, left and right, tilt, yaw, and rotation. It demands high eye-hand coordination, fine operation, and psychological quality.

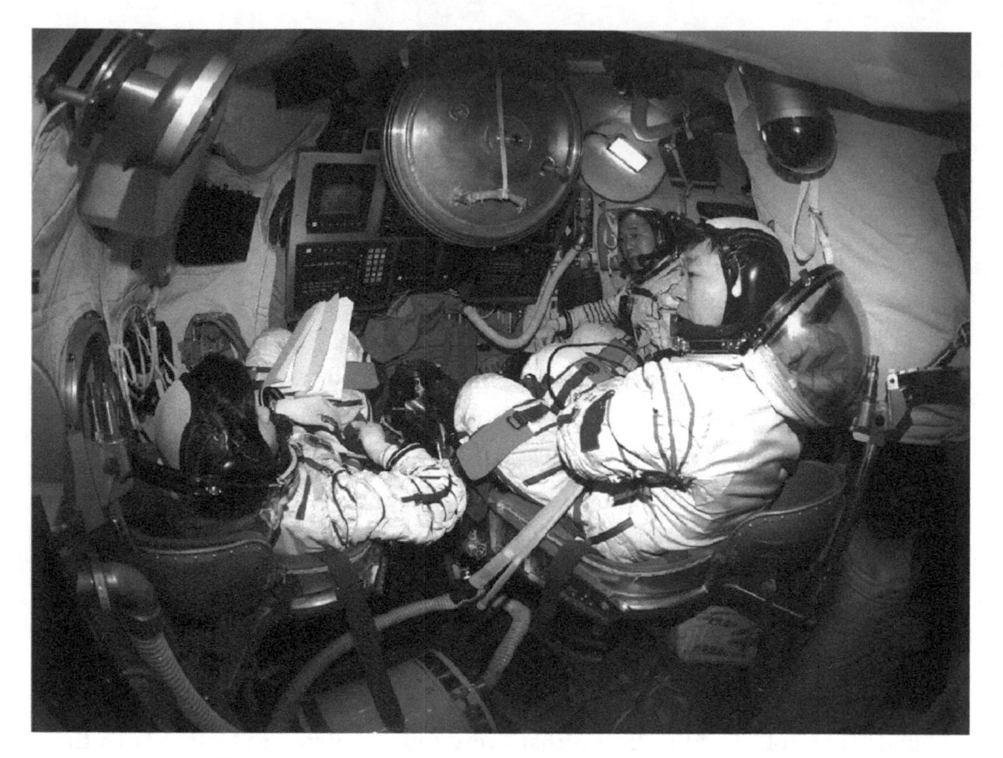

Jing Haipeng, Liu Wang and Liu Yang, astronauts in Shenzhou 9 task group, received manual docking training in the re-entry module. (Photographed by Qin Xian'an)

"Just like driving a car, only when you are skilled enough to turn an operation into a habit can one do as one wishes."Training expert Wu Bin explained, to master the docking technology, each astronaut's training volume reached about 1,500 times while each simulation docking took 30–60 minutes. It is the unimaginable training volume that helped Liu Wang, the astronaut, perfectly complete the task, thus realizing the first space docking of China.

Every action in space needed to be repeated over and over again on the ground. And so, did every detail. It was a test for an astronaut's intelligence, will, and perseverance.

Jing Haipeng said that in those days, he never delayed a day of training nor slept before midnight. Zhai Zhigang said that he contributed little to his family, and he hardly ever went shopping with his wife.

The research and development of the Shenzhou spacecraft and the training of astronauts commenced simultaneously. In the experiment, the spacecraft had been reformed several times and continuously improved. Yet, an unexpected difficulty came forward to the astronauts: despite the efforts made to get well around a large number of handles, buttons, and circuits, the improvement would surprise them on their next entry into the module with various changes.

To re-study, re-memorize, re-understand ... Faced with the torment, astronauts were in great anxiety. "In this case, when can we fly?"

Once, an old scientist came to the astronaut brigade, taking out a screw: "Screws, dozens or hundreds, were first used for destructive tests. Once they reach certain standard of knock resistance, compression resistance, and depletion resistance, the final data generated could be used to make a finished product."

Just as an item must go through challenges to qualify, so must an astronaut, who would break away from the gravity of the earth and fly into the vastness of space. The first step was to power through a hellish training of the human body.

There was a sentence that motivated 14 persistent flying hearts: as there was no shortcut on the ladder to space, to deliver the great mission of space flight assigned by the motherland, we must get down to earth and persevere in all hardship.

There was a cyclic climbing machine at the astronaut fitness training center. With the switch on, the rock surface would turn around and round without an end. Climbing it was equivalent to climbing Mt. Everest without a peak.

The 58 professional training sessions were 58 endless Everests, Chinese astronauts achieved it with extraordinary will and perseverance.

Harsh training allowed the duty to be shouldered. On July 3, 2003, the Manned Space Project Astronaut Selection Committee assessed that all 14 astronauts, certified third-class, were ready to independently carry out space missions and achieved zero elimination rate.

Wu Jie and Li Qinglong: Heroes on the Ground

In China's astronaut brigade, Wu Jie and Li Qinglong had dual identities: astronauts' coaches and astronauts.

On November 7, 1996, having passed rounds of selection, Wu Jie and Li Qinglong were sent to Russia's Gagarin Astronaut Training Center for one-year of training. In a way, they were the pioneers. With their studies complete, they returned and joined the Chinese astronaut brigade as astronauts' coaches/astronauts, playing a vital role in the exploration of astronaut training.

They used to be the ones closest to the space dream. However, it was a pity that

Wu Jie and Li Qinglong were not selected to pilot China's spacecraft into space, for several reasons. Today, a search of their names on the Internet would end up with merely a few short reports about their contribution. They were unknown to the public. There was no way for the world to deduce from a few words the glorious yet dull days.

During the Shenzhou 10 mission, I held an in-depth conversation with Wu and Li, and truly discovered how strong they were.

For Wu Jie, what happened 20 years ago was like yesterday.

On a sunny day, when he received the certificate of commander of the Soyuz Spacecraft from the director of the Gagarin Astronaut Training Center, he uttered words of excitement, "I just earned the certificate regarding the Soyuz spacecraft. Back in China, I will earn the certificate regarding a Chinese spacecraft, steer it to space, and dock it with the Russian space station."

Having earned the Commander's Certificate of the Soyuz spacecraft, Wu was qualified to drive any Soyuz spacecraft and realize the greatest ambition of all professional astronauts: to explore the vastness of space.

Over the years, Wu Jie spared no effort to achieve the dream.

In Shenzhou 6, Wu was only one step away from flying in space: he was selected as the backup crew. It was precisely this public appearance that gave people an opportunity to see this resolute face and catch a glimpse of his little-known experience.

In 1996, as excellent pilots of the army, he and Li Qinglong were selected as astronauts' coaches to be trained in Russia. "In a short year, in the face of language barriers, we must complete all training. It was of great difficulty."

"I must not disgrace the country!" With this belief, Wu powered through one hellish challenge after another.

An 8G load pressure on him when riding a centrifuge made it difficult to breathe; during the wild survival test in the Arctic circle, he only slept one hour in two days. "I am afraid a long nap would keep me from waking up by the cold."

"Before, I weighed 62 kilograms and came back weighing only 57 kilograms." In November 1997, Wu Jie graduated with a score of 4.98/5 in docking assessment, thus becoming the only foreigner in Russia who received the certificate of a Soyuz spacecraft commander.

In January 5, 1998, the astronauts brigade of China People's Liberation Army was established. Facing the national flag, as an astronaut / astronaut coach, Wu Jie was the first to take a solemn oath: "The interests of the motherland's manned space flight surpass all ..."

After returning to the motherland, Wu sorted and summarized the training courses that were taught. "I hope I can share my knowledge and ideas with my comrades-in-arms." Being an astronaut seems a glamorous profession but actually is tedious. "Basically, it was taking turns training and exams."

Over the years, Wu maintained a "battle mentality" on the training field: "As an astronaut, performing a flight mission is everyone's dream. However, when there were few missions at the beginning, everybody wanted to be selected. With little difference among the candidates, a simple detail would alter the result."

Off the training ground, Wu kept an "elder mentality": comrades-in-arms usually consulted him privately when stuck with technical difficulties. The first batch of astronauts referred him as Elder Wu, while the second batch called him Coach Wu.

Often, Wu enjoyed being a teacher. Before the launch of Shenzhou 9, he called Jing Haipeng, Liu Yang, and Liu Wang respectively to remind them of matters of attention onboard.

Losing the selection was rather disappointing. "Without a space flight in person, I had a feeling of an incomplete task" said Wu.

When he failed to make it to the cut of Shenzhou 7, frustration washed over Wu. He called his mother, who comforted him. "I would also like you to be selected. However, the failure must not discourage you. Look ahead and let go, so you can move on."

Wu quickly adjusted mentally. During the Shenzhou 7 mission, he provided technical support for his comrades-in-arms on the ground for two nights and a day in a row.

Without the passion for space flight, who would contribute as Wu did?

"Separate the satellite from the rocket." This was the first command Wu issued in the Russian Soyuz ground simulator module as the commander. Regrettably, he did not have the chance to issue a command in Shenzhou spacecraft. In June 2016, Wu retired from assistant director of science and technology department due to poor health.

From 1996 to 2016, Wu dedicated 20 years of effort. At the farewell party, he concluded emotionally, "We came because we dream, and we believe even though we were ignorant of worldly wisdom, interests, and dangers. Most of us were born in the 60s. We are influenced by traditional ideas. We, disciplined, were able to bear hardships. We experienced great changes, wanted to break through and innovate. The aim was not fame and profit, but the glory of the motherland, of the family, and of ourselves. We were lucky to be in the good times. Manned space flight may be the best way to show the value of life. Who could have imagined the luck we were blessed with since arriving in Beijing? I never imagined there would be the regret of not going to space.

The regret was also destined to accompany Li Qinglong for the rest of his life. Li was born in 1962, one year earlier than Wu.

Li Qinglong led a legendary life with two unusually early graduations.

The first early graduation was when he was specially recruited as the first batch of undergraduate pilots in the Air Force. Usually, the flight training took two to three years to complete. He nailed it in one year.

Shortly after he was granted permission to fly solo, he met his first life-and-death test. That day, nine fighter planes received special training. Li's aircraft sounded a fuel alarm. With excellent psychological quality and piloting skills, he landed the plane on the ground safely. Looking back on this scene, Li said: "The ground crew were stunned, thinking that a crash was going to happen."

The second was the "Chinese miracle" created by him and Wu at the Gagarin Astronaut Training Center in Russia. They finished four years of courses in only one.

Russia assigned a 70-year-old old lady to teach them Russian. "As she spoke no Chinese, classes were often taught in body languages." Li Qinglong said, "Regardless, after three months, we learned to speak Russian." When a delegation from China visited the Gagarin museum, he acted as the interpreter with perfect coherence. "One of my greatest achievements was managing to speak Russian," said Li.

The training in Russia was harsh beyond imagination. At one training, Li and Wu were taken to a snow field in the Arctic Circle for a 48-hour survival test at a low temperature of below 50 degrees Celsius. And merely two palm-sized compressed biscuits were made available. "I did not sleep at all for two days. A quick death would be better" concluded Li. The test led to Li losing two kilograms of weight.

A year later, Li earned the certificate of international astronaut, number 027, with unanimous approval of the Russian space experts' panel. As a result, he was proven qualified to perform duties on the Peace Space Station. Previously, only 26 trainees in the world made the cut.

As the first batch of 14 astronauts in China remembered, the Spring Festival in 2003 was much less festive: on the first day of the lunar year, tragic news came from the States that the US space shuttle crashed, killing all seven astronauts.

The date of the accident was merely 280 days away before the launch of Shenzhou 5. The Chinese Astronaut Center gathered astronauts for their opinions. Unexpectedly, the seminar turned into a heated meeting where solemn oaths were taken. Li's speech left no dry eyes in the room: "I am a pilot, long well-aware of the risk of death. I love this cause more than my life. No risk can affect my choice nor belief."

In 2003, the 14 astronauts were assessed capable of a space flight. But the density of manned space missions was not generous enough to grant the chance to them all.

Every flight mission involved a cruel elimination, should which be compared to a knock-out game, every roll of dice would have one candidate out. Although the astronauts often spoke of the acceptance of elimination, they desired nothing better than going to space.

"The defeat was not worn on their faces" said Li, "instead, the frustration was turned into more effort in the training for the next opportunity."

In the astronauts' brigade, Li and Jing Haipeng had a special relationship as both colleagues and mentor and apprentice. When Li was deputy brigade commander of the Air Force, Jing was the student pilot he mentored. "He was my pupil then, and he still is," said Li.

Communication was so easy that one eye signal sufficed. Before Jing's two flight missions, Li's exhortations were as short as "return safely."

"As an astronaut, we must be prepared in two respects, physical and psychological" said Li. "As a soldier, I would complete my mission regardless of perils if selected; I will continue to fulfill my duty if eliminated."

The road to space was a long journey. As Wu Jie and Li Qinglong, several of the first astronauts did not make the cut. They were still making contributions in obscurity. So far, their names have remained unknown.

A salute is well-deserved. Without going to space, they are still space heroes of the country.

Yang Liwei: The First Chinese Astronaut

Who is the first Chinese to fly into space?

Yang Liwei.

Among the Chinese astronauts, Yang was the most famous. "A round trip to space, a five thousand years of history." The feat of the first space flight was epoch-making.

Shenzhou 5 made China's first manned space flight. First meant worldwide attention and unprecedented risks. Undoubtedly, the road to space that no Chinese had walked was paved at the great risk of the life of astronauts.

The first Chinese astronaut—Yang Liwei. (Photographed by Qin Xian'an)

Yang was at home before the Shenzhou 5 mission. (Photographed by Qin Xian'an)

Who could shoulder the great responsibility?

Rumor has it that Gagarin stood out among astronauts in competitions because when he entered the spacecraft ground simulation module, he took off shoes while others kept them on. The Soviet Union chief designer noticed it. Regarding it as a respect for his work, he chose Gagarin as the first astronaut.

Over 10 experts and professors formed the selection evaluation committee to conduct a comprehensive assessment of 14 astronauts. In the 5 simulation tests, Yang scored 99.5, 99.7 and three 100s. Since the first round of selection in July 2003, Yang had been in the lead: he stood out in the 5/14 selection; he made it through the 3/5 elimination; and he ranked first in the secret ballot of the judging panel.

The Chinese nation chose Yang, whose excellent performance proved the choice correct.

With smile and a military salute. In October 15, 2003, when the rocket bottom spat flames in roars, faced with a dangerous journey, Chinese Space Hero Yang Liwei demonstrated incredible calm in the eyes of an audience of millions.

Will as strong as steel, heart as calm as water. Now when people talk about Yang Liwei, memory as follows has faded little:

At 6:45 on the 15th, Yang mounted the towering launch platform. The staff to escort him to the cabin agreed to tell a joke to help him relax, but none was able to utter a word. Suddenly, pointing to a big-nosed technician, Yang said, "Don't you resemble a Russian astronaut?"

The technician quickly said, "Yes, but that astronaut has been promoted to be the curator of the Russian Space Museum."

"Well, Mr. Curator, see you tomorrow." Yang's joke softened the mood.

"Five, four, three, two, one, ignition!" On the screen of the medical supervision post of Beijing Aerospace Command Center, a beating blue dot read Yang's heart rate: 76 bpm. Data tells that foreign astronauts at the ignition of a rocket normally had a 140 bpm heart rate. During the 21-hours flight, the spacecraft circled the earth 14 times, and Yang had a heart rate basically at 75–85 bpm.

The flight lasted 21 hours and 23 minutes, throughout which Yang remained calm and performed over 200 actions with great accuracy and no mistakes. Space scientists were overjoyed. Qi Faren, chief designer of Shenzhou spacecrafts, praised: "Yang was tremendously successful and perfect."

On June 21, 1965, Yang was born in a family of well-educated in Suizhou County of Liaoning Province.

His father was a college student who graduated in the early 1960s. He worked as a teacher, then as an administrative clerk in an agricultural byproduct company in a county. Mother was employed as a Chinese teacher in a middle school in the county until she retired. There was also a sister and a brother. The family of five lived in harmony and peace.

"Be practical and honest" was what Yang's parents expected of him. In his childhood, he was smart with a quick response. And he was also the head of a group of children. When he graduated from primary school, he entered the class selected at the county's key school with excellent grades. He took part in the mathematics competition of the county several times and won a considerable number of prizes.

In the summer of 1983, the 18-year-old Yang entered the Eighth Flight Academy of the Chinese People's Liberation Army Air Force. Throughout his four years, his academic and training performance were excellent.

In 1987, he graduated from the academy and became an attack plane pilot of a certain air force division. With innate intelligence and hard work, he soon grew into a topflight officer in his division, and later became an excellent fighter plane pilot. In 10 years, he has flown across China. The blue sky of the motherland remembers his name.

In the summer of 1992, the division came to a certain airport in Xinjiang for training. One day, he flew a plane at low altitude over Eden Lake Abruptly, it made a bang. Immediately, the instrument board showed a sharp rise of temperature at the air cylinder and a drop at the engine's rotation speed. Yang realized that he had encountered a serious "air parking" malfunction, that is, an engine of the aircraft stopped working. At the critical juncture, he was extremely calm: he must fly the plane back.

He held the joystick steadily, slowly closed the throttle, and steered the one-engine fighter plane up bit by bit. Five hundred meters, one thousand and fifteen hundred, the plane flew over the Tianshan Mountains and headed for the airport. As it approached the runway, the remaining engine stopped too. He resolutely released emergency landing gear and successfully landed the completely powerless aircraft.

When he exited the cabin, the flight suit was soaked in sweat. His comrades in arms surrounded and hugged him. The head of the regiment excitedly announced on the spot that Yang would be awarded Merit Citation Class III.

The correct handling of the air incident reflected excellent psychological quality of Yang.

Yang seemed to be born with the excellence to stand out but by no means was the result easily earned. The extraordinary experience of transforming from a top pilot to an astronaut has forged his valuable qualities: perseverance, meticulousness, and good habits.

He used to be weak at English. To memorize words and expressions, he called home every night from the astronaut's apartment, asking the wife, Zhang Yumei, to ask English questions on the phone. Over and over again, he passed the language test with full scores.

Yang fancies being creative. "Swivel chair" training was unbearable. Yet, Yang aced it. On a day off, his wife came home and found him whirling in the living room. In great surprise, she asked, "what are you doing?" He replied, "The swivel chair training and assessment is in two days. I am stimulating myself first."

An old expert who was very demanding on astronauts training took great pride in Yang. "He performed best in the swivel chair training. He is my favorite student."

At 6:23 a.m. October 16, 2003, the re-entry module of Shenzhou 5, bathed in the golden morning ray, landed on the vast grassland of Inner Mongolia. Looking at the jubilant crowd, he blurted out "I am proud of the motherland!"

He meant every word. Two years ago, it was the first time that Yang and his comrade-in-arms set foot in the Jiuquan Satellite Launch Center. At the foot of the soaring launch tower, an old scientist told the astronauts, "Shenzhou spacecraft is the crown of space science and technology, and astronauts are the gems on the crown. Now, there are 500,000 people involved in the development of the seven major systems of manned space project. 300,000 are working tirelessly on the front line so that you could go to space as soon as possible."

After the success of the first flight, Yang said to me that the astronauts were standing on the shoulders of countless scientists whose names were not widely known; and it was them that have allowed the astronauts to make a difference.

"There are 14 of us. I circled the earth 14 laps, which when divided is per lap per astronaut." added Yang.

The flight made the fame. But Yang kept a common heart. He said: "Engaged in a space mission, we need to be calm amidst the cheers. The road ahead is long, and the task will be more arduous and complex."

Today, Yang Liwei is deputy director of China's Manned Space Flight Office.

"Are you likely to take on another mission?" In the triumph of Shenzhou 11, Yang revealed his thoughts to the media.

"I, indeed, miss the yearning for another flight, especially after a long time without it. Each astronaut is very dedicated to space flight missions. Great efforts have been made and deep feelings were formed. John Glenn, aged at 77, became the oldest astronaut in space flight history. I think I will fly again should I be honored another chance."

Fei Junlong: a Chinese Dragon

Fei Junlong, Shenzhou 6 astronaut, is now the leader of China's astronaut brigade.

Today, speaking of Fei, people think of his signature move in space, the somersault. At an altitude of 343 kilometers, he made four somersaults in three minutes, each of which traversed 351 kilometers in space.

A Chinese dragon returned. One somersault of the Monkey King in the classic novel *Journey to the West* covered 54,000 kilometers. But after all, it is a mythical tale. But Fei was the first Chinese to make a somersault in space real.

On October 12, 2005, when Shenzhou 6 launched, Fei and Nie Haisheng jointly flew into space and triumphantly returned, completing a perfect flight widely recognized by the world's space industry.

For five days, they covered 3.25 million kilometers in space, four times the distance of a round trip between the Moon and earth; for five days, they experienced 76 sunrises and sunsets. Their entry to the orbital module for the first time marked that China entered the stage of human participation in the space experiments. For 5 days, they sent hundreds of instructions with perfect accuracy, and there were no mistakes in tens of thousands of operations.

In the vastness of space, Fei and Nie jotted down the words in their work journal: "Looking at the earth from afar, the heart never left home. This flight has made the motherland proud." Below were their signatures: "Fei Junlong, 13 October, in flight for 1 day 8 hours 32 minutes and 20 seconds. Nie Haisheng, in flight for 4 days 11 hours 2 minutes and 30 seconds."

Fei was born in Kunshan City, Jiangsu Province. He was a fan of painting at a young age. Out of expectation, Fei did not become a professional painter, but he took the works he painted 20 years ago to the sky and to space in this golden autumn.

Fei has gone through a series of unusual conquests from a pupil, who always carried a roll of Xuan paper (a type of paper used by artists) to school along a country road south of the Yangtze River to an astronaut, who shouldered the space mission of the Chinese nation.

In 1982, he graduated from high school. It happened that the Air Force came to the school to recruit pilots. He secretly registered for the pilot's physical examination without telling his family. His grandpa was the first to disagree upon learning that his only grandson was going to enlist as a pilot in the Air Force. "Premier Zhou left his hometown to participate in the revolution when he was young. I am already 17, able to take good care of myself." After two decades, Fei's father, Fei Changbao, still remembers these words.

"Now that you have made the choice, you must give it your all." His father's trust before he left for the army later became his life motto. Two years later, having graduated from the Air Force Flight Academy as a straight A student, he officially became an Air Force pilot. When acting as a flight instructor, he was rated in the Air Force as excellent; as a flight technical inspector, he classified and analyzed various accidents and compiled an investigation report titled "Keep the Alarm Ringing."

In July 1992, when there was an emergency of insufficient fuel during a test flight at high altitude, Fei made it safely back to the airport thanks to his superb flight skills. After the plane landed, the fuel in the tank was completely exhausted. Outstanding courage, excellent skills, and a calm mind made him a young super pilot at the age of 32.

In January 1998, he stood out from over 1,500 outstanding Air Force pilots and became a member of the Chinese People's Liberation Army's astronaut brigade. Owing to his meticulousness and perseverance, Fei performed excellently in every subject from basic theory learning to physique training, from professional skill learning to space environment endurance and adaptability training, from flight procedure and mission simulation training to psychological training, and lifesaving and survival training.

Shenzhou 6 was expected to fly in space for multiple days. Fei had the 400,000-word flight handbook engraved in his memory. He knew all the intricate flight procedures, operation keys, and contingency plans.

Fei's delicacy and persistence even surprised the trainers. At one training session, the trainer tossed a question: "How many failure modes can cause an abnormality in return?" The expected answer was five, but Fei answered six, which was proved correct.

For a long time, Fei kept a secret about his occupation to his parents. Once, mother could not help asking him, "So what is your job?"

Smiling, he answered, "I still fly, just higher than when in the Air Force."

On May 24, 2013, I again met Fei, head of the Chinese astronaut brigade, at the reception hall of the astronauts' apartment. As soon as I laid eyes on him, his style of rigor was clearly exhibited in a unique way. Clutching fists in both hands and standing one meter away, he politely said, "I'd better not shake hands when medical isolation is being carried out on the Shenzhou 10 astronauts crew before the mission. It is forbidden anyhow."

Rules must be obeyed; he saw it as "Beyond a personal matter, but instead a national matter." The conversation began with the classic space somersault eight years ago during Shenzhou 6's flight. By far, Fei, away from public spotlight, has grown to be the deputy director of the China Astronaut Center.

From an astronaut to a manager of astronauts, more responsibilities that came with the change of roles kept him extremely busy. "No matter how busy I am, I keep on training and often use holidays and evenings to make up for lost time." Fei Junlong said, "As an astronaut, I have to maintain my technical status."

"Treasure the noble honor and go deeper into space." Not long ago, he exchanged words with the second batch of astronauts: "The future mission is the space station. We need to be more demanding on ourselves."

Up to now, Fei has invested a massive effort in the training and management of astronauts, thus forming personal experience. "Management sometimes is similar to parenting. In additional to preventing "children" from falling, it should be taught to them where they may fall, so that they know better how to protect themselves the next time it happens."

In a recent foreign exchange, Fei told an ESA astronaut, "As a fellow astronaut, you can share my happiness, but I am proud because China is behind me.

In spare time, astronaut Fei Junlong with great joy helped his wife Wang Jie stir-fry vegetables at home. (Photographed by Qin Xian'an)

It was not only the voice of Fei, but of all Chinese astronauts fighting for the motherland.

Nie Haisheng: General's Expedition

In 2005, Nie Haisheng and comrade Fei Junlong executed the Shenzhou 7 mission. Nie traveled in Shenzhou 7 for five days. That year, aged at 41, he held the rank of colonel.

Eight years later in 2013, as commander of Shenzhou 10 flight crew, Nie led two comrades-in-arms to set a 15-day flight record. That year, at age 49, he held the rank of major general.

It was a record-setting flight. He became the first Chinese astronaut to enter space as a general.

In ancient times, "flying generals" garrisoned the border. In modern years a "flying general" went on an expedition in space for the motherland. The Chinese dream was perfectly described in the personal growth of Nie.

"When I herded cows during childhood, I had, on the back of a cow, a strange dream, in which I spread a pair of big wings and soared in the blue sky." Nie said, "After becoming an Air Force pilot, I thought my dream had come true. Beyond my expectations, I grew into an astronaut."

Nie was born in a remote village in the countryside. There were eight siblings, of which he was the sixth eldest. During his youth, poor as the family was, he, who fancied reading, was fully financed in his education. In 1980, when Nie turned 14, his father died of illness. He, who had just entered the sophomore year of middle school, had to drop out. His little brother was young. As the only male laborer in the family, he had to become the breadwinner.

In the pouring rain, the schoolteacher found Nie's mother: "The boy performed well at home. It's a pity for him to drop out. We are willing to help. Let's not ruin a bright future for him." Nie was working in the field. When he learnt that he could return to school, he threw himself into mother's arms and cried. From then on, he studied harder, was thus admitted to a key high school in the county. Considering his negative financial situation, the school subsidized 3 RMB monthly to him. Regardless, he had to work some part-time jobs during the holidays to finance the family and his tuition.

"Fortunately, the children of the poor families are better at survival. Slim as Nie was, he was healthy. From birth to adulthood, he was seldom struck by disease." Recalling the days, Zhang Jinxiu, his 74-year-old mother, could still taste the bitterness.

Dreams come true with hardships overcome. In 1984, Nie finally realized his dream by enrolling the Air Force Flight Academy.

Difficulties seemed to haunt the farm boy. On June 12, 1989, Nie faced a test of life and death in his first solo flight when the engine suddenly stopped at an altitude of 4,000 meters. The plane plummeted. The ground control instructor commanded him to abandon the plane and bail out. But he took risks again and again in an attempt to land the plane safely. One thousand meters, 800 meters … He did not press the button until the last minute. He bailed out and survived.

It was found out later that it was a mechanical failure of the engine that caused the accident. Considering Nie's bravery in trying every means to save the plane while setting his life aside, the army awarded him Merit Citation Class III.

His wife Nie Jielin disagreed with his decision to be an astronaut because the husband as a pilot was concerning enough for her. In the face of her husband's persistence, she finally caved in. "Nothing can change his decision when he has the mind set on it."

Nie has the inherent tenacity and perseverance of the rural. "He speaks little but is down-to-earth and cooperative. Once he has his mind set on something, he goes all at it. "Yang Liwei recalled that in 2002, Nie's zero fear in hardship in the harsh desert survival training left him an unforgettable impression.

Fifteen years as an astronaut, his heart was set on one thing – staying prepared for every mission. In his own words, "Age and rank are changing, but the identity of an astronaut remains, and the mission and dreams of an astronaut too."

For each mission, the principle of China's selection of astronauts is to "re-shuffle" the candidates and start from scratch. Shenzhou 6 performed a flight for five days, during which Nie and Fei Junlong were well-coordinated, thus creating a new leap in China's space flight. After the mission completed, he told himself that flowers and applause had become history. Restorative training, adaptive training, and new skills training followed. No matter what training he took, he placed himself as a novice to march forward step by step.

Returning to space is the aspiration of almost all astronauts in the world who have had a taste of it. Nie is no exception. "The universe is a big magnet. Once you have been there, you would like another trip." The words of a foreign colleague linger in his heart.

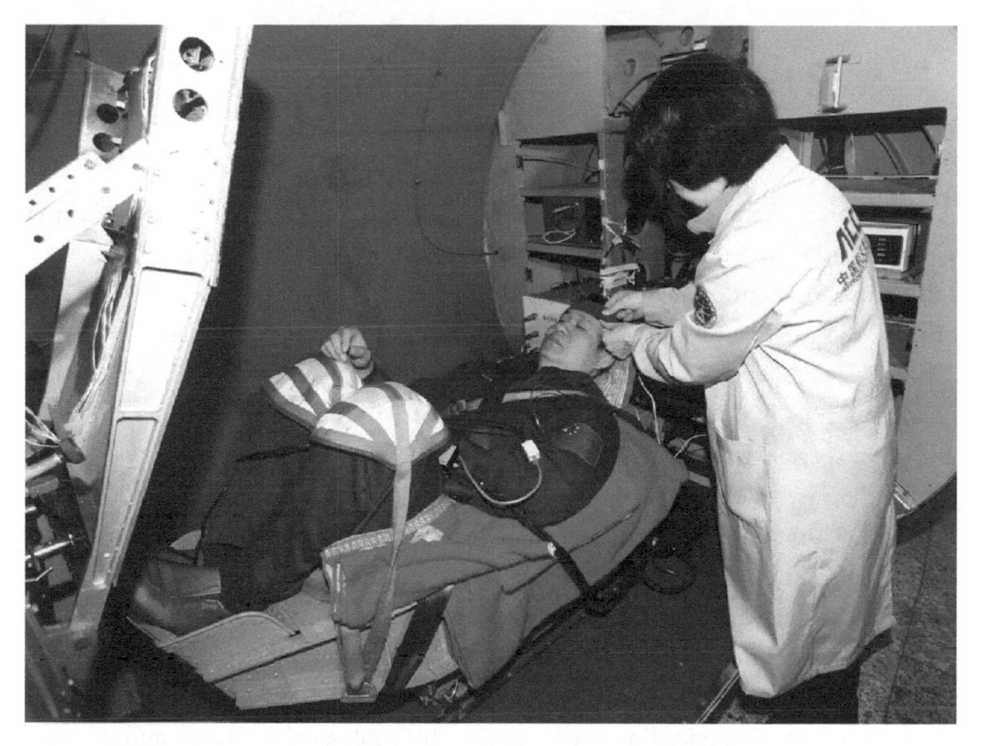

Nie Haisheng was preparing for centrifuge training. (Photographed by Qin Xian'an)

One Olympic champion said it was harder to keep a record than to break one. And Nie did a perfect job in keeping his record. The lifecycle was not a couple of days, months, years, but a full eight years.

For this reason, in the history of China's manned spaceflight, Nie left behind a legend which has never been duplicated.

From Shenzhou 5 to 10, he was either the first choice or backup in five manned missions. Such experience was unparalleled among Chinese astronauts.

"Very stable, very sturdy." In the evaluation of Nie, China's first flight astronaut Yang Liwei used two "very"s.

Shenzhou 10 took a 15-day space trip, where Nie led Zhang Xiaoguang and Wang Yaping to win the applause of the Chinese and the world with excellence performances.

"The country spent a great amount of money to train us. There are so many secrets in space waiting for us to explore. How can we slack off?" Today, Nie Haisheng, who is a general, remains as simple and unadorned as ever: "Should it be permitted, I hope I can fly into China's space station."

As the leader of the Chinese astronaut brigade, Nie sees no end in space flight.

Zhai Zhigang: A Chinese Footprint

At 16:34 on September 27, 2008, the glorious moment belonged to both the entire Chinese nation and the Shenzhou seventh astronaut Zhai Zhigang.

At this moment, Zhai, dressed in an extravehicular spacesuit, took the first step of a walk in the boundlessness of space. Up there, therefore, China has a footprint. The one small step of Zhai is a great step for the Chinese nation to explore the universe. China has since become the third country in the world to master the technology of space extravehicular activities.

In the first spacewalk of China that lasted 19 minutes and 35 seconds, Zhai "covered" 9,165 kilometers and became China's "fastest walker."

When he took the first step in space, a new English word "Taikonaut" (Chinese astronaut) was created with Chinese as its root. It appeared frequently in the reports of Shenzhou 7 around the world. Shortly after the creation, the word specifically indicating Chinese astronauts was accepted by the West and included in mainstream English dictionaries, reflecting the increasing technological influence and national strength of China on the world stage.

At 4:59 p.m. on September 27, 2008, Shenzhou 7 astronaut Zhai Zhigang successfully returned to the orbital module, marking the successful completion of the first spacewalk in Chinese history. (Photographed by Qin Xian'an)

Zhai made it happen. However, success did not come easily.

When Shenzhou 5 was launched, he did not make the cut; when it was Shenzhou 6, he failed again. He was the only one of the 14 astronauts to be selected three times in the flight echelon.

To seize the opportunity to fly to space, Zhai waited for ten years. In the past decade, he sat on the bench as a backup astronaut, seeing off teammates fly to space again and again, but he kept his mind at ease. In seeking improvement, he trained tirelessly, studied relentlessly to prepare himself and finally welcomed the precious chance, which he seized tightly and shone. There is no exaggeration in Yang Liwei's evaluation on him: his performance in Shenzhou 7's extravehicular mission was perfect. On a scale of 1–100, he is 100.

Zhai was born in October 1966 in Longjiang County of Heilongjiang Province. When he was an Air Force pilot, his good appearance had him become the cover of the Chinese Air Force magazine.

In his childhood, Zhai's family was in financial distress. His mother sold fried melon seeds to finance his education. Zhai sometimes felt sorry for his mother and wanted to drop out of school to work. The often-kind mother lost her temper: "I cannot read, nor speak any general principles, but I know one thing, which is you must attend school."

Xu Dongsheng, a former teacher in charge of Zhai's class, told reporters that during the senior year of high school, extra classes were given every evening. However, as total power failure happened often, the classroom could only be illuminated with oil lamps. Sitting in the back row, Zhai shared little illumination, so he carried a small candle in his pocket every day. Because of poor domestic finances, he only used it in class and blew it out immediately after class, wrapped it in paper and put it in his pocket.

He was also known as a filial son. After enrolling the Flight Academy, he kept only two RMB from a 12-RMB monthly allowance, the rest of which was mailed to his mother.

He used to be a test pilot instructor of subjects with high difficulty. He joked, "As a pilot, I'm lucky because I'm safe in whatever force I go to." In fact, the luck was not accidental.

When he was newly enlisted at the army, he had a dangerous flight experience. One day, Zhai and his comrades were training in a formation of airplanes. Suddenly, heavy fog struck. Emergency landing was demanded. And the commander was most worried about Zhai the new pilot. When everyone was in great fear of his safety, Zhai flew through the thick fog and landed safely.

In November 2003, the hometown of Zhai, located in the border of northeastern China, was clouded with grief while people were immersed in the joy of the successful first manned space flight in China.

Zhai hurried back to his hometown. Deep regret swallowed him as he failed to say goodbye to his mother before she passed away. He knelt long before her, wailing. His mother's greatest wish was to see her son fly into space. At the mourning hall, he solemnly promised, "I must make it. Should Shenzhou 6 pick someone else, I will still go for Shenzhou 7, 8 and so on."

After being passed over two times in the selections, Zhai received more and more attention. He was asked, "Two times in the echelon, but the final list. Now it should be your turn to be on Shenzhou 7?" He smiled indifferently: "Every selection starts from scratch. To travel in space, one must stand out with skills better than others."

Shenzhou 7 mission involved the first spacewalk. To this end, many more dangerous and challenging training subjects were added to better prepare astronauts. In order to adapt to weightlessness, Zhai put on more than 240 kilograms of training clothes to receive simulated-weightlessness tank training. Intent on experiencing as much as possible in the shortest possible time, he often performed a series of difficult actions in one breath. After training, he was exhausted. He laughed at himself, "I was over-experiencing it."

"Gold shines." With persistent pursuit, Zhai acquired the outstanding space skills to be in the Shenzhou 7 crew. The dream of walking in space was finally about to come true.

Walking in space consumes massive energy. To train to achieve super physical strength, every time physical fitness training was carried out, Zhai always added more to the prescribed load. After half-day of training, his sportswear was always soaking wet. It was common for him to lose a couple of kilograms of weight at once.

It was well-known that Soviet astronaut Leonov completed the first human spacewalk. However, few understood the hardships he bore and efforts he invested. The spacewalk lasted 12 minutes and 9 seconds, but his weight was reduced by 5.4 kilograms, and his shoes gathered 6 liters of sweat.

This time, Zhai also completed the first China spacewalk in great exhaustion of his physical strength, losing 3.5 kilograms in weight.

For Shenzhou 7 mission, I was there to see them off and welcome them back.

It had been five years since I saw Zhai there. He was in a blue training suit of the astronauts' brigade, and the chest mark on the suit was the symbol of the Shenzhou 7. At first second, I was given the impression that he had lost weight.

Zhai spoke frankly that after the mission various charity activities and popular science activities ensued, in the attendance of which he treated them seriously, especially in the face of primary and secondary school students. "I cherish every look of respect from people," he said. "It's for all astronauts, not only me."

"As long as we are in the team of astronauts, we will keep getting better. As long as there is still training and tasks, we will keep motivated," he said, "The personality of astronauts is to never give up. We are friends and comrades-in-arms at normal days and rivals at training. But the space mission is too arduous and glorious, so our competition is actually self-competition."

"Isn't it boring to train hard all the year round? Is it not lonely?" When Zhai heard it, he laughed heartily. "My secret is to practice calligraphy, imitating Yan Zhenqing's regular script in a single stroke. And the more I practice, the calmer I feel." During both space missions of Shenzhou 8 and 9, two pieces of Zhai's calligraphy were brought along.

"Practicing regular scripts takes efforts, and so does being an astronaut. One may have gained no progress after three or four months. But as long as one sticks longer to the fifth and sixth, a harvest awaits." Zhai said, "An astronaut is made step by step. The most effective way is to be down-to-earth but play tricks."

"How is your physical condition now?" Zhai Zhigang gave the answer with a smooth handstand in the physical training hall.

Liu Boming: The Biking Astronaut

Liu Boming, sincere and quiet, low-key but confident, is a Shenzhou 7 astronaut.

He was born in September 1966 in Yi'an County, Heilongjiang Province. He and Zhai Zhigang are from the same hometown.

Throughout Liu's life, he could not be more familiar with two routes: one is the bumpy mountain path from home to the county middle school, the other is the asphalt road from the Beijing Astronaut Apartment to the training center.

Twenty-six years ago, he biked back and forth on the bumpy mountain path from home to school every day. In slippery snow and the raging wind, he often slipped and fell. Often night had already fallen when he returned home.

When Liu was in high school, the family had to finance the education of five children at the same time. Father, who scraped a living by farming, was exhausted.

To spare the burden of accommodation expenses for the family, Liu chose to be a day school student, riding a dilapidated bicycle every day. The round trip was 20 kilometers, yet he was never late, regardless of harsh weather. Back then, Liu had hardly any new clothes. Corn bread was lunch, plus some pickles.

"A bicycle and hair full of frost." This is the impression on Liu Boming of Zhang Fulin, head teacher of senior students and deputy director of Yi'an County Education Bureau. However, it is precisely the route that has trained his physique and tempered his perseverance to endure hardships. Since then, Liu Boming believes that the endurance of hardships eventually would lead to greatness.

Today, the mountain path remains bumpy, but the student has embarked on the road of flying to space: the asphalt road from Beijing Astronaut Apartment to the training center.

Although the roads were different, Liu was still hardworking – physical training was conducted three times a week, which would consume a great amount of energy. Nonetheless, he pulled some extra hours; for centrifuge training, he always pushed himself to the longest time with maximum intensity; for swivel chair training, he powered through 3 to 6 minutes, and then 15.

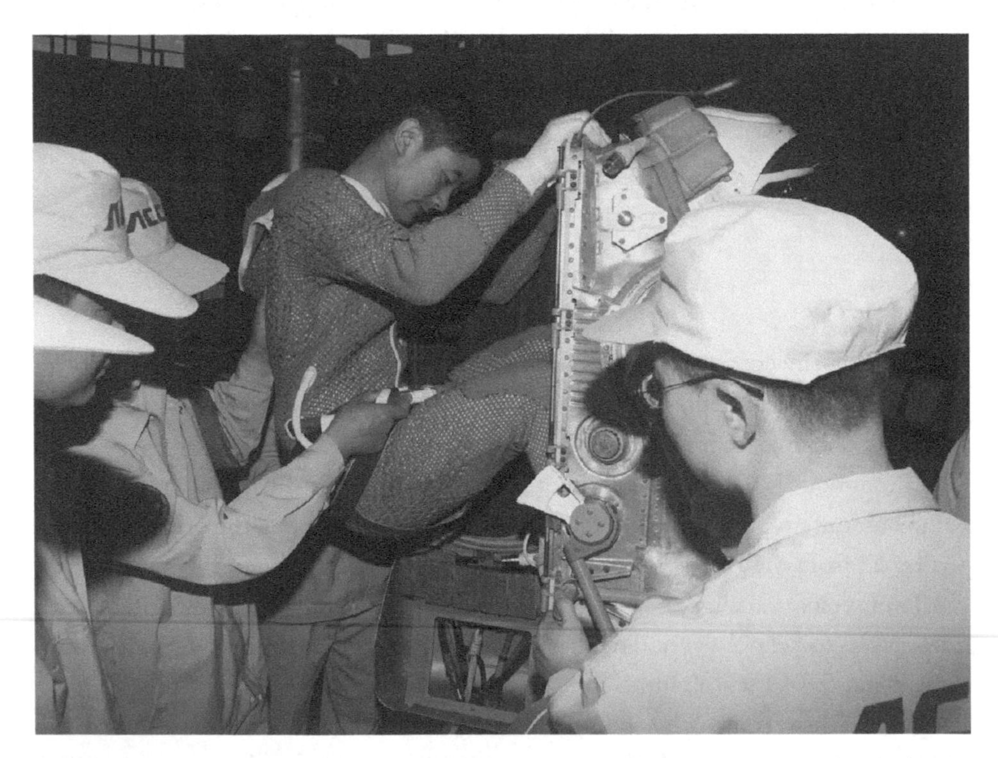

Liu Boming exited from an underwater simulated training suit. (Photographed by Qin Xian'an)

For a decade, he paved his road to space step by step.

He was a nobody when Shenzhou 5 was launched; he was selected as a flight echelon for Shenzhou 6 and stood behind heroic astronauts Fei Junlong and Nie Haisheng; when the it was Shenzhou 7, he stood out and completed the first Chinese space walk with Zhai Zhigang and Jing Haipeng.

Before the space dream, Liu aimed at steering China's best fighter aircraft.

In March 1985, the Air Force was selecting pilots. Liu was the only one selected in the county. From theoretical study to practical operation, from beginner plane to advanced plane, from the propeller type to the jet type, candidates were constantly eliminated. The first 125 candidates were reduced to 35. Liu's excellent performance all the way labeled himself the best of all.

In October 1990, he became a fighter pilot. To win a fierce competition was not the end of his pursuit. In his opinion, so as to continuously surpass oneself, the competition against himself was the ultimate goal. To achieve that, he continued to improve his flight skills with a silent effort. Two years later, as an excellent pilot, he was appointed to fly the best fighter plane in China at the time.

"Higher than pilot, there is astronaut." From the moment he entered the astronaut team, Liu set new goals. Strict standards were set for himself every time he trained to realize the space dream as soon as possible. After every training, he summed himself up. He never let loose of any part of the training. If one action was not mastered, he would apply for overtime training in the evening until it was.

In learning theory, he never stopped at the surface, but went straight to the root. For example, regarding the spacecraft, he went beyond remembering switches to understanding the circuit. Should there be a fault, an analysis over the circuit would often help him locate and solve it fast. Therefore, he earned the nickname Little Zhuge (after Zhuge Liang, who became a symbol of resourcefulness and wisdom in Chinese folklore) among his peers. Once a teacher famous for tossing tricky questions quizzed Liu more than 10 questions in a row. Liu answered them without a hitch, which enabled him to be exempt from subsequent quizzes.

Failure, for Liu, was the beginning of the next success.

Instead Liu was motivated when he failed to be selected for Shenzhou 5. In order to narrow the gap, he dug out the relevant books for a systematic analysis, adjustment of thinking patterns, and reorganization of his thought. After three months, he sorted out 700,000 words of notes and data. At the same time, according to personal physical defects, he formulated a targeted training plan.

When Shenzhou 6 made the selection, his great progress allowed him to be selected in the flight echelon. Although he failed to go to space as he wished, he was not discouraged at all. Silent preparations for Shenzhou 7 were made to improve himself instead.

Opportunities are for the prepared. To realize the space dream as soon as possible, Liu trained harder and studied harder. He is a good method-finder. Aiming at the Shenzhou 7 mission, he sorted out a set of training methods suitable for himself. For example, the extravehicular procedure, involving many steps, was very perplexing, so that cramming would not help much. He summed up the mental imagery memory method. Lying in bed every night, he played daytime training procedures in his head.

Liu was so meticulous that it impressed the teachers. During a training session, the trainer suddenly required him to give instructions, off manual, for the whole process from exiting the cabin to returning to the gate. For more than an hour, he gave instructions while operating, all of which were correct.

The first spacewalk has been deeply remembered. However, behind this glory, Liu and his comrades Zhai Zhigang and Jing Haipeng experienced a thrilling life-and-death test, little of which was previously known.

When exiting the cabin, Zhai pulled the door forcefully three times, none of which worked at all. "No matter what, we must not let the whole nation down." Liu handed Zhai auxiliary tools. Zhai maximized efforts to pull, and the door to the vastness of space was open at last.

The moment the door opened, a fire alarm went off. What next? In that instant, Liu said to Zhai "If there is a fire, there may not be a return. Cast everything else aside. Go ahead with the procedure and finish what we have to do." Zhai did not hesitate to exit the cabin. Liu adjusted the steps and handed him the national flag. "Even if we cannot go back, we shall place the five-star red flag in space."

Deep space, in a unique way, had tested the three astronauts for life and death. Afterwards, Liu said, "We thought nothing of it."

Since the return of Shenzhou 7, Liu, a low-key figure, has rarely appeared in public media.

Liu provided technical support for Shenzhou 9 and 10. As one of the members of the Shenzhou 10 Teaching Expert Group, he discussed with other members about the curriculum. "After all, I've been there and advice about weightlessness is within my ability."

Over the years, Liu Boming's weight change has been kept within one kilogram. Weight control is a compulsory course for astronauts. In addition to those, Liu took the initiative to talk about his optional course, English.

He admitted that he had a weak foundation. Although he learned a lot in recent years, thus forming a good vocabulary, his listening was still poor. Increasing exchanges and cooperation between Chinese and foreign astronauts made improving English necessary. "My goal is to achieve barrier-free communication with foreign counterparts without an interpreter." Liu admitted that he spent three to four hours practicing oral English every night.

One of Liu's biggest feelings after the Shenzhou 7 flight was the strength of the team. "In the vast universe, the spacecraft was floating alone. To tell the truth, we felt small and lonely. After the experience, you will feel strongly that the team is really a strength behind."

Perhaps it was this feeling that made him write down the words: "Dedicate loyalty to the Party, perseverance to the cause, enthusiasm to partners, and confidence in yourself."

Liu said, "With these words in mind, I think I will be forever invincible."

Jing Haipeng: A Soul with Dreams

Among the Chinese astronauts, Jing Haipeng's face is probably the most familiar to the public because he has been the astronaut of Shenzhou 7, 9, and 11.

Up to now, Jing is the only Chinese astronaut to fly three times to space.

On Shenzhou 7 in 2008, he and comrades Zhai Zhigang and Liu Boming completed China's first spacewalk.

On Shenzhou 9 in 2012, he and comrades-in-arms Liu Wang and Liu Yang completed China's first space docking, and first checked in the space "China home."

On Shenzhou 11 in 2016, he and comrade Chen Dong lived in space for 33 days, creating another new record for manned space flight in China.

Nevertheless, Jing has not changed in these years, not only the face, but also the inner state of mind.

He is still modest: "Every time we fly, we are greatly supported. Behind us are the motherland, family, comrades-in-arms, organizations at all levels, countless astronauts and all Chinese people. The unknown scientists are the real heroes."

His first desire remains unchanged: "Feel on site when given the chance. Under our buttocks was the second-stage rocket. When you hear the huge roar after ignition, you will understand what a high-risk occupation is. But it is worth our lives. An astronaut's duty is to always be ready for flying missions. It is more than accomplishing a feat only. The country spent much to train us. Only when we are prepared to be selected by the motherland and fly at all times can we repay the country.

His pursuit remains the same: "If I say that I dreamed of being an astronaut since I was a child, it must be a lie. At childhood, becoming the main player of the basketball team was the ambition. The aspiration to be a pilot, an astronaut and flying came later. When a dream comes true, one must have another, otherwise one will collapse. Like people in space, they feel zero gravity. Back on the ground, if there is no dream, the soul will be weightless. Always having a dream let me feel down-to-earth."

When introducing herself, Jing Haipeng still goes like, "Jing means scenery, Hai the sea, and Peng the bird that soars across the sky."

Should the talented be compared to beautiful jade, Jing was only raw at first. From a remote village to the blue sky and space, he gave persistent efforts with sweat allowed him to polish himself into a shining jade.

Jing Haipeng, a native of Yuncheng, Shanxi Province, was born in October 1966 and enrolled in the army in June 1985.

In primary school, Jing used to like playing basketball, but he was short and thin. At the beginning, his was assigned the task to carry bags for the players. He was not content and tried his best practice shooting. For countless nights, in the dim moonlight, he ran alone on the court.

In one match, one main player was injured, thus a chance for him to play. He jumped up and shot. In!

"I was wearing a vest my father bought, with an image of a seagull on the front and a large '5' on the back. Every time I scored, the audience shouted, "Come on, Seagull 5." Looking back, Jing exhibited deep pride.

The match made his name. "Seagull 5" became the nickname of Jing. In middle school, high school, and the army, he was the main force of the basketball team. Today, he is still the vanguard of the astronauts' basketball team. Jing admitted, "Playing basketball allowed me to understand that as long as I keep at it, I can make a difference."

Every step toward space was tremendously challenging. There is physical training that challenges physiological limits, desert survival training in the wilderness, skydiving training, etc. Every time in the face of difficulties, Jing encouraged himself: "I'm Seagull 5. Hold on, I can do it."

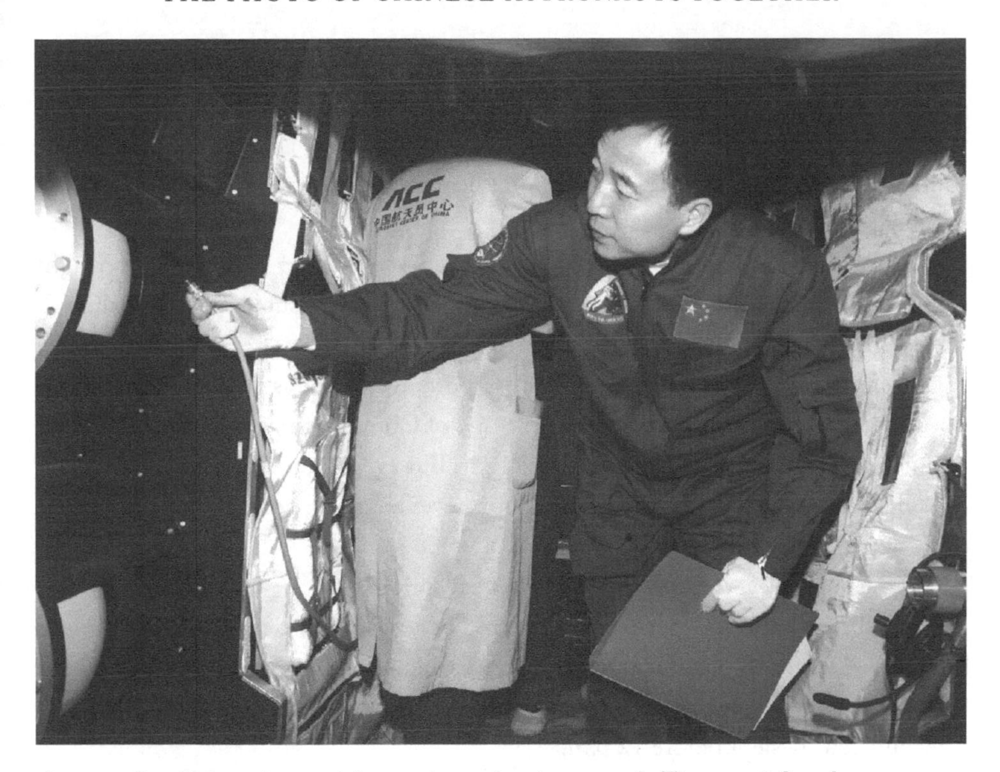

Astronaut Jing Haipeng inspected the experimental equipment at the Tiangong-1 Complex. (Photographed by Qin Xian'an)

Basketball shooting seems to be a metaphor of life. The dream has been constantly upgraded to a bigger one. It was the efforts and persistence that helped Jing realize his dreams.

In his junior year at high school, he was fascinated by a picture in the newspaper: a pilot standing radiantly beside a fighter plane with an oxygen mask at hand. He remembered clearly that the pilot was named Zhang Haipeng. The double strike of the same first name and envy motivated him to be a pilot.

In 1985, when 19-year-old this ordinary farm boy ran in the fields waving the acceptance notice of the navigation school, who could have thought that he would become make China proud? That day, his parents were still harvesting corn in the field.

The moment the journey to space began, obstacles were there to impede him.

The swimming training required trainees to swim continuously for 50 meters. Otherwise they would be eliminated. Jing was unable to swim. He either fluttered or choked in the water. As the test was approaching, he still sank like a brick.

"I shall not allow elimination to happen to me." Undiscouraged, he still insisted going to the water. The night before the test, he managed to float. The next day, with teeth gritted, he plunged into water and swam 50 meters across the pool. In the sound of cheers, he swam back. He not only learned to swim overnight but achieved excellent results.

With perseverance, Jing created a miracle of excellence in swimming overnight and conquered the most intensive and dangerous training subject of Shenzhou 7: the weightlessness buoyancy tank.

There were many ways to exploit Jing's potential. Time is undoubtedly an important key. From the moment he became one of China's first astronauts in 1998, he had held on for 10 years before his first flight. But ever since, time seemed to be on his side. Astronaut Liu Yang said: "His rigor never changed, nor did meticulousness, diligence, and even face and weight."

Time defeats everyone. Why is Jing an exception?

"Persist, persist, and persist!" The secret of his victory over time was simple, but hard. Despite his understatement, it was illuminating: "When preparing for Shenzhou 11, for about a year, I hardly ever went out of the space city, hardly had a weekend, hardly slept before 12 p.m."

Every flight is a lonely journey. Jing said, "My comrades-in-arms and I have enjoyed this journey and are willing to dedicate everything for it year after year, day after day. If to say at first, I just liked the profession, now I am loving it."

Successor Liu Yang was still deeply shocked during an occasional visit. Shortly after the return of Shenzhou 9, astronauts had just broken away from intense tasks and not yet had time to rest. Once she knocked opened Jing's door and saw a room full of learning materials. Liu asked in surprise, "Bro, why are you filling the room with books?" Grinning, he replied, "Preparing for the next mission." "But we can't participate in the selection according to the regulations for Shenzhou 10? "Well, there are Shenzhou 11 and 12, aren't there?" Extraordinary perseverance enabled him to defeat time and time rewarded him in return. The flying time was getting longer and longer, from three days of Shenzhou 7 to 13 of Shenzhou 9 and to 33 days of Shenzhou 10; the altitude was getting higher and higher: the flying height of the Shenzhou-Tiangong complex was 50 kilometers higher than before, which made him one of the two Chinese with the highest flying altitude so far.

"No matter how high or far I fly, my gratitude remains." Jing said passionately that as a farmer's son, what he achieved today cannot be separated from the cultivation of the country and the nurturing of his parents; every footprint of his growth marks the help from everyone.

An instructor trained astronaut Jing Haipeng to perform urine bag sampling. (Photographed by Qin Xian'an)

Jing's original intention has never been forgotten regardless of the incredible experiences with three flights. April 24, 2016 was China's first space day. In front of the national flag, Jing and his comrades recalled the oath they took when they joined the astronauts brigade 18 years ago.

"The last sentence, 'Willing to strive for the motherland's manned space cause for life' perfectly elaborates our original intention" said Jing. "There is only one reason behind the choice to continue flying: to fulfill our promise to the motherland."

Liu Wang: Sing a Ballad from Space

Liu Wang, Shenzhou 9 astronaut, opened the history of China's manual spacecraft flying in space. He completed China's first manual space docking with perfect accuracy.

Among Chinese astronauts, Liu Wang is the most romantic.

On June 21, 2012, Liu's melodious harmonica sounded on Tiangong-1, China's space home. It was his wife Wang Wei's birthday. When the wonderful harmonica sound of "On the Water Side" passed through thousands of kilometers to the ground, immediately Wang burst into tears.

This was an elaborate birthday gift for his wife, to express gratitude for her silent support for many years.

When Liu Wang became one of the first Chinese astronauts in 1998, he was only 29 years old, the youngest of all. He had turned 43 when Shenzhou 9 was launched in 2012.

To fly in space, he had made 14 years of efforts.

For 14 years, failure in the selection of astronauts struck repeatedly.

For 14 years, he led a life where quick adjustment had to be made for the next round of selection after losing one.

Many asked him: "It must be exhausting, right?"

Liu Wang answered, "When a person likes a thing, he will have no exhaustion in going all out to do it. For instance, mountaineering is exhausting in the eyes of the ordinary, but for the hikers, they see joy. Without devoting great efforts, an easy outcome is less satisfying."

Liu said little. Yet, in the few words he spoke, he exhibited an inherent optimism.

Speaking about the failure of previous selections, he said, "It's alright. After all, manned space flight was not a train that can take a great many passengers at once." Hearing me praise the female astronaut Liu Yang's literary talent, he quipped: "Hire her as Tiangong-1 special correspondent."

Because Liu Wang is from Shanxi, Jing Haipeng joked that he shared the similarity with Shanxi old vinegar. The longer it brews, the more fragrant it becomes. When the hard journey ferments into happiness, the destination is close.

Liu Wang was born March 1969 in Pingyao, Shanxi Province. He inherited his father's ingenuity and his mother's perseverance. Despite failing to graduate from elementary school, father learned all the knowledge of electricians, thus was able to fix neighbors' sewing machines, radios, and pumps. Mother's words were what he lives by: "Be true to yourself as you cannot control what others think of you."

Liu Wang never forgot walking seven kilometers each way to attend middle school. On the dirt road leading to the middle school in the city, he rode happily every week on the bicycle the family had bought with every dime saved. After three years, he was admitted to the pilots' school.

In 1988, dreaming of becoming a doctor, he instead seized the opportunity to enlist in the Air Force that was conscripting in his town. As a pilot, there was an elimination rate of 67% percent. From then on, he learned to calmly face selection and elimination.

Ten years later, Liu Wang, with a record of 1,000 hours of safe flight, after strict selection, became one of the first astronauts in China. However, every time Shenzhou manned spacecraft flew, he regretfully missed the opportunity.

Looking back, he said, "Often one waits to be chosen, but before that, one shall be proactive and must do the job well."

For 14 years, Liu Wang has always been ready with a positive attitude to be chosen. For example, when he first became a pilot, he weighed 62 kilograms. Twenty years later, his weight was controlled at 64 kilograms. From a pilot to an astronaut, Liu Wang kept at it day by day with a strong will of steel, and finally achieved the most perfect result.

Liu Wang said: "There is a kind of dedication called waiting, and a kind of gain called the same. In waiting, I have accumulated and enriched y my experience. For the sake of ideals and pursuit, some waiting is necessary."

After the flight crew list was confirmed, his wife asked Liu, "Are you excited?"

He said, "There was only a moment of relaxation." In the view of Chen Shanguang, director of the China Astronaut Center, more often than not, Liu Wang's state was unrelaxed: never give up and never shrink back.

According to international practice, the hand-controlled docking can be mastered after 1,000 times of training. But Liu conducted over 1,500 times of training on the ground. With tenacity, Liu developed self-confidence for the task.

On the eve of the expedition, Liu Wang, who spoke little, was asked how sure he was of hand-controlled docking. He answered decisively: "100%, I believe not only in my own strength, but also in the strength of our aerospace scientists."

140 meters, 30 meters, 5 meters ... On June 24, 2012, Liu Wang calmly piloted the Shenzhou spacecraft. It took seven minutes, with an error of 1.8 cm, to have a job done perfectly in the vastness of space, far superior to automatic docking.

Chinese precision won world acclaim. In this regard, Liu Wang remained modest, "A little worse than my own standards."

Before the Shenzhou 9 mission, Liu Wang's problem was how to break the news to his aged mother, who was not in good health. After he was selected, he once tried to ask her whether she would like her son to be selected. After a moment of silence, she always replied that she would not.

Liu Wang and Jing Haipeng conducted manual docking training. (Photographed by Qin Xian'an)

As the eldest son in the family, he knows his mother well. Children's safety is the wish of mothers all over the world, but his mother must also like it if he could realize the dream of the years and soar higher.

In the days of traveling in space, he always thought about mother. When he returned from space and stepped on the earth, a voice in his heart yelled: "Mom, I'm back."

"I personally believe that our manned space project is a fine art created by experts in various fields. The role of astronauts is to show the art." It had only been eight months when he was awarded the title of hero astronaut.

The space flight made him famous. Liu Wang's life has experienced many changes. "As a soldier, I used to have a single battlefield, the training ground; after the space mission, I had to take on more roles."

Faced with various new challenges, Liu Wang chooses to behave by his nature and do things according to the role. In September 2012, the International Astronautical Federation (IAF) flag exchange in Berlin, Germany, left him a deep memory.

"At that time, as soon as the short promotion film of our space docking was broadcast, all the colleagues and experts present spontaneously applauded. Their recognition and praise prove that China's space technology is first-class, and the strength of Chinese astronauts too. At that moment, I was very proud."

Liu Wang is the first Chinese astronaut to pilot a spaceship. Recalling it, he became calm and exciting: "It was spectacular, the rhythmic voice of the spacecraft engine was intoxicating."

Liu Yang: First Female Astronaut in China

Liu Yang, a Shenzhou 9 astronaut, is China's first female astronaut to go to space.

In the spotlight, Liu Yang has neat, short hair, shallow dimples, and an outgoing face. The face has become ever-beautiful scenery in the history of manned space flight in China.

Liu Yang said more than once: "I am lucky to be in the great era of China's manned space development and become one of the first batch of female astronauts, and I am very happy to represent Chinese women in space and show the image of us to the world."

The greatest imagination would not guess astronaut Liu Yang's childhood dream.

"When I was a child, I went to the park by bus with my mother. I felt that the bus was amazing. And I dreamed of becoming a bus conductor when I grew up, so I could take it every day."

As every other girl, her dream has changed with age: in middle school, she wanted to be a teacher because she enjoyed explaining intricate questions to classmates; in high school, she aspired to become a lawyer because the heroine of the popular TV drama *Fa Wang Rou Qing* was a beautiful and just barrister.

Nevertheless, she never thought that a flight to space would happen to her.

This is the drama of life. Destiny sometimes changes in an instant.

When she was in her senior year, the Air Force happened to recruit female pilots at her high school. Back then, every seven years a group of female pilots were recruited. Parents did not expect that the daughter, who used to be thin as a chopstick, eventually stood out as the only candidate from the region.

A few years after becoming a pilot, in 2009 the state started to select the first batch of female astronauts; and she did not expect that the chance of participation in the Shenzhou 9 mission would come soon after joining the astronaut team.

Today, Liu Yang, in a spacesuit, looked back at her growth footprint and sighed with deep feelings: "I have been really lucky that every key step of my life has just caught up with the good times of national development and of manned space

Astronaut Liu Yang checked laboratory equipment on the Tiangong-1 complex. (Photographed by Qin Xian'an)

development."

Recalling the past, she confided in me: "When they were recruiting pilots, the teacher signed me up for it without telling me."

I tossed another question "How about this time? You signed up for yourself, didn't you?" She answered, "Of course, I did it myself because I really wanted it."

As a pilot, Liu Yang suffered a lot: "For the first camp and field training, my feet were blistered, and I cried as I hiked." After reporting to the astronauts' brigade, despite psychological preparation, Liu Yang did not expect that the training of astronauts would be so harsh and the standard so high.

Swivel chair training takes four minutes. But at the astronaut brigade, the duration was set at 15 minutes.

At first, five minutes was her limit. She used to sweat massively, totally on the verge of collapse. "What should I do?" Liu Yang figured out her own trick: "At the moment, I fantasized about myself at the beautiful seaside, shifting my attention away from the pain."

Every training is a challenge to astronauts' physiological limits. The only way to overcome them is to hold on with extraordinary perseverance.

In the first centrifuge training, Liu Yang's legs started to tremble in less than three minutes as if they have run a marathon. Today, she can bear the same standard load of 6G as a male astronaut. She said, "For the space cause, the more you suffer for it, the more you will love it."

It usually takes about four years to nurture a mature astronaut. But Liu Yang did it in two. Her diligence paid off. A reporter asked: "Will there be any shortcomings in skills since it was a crash course"? Unexpectedly, she said with confidence, "I am sure of myself, you know, the first female cosmonaut in the Soviet Union, Valentina Tereshkova, only trained for half a year."

For laymen, astronauts lead a mysterious life. But for Liu Yang, it was only intense and monotonous. She just kept learning and training. For more than two years, Liu Yang did not go shopping or see a movie.

As a woman, Liu Yang believed: "It's a kind of happiness to marry and lead a stable life; it's also a kind of happiness to fight for a cause that one likes. As an astronaut, the happiness I experience is beyond what one could imagine."

In the eyes of mother Niu Xiyun, Liu Yang is a filial and sensible child. "When she was in middle school, the family was poor. She did well in an exam during sophomore year, for which the school awarded her 10 RMB. Without leaking any information, she asked her father about his shoe size. A few days later, she bought a pair of sneakers for him, which were priced at 10 RMB. She spent it all on the gift with nothing for herself. In senior year of middle school, she received another 10 RMB scholarship and bought me a pair of sneakers too. She suggested us to exercise. As a pilot, she told us nothing because she feared that we would worry. Once the plane was hit by a flock of birds. She said nothing. I read it in the newspaper instead. And it was terrifying for me and her dad."

In the eyes of Huang Weifen, her trainer, Liu Yang is an outgoing and diligent girl with many talents and strong affinity. "In the astronaut center, she hosted all kinds of recreational activities. Some time ago, she visited the martyrs' cemetery in the Jiuquan Satellite Launch Center. Her speech after seeing the diversiform-leaved poplar (an ancient species in Xinjiang) and the desert moved us all. At training, she is very diligent and conscientious, paying attention to every detail and every movement. She is good at communicating with teachers and peers. She would not stop until she got a proper answer to her question. Once when we rehearsed the response drill to a fault, there was slightly difference from the manual. She immediately came to ask about the relevant details."

During four years as a pilot, she flew 1,680 hours. In rescue and relief sites, in the blue sky of the motherland, she has actively participated—

One year, Liu Yang and her comrades-in-arms piloted transport planes at great risk for more than three hours continuously to transport essential supplies to a certain destination. Once, she and her comrades-in-arms encountered a flock of birds, which hit the plane, during a rescue mission.

Despite the experience of tests of life and death, Liu Yang never turned her back in a call from the motherland. Today, in a golden astronaut logo, her options do not change. She laughed and said, "I used to fly an airplane and now a rocket. The risk gets higher. Regardless, I would like to fly in space on behalf of the motherland."

Somersaulting, displaying panda toys, bicycling, showing Chinese Kung Fu ... As China's first female astronaut, Liu Yang's every move in space has been widely talked about. She is known as the Shenzhou woman: a Chinese woman in Shenzhou spacecraft.

However, what impressed the reporters most was this:

Before going to bed on June 19, 2012, she held up a board in front of the camera and wrote "All the staff, I am grateful for your hard work. Good night."

The simple words also reflected a valuable quality of the female astronaut. The fame did not blind her eyes to see who it was that lifted her to space.

Now, Liu Yang has a new title: deputy to the National People's Congress. "I used to watch the National People's Congress and the Chinese Political Consultative Conference on TV. This year, as a deputy to the National People's Congress, I am deeply honored to listen to and review the work reports of the government," said Liu Yang.

"The folks moved me the most." Liu Yang told me that recently she received two special gifts:

The first was a heart-shaped cardboard covered with colored post-it notes. Not long ago, she was invited to make a report at a school. Primary school students had written down the words they most wanted to say to her. Thus, there was this message board. She cherished the gift much and put it in the middle of her desk at home. "Reading the words full of child innocence warms my heart."

The second was a group photo. On the back of it, there were some words: pay homage to those who have made contributions to the aerospace cause. Liu Yang attended the 60th anniversary of the Beijing University of Aeronautics and Astronautics. A gray-haired aunt made her way through the crowd to take a picture with her. Unexpectedly, when she went in May to the Beijing Airlines Affiliated Middle School to have a seminar with students, a young boy came over with an envelope and said, "Sister Liu, this is for you, from my grandma. When I grow up, I will be an astronaut too."

"I was particularly touched when I received the photo. That grandmother is a retired professor of Beijing University of Aeronautics and Astronautics. She has been cultivating talents for the cause of space flight of her motherland all her life." Tears were circling in her eyes when she told the two stories to me.

"Every astronaut dreams of flying higher. I will pursue my dream and keep working for it," said Liu Yang, "As long as I maintain my physical condition, training results, and psychological quality, standard requirements will be met in all aspects, and I will stand out in the selection for another chance of a space journey."

Zhang Xiaoguang: the First Chinese Space Classroom Cameraman

Zhang Xiaoguang, a Shenzhou 10 astronaut, is China's first space classroom cameraman.

For the chance to fly, Zhang has been working hard for 15 years. His life truly elaborates the determination to full one's aspiration.

"Success is a part of life, so is failure; winners are not only those who never fail, but those who never give up." Zhang compared his failures to a few clouds when the sun rises. "When I find my own shortcomings and make a plan for improvement, the sun warmly and hopefully shines on me."

Zhang Xiaoguang was born in May 1966 in a village in Jinzhou, Liaoning Province.

Teenage life, as described in his favorite book *The Ordinary World*, involved a 5 kilometer walk every day to school with pickles and rice. Occasionally, mother would add an egg in his lunch box without telling him.

Zhang's father, who worked in the countryside in Zhang's freshman year at middle school, brought back a prize, a golden Hero-brand pen made in Shanghai. Father held up the pen, and said to the children, "Whoever performs best in school gets the pen."

The golden pen was so tempting to a teenager that he blurted out to his elder brother and sister, "Give it up, it will be mine."

When the words came out, he, who was often lost watching planes spit smoke, did not see that in six years, he would be admitted to the Air Force Flight Academy and get the pen as he had wished.

On that day, the family celebrated for him, and father shed tears when he handed him the pen. That night, he experienced the first insomnia in his life. The night was destined to be Zhang's most unforgettable memory.

The blood of competition seems to flow in him. Zhang Xiaoguang, who won the golden pen, soon realized that more "golden pens" were waiting ahead.

The competition in the flight academy was extremely fierce. From enrollment to graduation, only 20% of the students were able to make it to the combat forces to fly a plane, while the rest were eliminated. As soon as he got on the plane, Zhang was identified by the teacher as being cut out for the job. Thus, he started to fly solo early. In 1990, he was assigned to the combat forces and became the fighter plane 7 pilot.

Six years later, the opportunity came again. Zhang Xiaoguang, already a first-class pilot, participated in the selection of the first batch of astronauts in China.

And the results came out on January 5, 1998 with Zhang becoming one of the first Chinese astronauts.

Many years later, he poetically described the unforgettable moment: "It was a sunny day. Facing the bright red party flag and the national flag, I raised my right fist to swear the oath of astronauts and my heart was filled with elation. The responsibility and strength that came with the word astronauts were so great."

Fifteen years later, his persistence earned him another "golden pen" to be selected into the Shenzhou 10 crew to fly into space. "It's a dream I have been chasing, and I know I can reach it," he said.

Despite not having seen the book for many years, Zhang still clearly remembers every detail in *The Ordinary World*. "I share some similarity with the protagonist Sun Shaoping" said Zhang. "His growth, simplicity, positivity, and especially great love, deeply attracted me."

Love and effort were the themes of his words. As his favorite motto goes, life is filled with love and efforts.

"I had not even gone to the county center until I graduated from middle school. It's flying that lifted me high enough to see the world. "Zhang likes to express love for flying in the words of "soaring in the blue sky of the motherland." "Especially in an ultra-low altitude flight, flying over a village or a town gives the view of the school and the playground, where there are the motionless students. When gazing into the sky that I was flying across, a sense of responsibility and pride rose."

Zhang depicted a most romantic moment: "It was a night flight, and the Moon hung in front of me on the left. It was my wife's birthday, and I flew back and forth at the Moon two or three times. In my head, I compared her to the Moon."

It's hard to imagine that the moment is still a treasure of his own. I asked out of curiosity, "Why not share it with her?" He answered humorously, "Loving someone takes a lifetime. It is easy to call her my sweetheart, my darling, but very difficult to call her my old lady once."

"He is the center of joy in the team" said Shenzhou 10 Commander Nie Haisheng. Zhang is labelled as warm hearted, responsive, and helpful, which was the unanimous evaluation by the astronauts we interviewed. In addition to singing and playing *go*, one of the happiest things for him is to share the stories he has heard about the history of the party and the army.

Zhang attaches great importance to love and righteousness. For example, in 2003 he went back to his hometown to find Liu Yingfu. Had it not for Liu Yingfu who insisted on taking him to Jinzhou for the physical examination for the astronaut selection, Zhang, who missed the car, would have probably given up. "Without you, I would not have become a pilot, let alone an astronaut. Brother, you have changed my life." At a party, he drank a large glass of wine to express his thanks.

The Shenzhou 9 team received a grand farewell in the Wentian Pavilion before the expedition. Zhang said to Liu Wang, "Difficulties make greatness, hardship lies everywhere; you may relate to the message delivered in this verse. Congratulations on your dream of flying, I look forward to your triumph."

A year later, Zhang dedicated the words to himself. He admitted that he could finally relate to what Liu Wang must have been feeling. "About the verse, only after you have experienced it, will you find it unforgettable."

Over the years, he has witnessed such perfect moments where he saw off comrades to space in batches. Over the years, he has participated in the selection, failed, shut himself up in the office and sat alone in silence again and again. He wept, dried the tears, and returned to the training ground.

"The pursuit of perfection is to make up for the defect." Zhang firmly believes these words and knows how to act on them.

In the early years as an air cadet, he developed the habit of polishing. At night, when the lights were off, he used a flashlight to illuminate the textbooks while having a flash of the essentials of the flying movements instructed by the trainer during the day in his head.

Now, a flashlight is no longer necessary, but the habit has been kept. He reached 2,000 times of docking training. Plus, with the training in his head, the number would be even greater.

One night when Shenzhou 9 was on a mission, Zhang did push-ups in the dormitory alone, and there was a figure of sweat under him. Jia Zhongyuan, political

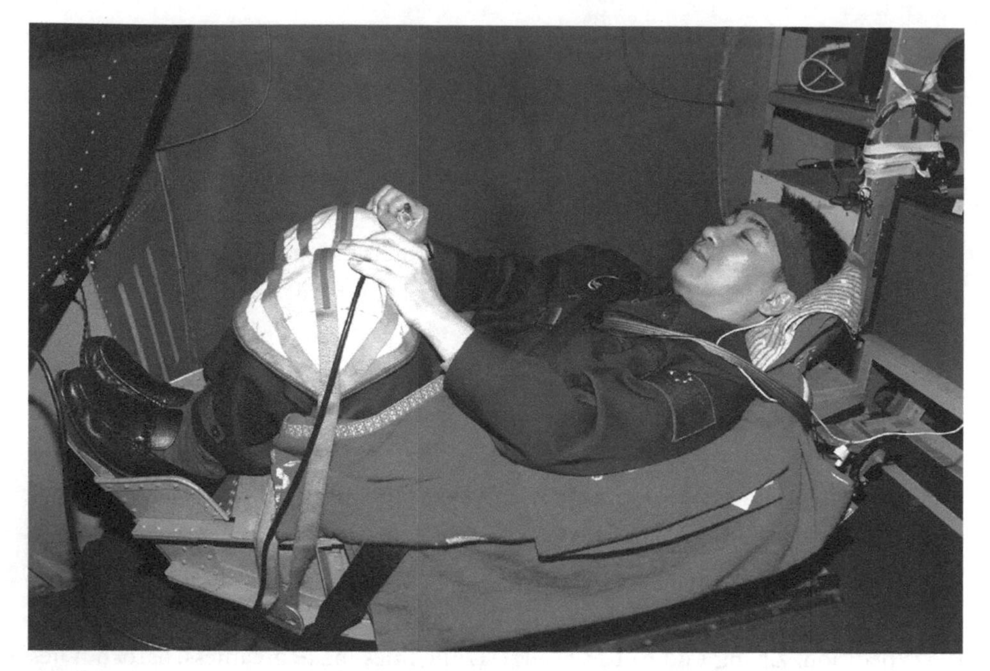

Zhang Xiaoguang prepared for test in the centrifuge. (Photographed by Qin Xian'an)

commissar of the astronaut brigade, who happened to bump into this scene, sighed: "What tenacity and perseverance! No wonder his physique is getting better, his condition better, and his performance better!"

After he was officially selected into the Shenzhou 10 crew, he thought about every angle of the hands and feet in a weightlessness environment to complete the recording task of space teaching. "The better preparation made on the ground, the more comfortable in space," he said.

He also had the courage to face space risks at all times. On February 2, 2003, the second day after the crash of the U.S. space shuttle Columbia, Hu Shixiang, then deputy commander-in-chief of China's manned space project, came to the astronaut center to hear out their feelings. Zhang Xiaoguang was the first one to stand up and say, "Who but me can?"

Ten years later, he still blurted out: "Who but me can?"

It was a kind of courage to say that in those days, but now a kind of confidence.

In the space teaching process, Zhang acted as a cameraman, recording the teaching with a hand-held camera and transmitting it back to the ground in real time. "The basics of recording a video are not difficult to learn. But to tell a complete 40-minute story in one shot would require some careful thinking." He repeatedly practiced shooting skills: where to give a close-up, when to change the angle, whether a follow-up scene? The amateur cameraman who was assigned the job at last minute turned out to be not an amateur at all. "This is the responsibility given to me by the task, so I must give my all."

While completing tasks, Zhang had fun in the 15-day space tour. He recorded the flight with a video camera and a diary, the stories of which were told to family and friends when he returned. And he also detailed the stories to the comrades who shared the same space dream.

The flight of Shenzhou 10 was pure perfection. From take-off of the rocket to the flight of spacecraft, from docking to entry into Tiangong, from return braking to the safe landing, Shenzhou 10 astronauts Nie Haisheng, Zhang Xiaoguang, and Wang Yaping accomplished a flawless flight. They deserved to be named the Dream Team.

"A rainbow comes after a storm. And the storm is the experience that I should cherish the most," said Zhang, "Success does not last after all. Diligence shines forever."

When it was time to leave Tiangong-1, Zhang and two comrades showed reluctance. They tried to leave a trace left and right in every corner of the space home. And they saluted in military manners. And a message was left in the Tiangong-1 computer: precisely control every flight, and China's aerospace industry will continue to thrive; bear in mind every progress, and the great motherland will prosper.

"The greatest happiness in life is doing what one likes." Zhang Xiaoguang, who loves reading, described what he felt during each flight eloquently: "We are the ones who pursue dreams and realize them. We stick to an idea that one shall not seek the best but better as the space dream has no end."

Wang Yaping: The First Chinese Space Teacher

"The change of life often goes beyond one's imagination," Wang Yaping sighed to me.

In the golden autumn of 1992, China's manned space program had just been launched. In a small village in Jiaodong, the 13-year-old girl helped her parents weed in the field under the scorching sun. Her face was sun-burned red.

In the golden autumn of 2003, when Yang Liwei, astronaut of Shenzhou 5, flew to space for the first time, the little dark-skinned, slim girl had grown into an excellent female pilot in the army. Seeing the rocket liftoff, she wondered: "Now that there is a male astronaut in China, when will there be a female one?"

In the midsummer of 2013, in a ponytail, she flew into the vastness of space: Wang Yaping, the third member of the Shenzhou 10 flight crew, became China's first space teacher.

Wang is eager to excel. The seed of the eagerness took root when she was little. Searching her deep memory, she remembered:

"When I was a pupil, I went to the field to help my parents with farming. To plant peanuts, they dug a hole, in which I planted the seed. Neighbors teased me: Aren't you too young to know how? I pouted and said of course."

In 1980, Wang was born to a farming family in Yantai, Shandong Province. "In my childhood, my family was poor. I had only one toy. It was the doll my grandmother sewed for me. The hair of it was made of red wool" she said.

From a young age, the hardships of her childhood taught her to be grateful to her parents. Her first experience of helping mother with the laundry remains fresh in her memory.

When mother was not home, the eight-year-old secretly took a basin of dirty clothes to the brook in front of her home to wash. The result turned out that the laundry was not done well despite that a large chunk of soap was used up. Mother came back and waved the hand in disapproval: "My dear daughter, you are not helping, but otherwise."

Parents are the first teachers of children. As a child, father often taught her "Honesty comes first, and then kindness." And he always saw her as an outstanding daughter. Once, Wang "failed" and came in the second place in an exam. She lingered at the door for over an hour with her result. And the moment she laid eyes on her parents, she burst into tears. Mother asked about what happened, and she answered: "I failed. Only the second this time."

Hearing this, dad comforted: "It's OK. I'm sure you will get it next time."

Since childhood, Wang has continued to strive for self-improvement. The teacher who taught her shared the same comment on her: "This child could not study any harder."

As a child, Wang had a simple dream. "I hope that one day I can go beyond the small village, enter university, and become a doctor or lawyer."

However, the change of one's life often happens without advance notice.

In her senior year of high school, Wang was just in time for female pilots' recruitment. "Of the 20 girls in the class, I'm the only one without glasses. Everyone encouraged me to give it a try." Unexpectedly, she passed all the tests and was admitted. Hereafter, the village girl stepped on the stage of blue sky.

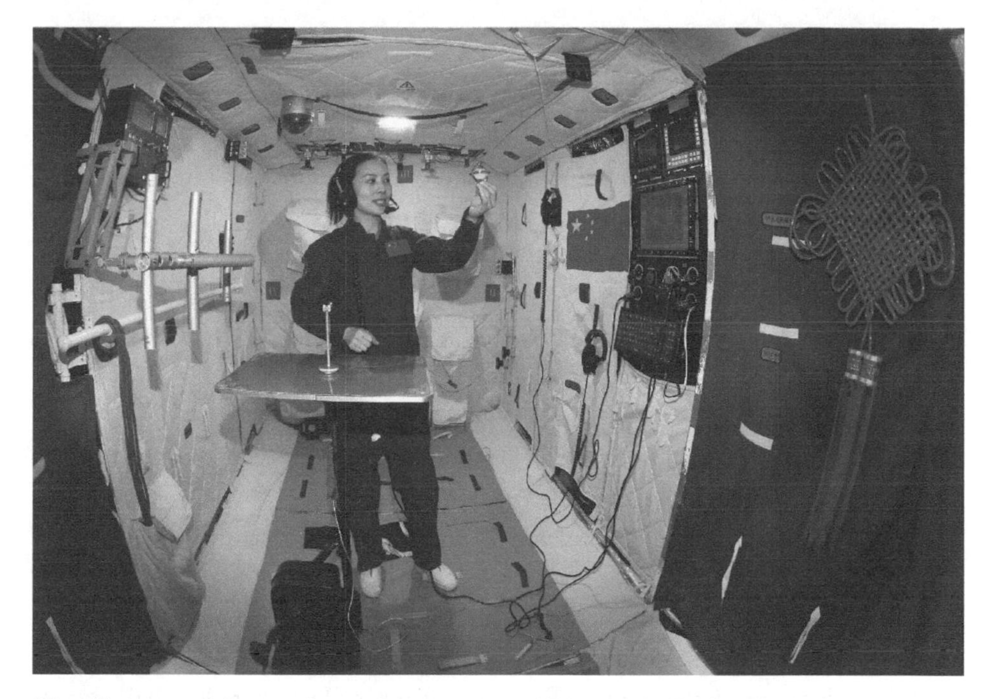

Wang Yaping tested the air tightness of the pressure suit. (Photographed by Qin Xian'an)

The hard life in the aviation school in northeast China far exceeded her expectation. "It's freezing in the temperature of minus 20 degrees C. Every time I jogged, the inside of my clothes would be sweaty. Frost would form on the outside. And my eyelashes would be with some little icy frost too." She considered the four years "A competitive long-distance race, for which if maximum efforts were not given, one would be eliminated."

Wang has always been adept at middle- and long-distance running. She won the first prize in a 7,000-meter race. Schoolmates asked her the secret. She disclosed: "I persevere even though I am exhausted."

With persistence, Wang eventually fell in love with the blue sky. "At the first 800-meter parachute jump, I did it without hesitation; at the first solo flight, when taking off, I subconsciously checked around. And I realized the trainer was absent. It made me so excited that in the cockpit I screamed." When she graduated, she was assigned to an aviation division with the second-best performance.

As a Shenzhou 10 astronaut, Wang made her public debut. But as a pilot, people have known her for a long time.

In recent years, she has been called to where there was a drought that needed artificial precipitation. The media referred to her as the rain maker.

In 11 years, Wang had flown over 1,600 hours. The most unforgettable flight happened during the Wenchuan earthquake relief:

At noon on May 18, 2008, she was ordered to deliver relief materials to the Chengdu airport. "From 11 a.m. to 11 p.m., there was no time for a bite of food. Yet, it made me happy to be able to do something for the disaster area."

That day, Wang and her husband had a romantic meeting at the airport.

The husband's plane landed a second before Wang's. The urgency cut their meeting short. They soon embarked on the next flight mission. Before taking off, she received a message: "Sweetheart, concentrate on the flight. I got your back." A sweet rush washed over her.

In full support of her career, the husband quipped, "Babe, we used to fly planes separately, but now we are together on the same spacecraft."

The opportunity once again favored Wang, who stood out in the selection of female astronauts in 2010.

Hearing the news, her father said, "Being a pilot is dangerous enough. An astronaut's job must be more dangerous, as it involves a mission outside the earth."

She was aware of the danger. But she did not expect that the challenge of evolving from a pilot to an astronaut was extremely arduous. Over 50 courses were to be learned, and zero error was acceptable at assessments. She said, "A correct action cannot determine the success of a task; however, a wrong one is likely to have it fail."

In fact, the life of astronauts is monotonous. "It was just training and assessment." For three years, Wang had never gone shopping or seen a movie. But on the training ground, she was delighted every day. She said, "Nothing to be unhappy about is a happy thing itself."

The personal item Wang brought to the space was simple: a group photo of the two batches of 21 astronauts in the astronaut brigade.

"When one crew of astronauts flew in space, all the astronauts of the brigade were there, and all the Chinese aerospace workers," said Wang. "With a sense of mission, I shall overcome all difficulties."

In the Shenzhou 10 mission, Wang was the caretaker of the flight crew. Gentle and delicate, she showed female superiority managing daily life in space. During the 15-day flight, she was responsible for heating food, sanitation, and sending greetings to the ground duty personnel via computer every day.

At 10:11 on June 20, 2013, China's first class in space began: the highest platform was located in Tiangong, 1,300 kilometers away from the earth; the largest audience included 60 million primary and secondary school students from 80,000 schools in China.

Sitting in the middle of space, magic pendulum, magic water polo ... In 40 minutes, Wang, a female astronaut, gave a wonderful space lecture. It planted the seeds of science and dreams in the hearts of millions of young people. And it left people an unforgettable impression of the first Chinese space teacher.

Hereafter, Wang has earned a new title: Teacher Wang. Wang could not have been happier. "Teaching is a sacred occupation, and I am happy to be seen as one."

After returning from space, Wang received many letters from students.

A student wrote in the letter, "Miss Wang, it's amazing to see three of you floating around in space, as if birds that fly and kangaroos that jump. You taught me to dream. Like a caterpillar, I will slowly climb to the top of the tree, become a butterfly, fly, and realize my dream."

June 24 is Wang's mother's birthday. Wang had not been at her mother's birthday since she left home as a soldier at the age of 17. On this day, far away in Tiangong-1, she sang mother a loving song, "Mother with Grey Hair."

"In space, I realized that kinship, friendship, and patriotism are intertwined," said Wang, "On the ground, the motherland is home, and space is a dream; in space, space is home, and motherland is a dream. And one more flight to space would make my biggest dream."

On the eve of Shenzhou 10's expedition, I asked Wang, "What would you like to do the most when the mission is complete?"

Wang replied, "I would like to accompany mom and dad for a walk and perhaps go shopping."

A simple and joyful wish it is.

Chen Dong: Setting the Record for Longest Space Flight in China

Chen Dong, an astronaut of Shenzhou 11, spent 33 days in space with comrade-in-arms Jing Haipeng, setting the record for longest space flight in China.

In the autumn of 2016, Chen Dong, who was about go on the mission, made his first appearance in public. At the age of 38, he was the first selected astronaut among the second batch of astronauts in China.

"We are lucky to be in this era. Every major change of my personal life resonates with that of the nation." Chen Dong owed it to luck, "For the opportunity, Jing Haipeng has waited 10 years, Liu Wang 14 years, Zhang Xiaoguang 15 years, and me six years."

Chen Dong was born in Zhengzhou, Henan Province in December 1978. He, who grew up in a factory, has always been the leader child at school. Chen, once a naughty troublemaker, did not expect to become an astronaut with a calm personality and strong sense of responsibility.

It takes an opportunity for the dream to come true. The transformation of Chen's life was waiting for it. And the chance came when he was in the third grade, when he caused some trouble.

The school served the teaching faculty vegetables, which were all piled on the small playground. After school, Chen led a group of children to destroy the pile, trampling the vegetables to pieces. "Darn! I think I will be in trouble." The more Chen thought about it, the more afraid he became. But to his surprise, Mr. Yang, the teacher in charge of the class, did not punish him, but gently said to him, "I know that you did not mean it, but in this way, teachers would have nothing to eat. Think of others next time." In the end, Mr. Yang took the blame.

"Henceforth, I have always wanted to behave well and make my teacher proud." His good performance soon had him elected class monitor. And he managed to stay that position in the following schools. Over the years, Chen has been grateful to Mr. Yang as the tolerance and responsibility he learned from him constantly nurturing his growth.

One day in middle school, with several friends, he was lying on the ground observing the night sky and noticed a moving bright spot. Curiously, he asked his brother, "Is that a meteor?"

"It's a satellite," Brother told him. Curiosity motivated Chen to the library to search for books about satellites and space.

This was Chen's first contact with space, and the first time space entered his world.

From then on, born in a family of ordinary workers in Luoyang, Henan Province, the young man who once aspired to be a teacher, began to shift more attention to the blue skies.

At the college entrance examination of 1997, Changchun Aviation Academy was his only application. Finally, Chen was admitted by the Academy with flying colors. Here, he would gradually adapt to the shaping of another kind of character. At the freshman training, his quilt was thrown away as it was not folded as required.

At first, Chen showed resistance as he thought it was unnecessary to fold it into a tofu-shaped piece. He did not change until the trainer told him that folding quilts was actually to nurture a kind of military temperament and behavior.

"Being a monitor has enlightened me about what responsibility is; in military schools, I have understood what responsibility a soldier shoulders." As time went on, the weight on his shoulders became increasingly heavy.

In a twinkling of an eye, he graduated from the military academy, and became a pilot, squadron leader, and captain. He once thought that life would only revolve around the blue skies. Until one day, his attention was firmly attracted by astronaut Yang Liwei's first flight. His eyes were lit by the rocket's tail flame. Chen, aged at 25, seemed to sense the scent of space.

At that time, Chen was a fighter plane pilot, mainly flying at low altitude. "I want to fly higher beyond the atmosphere to see what it looks like." This desire grew stronger and stronger with the successive triumphant flights of Shenzhou 5 and 6. When he found out that all astronauts were selected among the pilots, and Yang Liwei, like him, was also a fighter plane pilot, he heard the heartbeat of his dream. "Perhaps, my chance is coming."

And soon a chance came, unannounced. In 2009, the selection of the second batch of astronauts began. Unfortunately, Chen Dong, then the leader of the team, took the team to carry out maneuvers in northwest China. The chance mercilessly slipped out of his fingers. In chagrin, he received the order to take part in the physical examination again after the maneuver.

Chen was called for the interview soon. To his surprise, the interviewer turned out to be Yang Liwei, the hero he idolizes. "After an hour's talk, my dream to become an astronaut has never been surer." Speaking of the interview, Chen has always enjoyed sharing the process.

Jing Haipeng, who has been to space twice, is also Chen's idol. "What could be luckier and happier than being brought into the team by my own idol and then going to space with him?" he said.

It took six years for Chen Dong to stand on the starting line of space exploration. During the six years, Chen seemingly "vanished."

In 2012, Yao Zhiqiang, the head teacher of Chen's class in high school in Luoyang, was concerned about this beloved student. In the Luoyang Evening Newspaper, he published a notice titled "We Miss You." Even now, the notice is still available on the Internet. Mr. Yao did not know that in order to realize the space dream, Chen was in the space city devoting himself to training. For six years, he practically isolated himself from the outside.

Six years was neither short nor long. "It's both a marathon and a sprint. Therefore, stamina and explosive force are needed." Chen frankly disclosed that he felt as if it was senior year in high school during which there was intensive learning every day. Back in the classroom, he picked up the textbooks that had been set aside for ten years. For Chen, it was not easy. Faced with astronomy, space technology, and space inertial coordinates, Chen suffered headaches. "When I first sat in the classroom, I found it difficult to keep my head cool. So, I put some cool oil on and stood in the back for a while to keep myself awake" said he.

Fortunately, learning is his strong suit. In addition, he never gave up should he fail. In two years, Chen once again found his learning mojo: he successfully passed all the examinations of 58 courses in eight categories.

"It is inevitable that he would make the cut." In the eyes of Huang Weifen, deputy chief designer of the manned space astronaut system, Chen excelled in all aspects without any obvious shortcomings. From the primary selection to the final, Chen always ranked first among the same group of astronauts. His comprehensive understanding sometimes surprised trainers, and his pursuit of perfection was admired by Jing Haipeng.

At 4:40 on October 17, 2016, the seeing-off ceremony was held for the Shenzhou 11 astronauts Jing Haipeng and Chen Dong in the Jiuquan Satellite Launch Center. (Photographed by Wang Sijiang)

Chen has twins, aged five years old. Before the expedition, the father, who occasionally told the story of Astro Boy, said to the twin boys, who fancy observing the stars: "Daddy is going for a long journey. Look at the night sky and search for the brighter stars. Maybe daddy will be there so you boys can say hello."

During the space flight, the five-year-old twins once had a video call with him. "I'm doing somersaults for them up here, and they're so happy down there." Chen asked the children what they wanted as gifts, and answer was stars. He promised them, yet back to the ground he found out that the sons had promised to give friend A five stars, friend B three, and the kindergarten teacher the Moon. "I was clueless for a moment," said Chen. "Fortunately, it dawned on me that I took many pictures of stars in space, which can be given as gifts as promised."

For Chen, who was on his first mission, there were countless unforgettable experiences, the first space drift, the first vegetable planting in space, the first silkworm rearing in space, the first viewing of the magnificent mountains and rivers of China in space, the sunrise and sunset in space, etc.

The first real sense of weightlessness is still fresh in his memory. "I didn't know how to exert force. A little push can result in too much." He said that at first, they needed to use shackles at every step. Later, they gradually adapted to weightlessness, so there was no longer the reliance on shackles for moving. "Instead, I began to enjoy it, and created a lot of postures myself."

During the 33-day space trip, astronauts participated in more on-orbit trials and experiments than ever. Each was a new challenge.

Technical verification of running a platform binding system was one of the challenges. The space running platform, unlike the treadmill on the ground, required to be verified by astronauts in space. At first try, they could not even walk on it, let alone run. That night, Jing Haipeng said to Chen Dong, "Should we fail, I will find myself ashamed to face the researchers." They started to observe each other in search of the trick, and finally succeeded. They could walk or jog with arms outstretched, hands wrapped on their heads or on the back, and in other various positions. "We can run as easily as we can on a treadmill on earth," said Chen.

"Chen performed perfectly. On a scale of 1–100, I will give him 100." Jing Haipeng said that Chen has put effective use of what trainers on the ground have taught without any mistakes.

At 13:59 on November 18, 2016, the re-entry capsule of Shenzhou 11 landed. Jing Haipeng and Chen Dong exited the capsule with a smile.

The 33-day live broadcast in space left the general public familiar with this outgoing and handsome face, Chen Dong, an astronaut of Shenzhou 11.

"I want to fly higher and higher!" For lucky Chen Dong, there may be more luck waiting for him in the future, because he happens to be in the good era of China's aerospace industry.

A Military Salute, the Unique Code of Chinese Astronauts

In May 2009, the selection of the second batch of astronauts was initiated. After strict evaluation in three stages of preliminary selection, call back, and final selection, seven astronauts (5 men and 2 women) made the cut to the second batch of Chinese astronauts.

The seven astronauts all hold a bachelor's degree, range from 30-35 years old with an average age of 32.4, and are married. Five male astronauts are active fighter plane pilots, and two female astronauts are active transport pilots. Their average flight time was 1,270.7 hours. Most of them have engaged in major missions such as Wenchuan earthquake relief and military maneuvers.

On May 17, 2010, the training of the second batch of astronauts was fully initiated. The pace to train the second batch of astronauts was obviously faster than that of the first.

On July 20, 2016, Ye Guangfu, a second batch of the Chinese astronauts who successfully completed the ESA cave training, made a public appearance. Ye obtained the qualification of manned space flight in 2014 and is now a class III astronaut. Huang Weifen, deputy chief designer of the China astronaut center, said, "He has made outstanding achievements and fully demonstrated the good spiritual outlook and personal quality of China's astronauts."

Shenzhou 5, 6, and 7 missions were all carried out by the first batch of Chinese astronauts. From Shenzhou 9 in 2012 to Shenzhou 11 in 2016, the second batch of Chinese astronauts Liu Yang, Wang Yaping, and Chen Dong took the baton.

In 2017, the selection of the third batch of astronauts in China was fully launched. Aimed at the construction of a future space station, the quantity and types of astronauts selected will be further expanded.

Chinese astronauts are the cover of China's aerospace industry.

Over the years, China's astronauts have made wonderful achievements that made the world applaud. Looking back on China's road to space, unforgettable scenes flash back:

When Shenzhou 5 returned safely, astronaut Yang Liwei shouted to the cheering crowd, "I'm proud of our motherland."

During the Shenzhou 6 space flight, astronaut Fei Junlong made three somersaults in space.

On Shenzhou 7, Zhai Zhigang, dressed in an extravehicular spacesuit, left the first Chinese footprints in space, when waving the national flag.

Shenzhou 9 astronaut Liu Wang calmly piloted the Shenzhou spacecraft, performing a smooth docking in the vast space.

The sweet smile of Wang Yaping, China's first space teacher and astronaut;

Shenzhou 11 astronaut Jing Haipeng's birthday in space …

All the images recorded the historic moments of China's manned space flight and ignites the pride and confidence of the Chinese people.

Over the years, I have witnessed all the occurrences. With every Chinese, I have shared the excitement.

However, what impressed me the most was the common action of astronauts when they went on the expedition: a military salute.

A hug speaks louder than thousands of words to a spouse. To the motherland, what action outweighs thousands of words? Chinese astronauts gave their own answer: a military salute.

Yet, in the flight manual for astronauts, there were thousands of words depicting flying procedures but not for making a military salute.

Foreign astronauts used to compare themselves: "An astronaut is a person who flies at the top of a bomb." Shenzhou 5 was China's first manned space flight. There was the concern over whether astronauts could return safely.

When Yang Liwei took Shenzhou 5 into space, he saluted in a military manner for the first time when the rocket was ignited. Since then, it has become a tradition for Chinese astronauts.

Over the years, I have seen the astronauts' expeditions many times, each of which gave a vibe of worry over the safety of the astronauts. After all, risks lurk throughout the space journey.

However, every time, astronauts demonstrated a confidence of success with a military salute.

At the launching ceremony, the astronauts' military salute left no dry eyes among the audience who came to see them off; at the instant when the rocket took off, astronauts paid a military salute to the motherland in front of the camera without prior consultation; the moment when they first entered Tiangong-1, they made the same gesture; when the leaders of the party and the state extended their regards, they did it again, reassuring the whole nation; when they exited the re-entry capsule, they responded to the audience's cheers with another military salute.

A military ceremony reflected the calmness they held in front of the test of life and death:

"Soldiers usually write wills before going to war. But we astronauts have never made one before an expedition." Yang Liwei said that he went home the night before. He usually adjusted the electronic alarm clock at home, so he picked it up and said to his wife, "When I am absent, you should know how to adjust it. Let me show you." When his wife heard it, she grabbed it from him and said firmly, "No, you will do it when you return."

Astronaut Jing Haipeng visited Hong Kong to meet with some students in a stadium. A pupil asked: "Uncle Jing, are you not afraid of death?"

He replied: "Like every ordinary person, not Iron Man, I fear death. But there are many things in the world worth devoting one's life, such as being an astronaut. From this point of view, I fear not." He suggested that students see the launch site for real should there be a chance. "We are sitting in the bottom of the second stage rocket. When it roars after ignition, one will understand the definition of a high-risk job. But this job is worth our lives."

A military salute is their commitment to the motherland as a soldier.

On January 5, 1998, Zhang Xiaoguang became one of the first astronauts in China. Many years later, he described the unforgettable moment, "It was a sunny day. Facing the bright red party flag and national flag, when I raised my right fist to swear the oath of astronauts, my heart was pounding. The responsibility and strength that came with the title astronaut could not be greater." On February 2, 2003, the second day after the crash of the U.S. space shuttle Columbia, Hu Shixiang, then the deputy commander-in-chief of China's manned space project, came to the astronaut center to inquire the opinions of astronauts. Zhang Xiaoguang was the first one to stand up and said, "Who but I can?"

Little was it known that at 16:30 on September 27, 2008, when Zhai Zhigang was going to display the national flag the moment he exited Shenzhou 7, the fire alarm of the orbital module went off in their earphones. Should the danger happen, three astronauts would not be able to return to earth.

At the critical moment of life and death, Zhai Zhigang did not panic, and Liu Boming and Jing Haipeng, who remained in the cabin, were also calm. Fortunately, it was just a false alarm. Liu Boming thought nothing else but: "Even if we can't go back, we shall leave the five-star red flag in space."

Looking at the military salute, I think of astronaut Liu Wang's words: "I am not only an astronaut, but also a soldier and communist."

Looking at the military salute, I think of astronaut Liu Yang's words: "A military salute represents our oath and lofty commitment to the motherland and the people."

A military salute in space is the unique code of Chinese astronauts, the oath of the Chinese astronaut team: "To contribute it all to the aerospace cause of the motherland."

CHAPTER 6

Chang'e Lunar Dream

In the spring of 2017, two pieces of intriguing news put the distant Moon under a heated discussion.

One piece took place in Los Angeles, in the United States in the Western Hemisphere. In the cold wind of early spring, American astronaut Eugene Cernan, the last to leave a footprint on the Moon, died at the age of 82.

In the summer of 2012, American astronaut Neil Armstrong, the first man to set foot on the Moon, also died quietly at the age of 82. Then-President Barack Obama said, "Goodbye, Mr. Neil Armstrong, thank you for showing us the power of a small step."

In December 1972, Cernan, as commander of Apollo 17, was the last to return to the lunar module, becoming the last person to leave a footprint on the Moon.

Upon return to the lunar module, Cernan said to the ground command center, "Bob, this is Gene, and I'm on the surface; and, as I take man's last step from the surface, back home for some time to come – but we believe not too long into the future…"

Forty-five years passed in a flash. Now, Cernan has left this world, regretfully without seeing the not-so-far future.

Since Cernan left the Moon, no human has ever been there again. "The death of astronaut Eugene Cernan marks the end of a great era and the disconnection of the once-close contact between man and the Moon," said a netizen.

When can humans return to the Moon? Who will be the next great Moon hero?

Cernan once predicted that China would be the next country to land a man on the Moon, "It's a good idea to explore the Moon with China."

Another piece happened in Beijing, China, in the eastern hemisphere. During this year's National People's Congress and the Chinese Political Consultative Conference, academician Ye Peijian, general director consultant and chief designer of lunar exploration satellite of the Fifth Academy of China Aerospace Science and Technology Corporation, took the initiative to disclose: China would launch Chang'e-5 lunar probe around the end of November 2017 to realize a soft landing and Moon sampling, and China's lunar exploration project would complete the last step of "circle, land, and return" scheme.

In 2017, Chang'e-5 was the biggest anticipation of China's aerospace industry. In this mission, China's lunar exploration capability achieved four firsts: the first automatic sampling on the Moon's surface, the first take-off from the Moon's surface, the first unmanned docking in the Moon's orbit beyond 380,000 kilometers, and the first return to earth with the Moon's soil at a velocity close to the speed of escape.

"If Chang'e-5 goes well, the whole project will be completed in one month, and it can pack 2 kg of lunar soil back to earth," said academician Ye Peijian.

What a 2 kilograms of lunar soil!

The weight, in the eyes of Chinese lunar researchers, was beyond imagination.

In 1978, before the establishment of diplomatic relations between China and the United States, President Carter sent a precious gift to China: a piece of rock collected by American astronauts from the Moon.

The Moon rock, the size of a fingertip, weighed only 1 g.

This one gram Moon rock was the starting point of Chinese lunar exploration research.

From 1 gram to 2 kilograms, the change of weight reflected the rapid development of China's lunar exploration over the years.

In 2004, China's lunar exploration project was officially initiated. In 2007, Chang'e-1 was launched. In 2010, Chang'e-2 took off. In 2013, Chang'e-3 successfully landed on the Moon. In 2014, as a pathfinder of Chang'e-5, the "Xiaofei" test was successfully completed. Chinese aerospace workers flew to the Moon with incredible Chinese speed.

"Moon, the Chinese are coming." Luan Enjie, the first commander in chief of China's lunar exploration project, said, "Regarding space exploration, no one knows how far and how fast we can go."

Two thousand years ago, *Huainanzi*, written by Liu An, the king of Huainan in the Western Han Dynasty, recorded the myth of Chang'e ascending to the Moon.

We believe that in the near future, the myth will come true.

One Small Step, One Giant Leap

The Moon is the nearest celestial body to the earth. It has been the satellite of the blue planet since creation. In the course of human's journey to the universe, flying to the Moon has always been a dream of mankind.

In 1988, American astronaut Neil Armstrong, the first man to set foot on the Moon, was invited to visit China. At the welcoming ceremony, he joked in a jovial mood: "The first to fly to the Moon is Chang'e, a beautiful woman in Chinese legend. The first to set foot on the Moon is an American. That's me."

Off the stage, Chinese scientists stood in silence, like a serene forest.

This joke, for Chinese aerospace workers, was greatly ironic; the myth of Chang'e flying to the Moon belongs to China, while the real Moon landing has nothing to do with China.

However, it is a fact that China was absent in the first wave of Moon exploration in the 20th century.

During the expeditions to the Moon, the Soviet Union took the lead. On January 2, 1959, a carrier rocket refitted with a strategic missile sent the first human lunar probe "Luna 1," designed and manufactured by the Soviet Union, into space and flew by the Moon. The news astonished the he world.

Eight months later, the Soviet Union sent the lunar probe to the Moon again, hitting the surface. For the first time, it left human marks on the Moon. Twenty-two days later, the Soviet Union launched another new lunar probe and transmitted back pictures of the dark side of the Moon. For the first time in human history, the mysterious side of the Moon was seen.

From 1959 to 1968, the Soviet probes took the lead in a series of lunar exploration feats, such as lunar overseeing exploration, lunar soft landing, lunar circling exploration, lunar sample collection, and lunar surface inspection. It opened the gate to successful lunar exploration by humans.

The brilliant achievements of the lunar exploration motivated the Soviet Union to realize a manned landing before the Apollo spacecraft of the United States. It planned to use the N-1 super heavy carrier rocket as the launch vehicle to send two astronauts to the Moon.

Haste makes waste. In order to have astronauts go to the Moon earlier than the United States, the engine of the N-1 rocket never went through any ground testing. From 1967 to 1972, launches of N-1 rocket failed one after another. In 1969, the U.S. astronauts beat it first, crashing the Soviet vision.

Instead, the United States took the lead in Moon exploration. At 3:51 GMT on July 21, 1969, Neil Armstrong, commander of Apollo 11, stepped out of the lunar module.

For the first time, human footprints appeared on the Moon. "That's one small step for man, one giant leap for mankind," announced Armstrong to the world.

The whole world was in awe of this amazing feat: the Apollo landed on the Moon.

There were three plans in the Apollo Moon landing project. Eventually, the US decided that the plan would be the lunar orbit docking method, which enabled Apollo's Moon landing to be at least two years ahead of schedule.

Apollo 10 launched on May 18, 1969 and carried out the trial flight for a full process Moon landing, circling the Moon 31 times.

On July 16, 1969, Apollo 11 completed the first human mission to the Moon after three days' flight. Astronauts Armstrong and Aldrin placed a metal plaque on the Moon with an inscription: "Here men from the planet earth, first set foot upon the Moon, July 1969, A.D. We came in peace for all mankind."

From 1969 to the end of 1972, the United States launched seven manned spacecraft to the Moon, and 12 astronauts landed on the Moon. They spent 302 hours and 20 minutes on the surface, traveled 90.6 kilometers, and brought back 381 kilograms of lunar soil and rock samples, thus initially revealing the real face of the Moon.

The Apollo Moon landing project lasted about 11 years at a cost about US$25.5 billion, with a total number of more than 300,000 personnel participating in the research. It has driven the development of science and technology in the United States, especially in jet propulsion, guidance, structural materials, electronics, and management science.

The United States and the Soviet Union shared a special enthusiasm for the Moon. In addition to hegemony, there were other practical considerations. Moon landing and exploration are the high-end test and promotion of technological development, as well as the chance for exploring further celestial bodies. The dream of human beings to understand the universe and exploit space began to experience "small steps" and "giant leaps."

Since December 1972, when Apollo 17 made its last expedition to the Moon, the exploration of the Moon has fallen into a sluggish period.

The Beginning of China's Lunar Exploration

China's lunar exploration commenced nearly 40 years later than that of developed countries.

Looking back on the history of China's lunar exploration, we should start with a gram of Moon rocks:

In 1978, U.S. President Carter's national security advisor Zbigniev Brzezinski came to China with a special gift: a stone cast in plexiglass and collected by American astronauts from the Moon.

This stone, only the size of a fingertip, weighed only one gram. The Americans did not elaborate its origin, as if it deliberately left a mystery.

Chinese have chanted about the bright Moon for thousands of years in poetry and lyrics. However, it was 39 years ago that the Chinese first laid a finger on it.

Who could read it? Ouyang Ziyuan, a researcher at the Institute of Geochemistry, Chinese Academy of Sciences, carefully chiseled a corner from the Moon rock with a little hammer. From this moment on, lunar exploration has become his lifelong pursuit. In that second, he thought to himself, "When can we Chinese study the samples we take back from the Moon?"

The Moon stone was the starting point of China's lunar exploration. In the following two years, Ouyang published 12 papers, which proved that the Moon stone was a high titanium Moon sea basalt sample collected by Apollo 17 spacecraft of the United States. The main components are: 51.5% pyroxene, 25.7% plagioclase, and 21.4% ilmenite.

Thirty-nine years later, Ouyang became the chief scientist of China's lunar exploration project. His dream to use the lunar soil he retrieves for research will soon come true.

It is from this piece of Moon stone that Chinese scientists have realized that to explore the Moon is for the glory of mankind, but only powerful countries can share it.

Project 863 ignited the dream of lunar exploration hidden in the hearts of Chinese aerospace researchers for many years. In the early 1990s, at the advocacy of Min Guirong, chief scientist of the National Project 863 Aerospace Experts Committee, the lunar exploration group was established.

At that time, the manned space project was at full speed. In this circumstance, the proposal of lunar exploration did not stir up more waves.

"At that time, our rockets and launch sites were ready to explore the Moon. It was no problem to aim directly at it." Long Lehao, a Long March rocket expert, recalled that before the first test flight of the Long March 3A rocket in 1994, scientists had proposed to carry a simulation satellite and "smash" it to the Moon in the way of hard landing. However, a poor economy, poor science and technology, and especially the lack of a clear scientific goal meant that the plan had to be given up.

Although the bold vision failed to come true, it made aerospace workers and scientists understand that to carry out the lunar exploration, we must figure out the scientific goal of researching and designing the lunar exploration as soon as possible.

In 1995, Ouyang Ziyuan and Ye Zili put forward the plan of developing the first Moon satellite. In 1997, three academicians, Yang Jiagong, Wang Daheng, and Chen Fangyun, published a proposal on the development of lunar exploration technology in China. In 1998, the expert group passed the project of "overall scheme design and key technology decomposition of lunar exploration robot," which started the research of a lunar rover in China.

In 1998, the State Council established a new National Defense Science and Engineering Commission, with the National Space Administration affiliated. At the beginning of this establishment, the Commission began to study the long-term planning of national space exploration, including lunar exploration.

On November 22, 2000, the Chinese government published a white paper on space projects for the first time, proposing to carry out the pre-research of deep space exploration focusing on lunar exploration. Subsequently, the National Defense Science and Engineering Commission entrusted academician Sun Jiadong, an aerospace expert, to be responsible for organizing the project approval demonstration.

"After the manned space project, China aimed to explore the Moon on the next move." On March 1, 2003, Luan Enjie, director of the National Space Administration, announced that the lunar exploration project would be carried out in three phases. The first involves the obtainment of three-dimensional images of the lunar surface by launching a lunar exploration satellite, analysis of the distribution characteristics of the content of useful elements and material types on the lunar surface, and the exploration of the thickness of the lunar soil and the environment between the Earth and the Moon. The second phase aims to have a lunar probe launched and land on the Moon. In the third phase, the probe performs surface patrol and sampling.

The engineering objectives and science objectives share the same core – not only standing at the international forefront, but also combining China's national conditions. The scientific ideals shall be combined with the country's actual technological capabilities. An entirely innovative Chinese lunar project shall be built on independent research, manufacturing, and testing.

Presence in the Era of Space Exploration

On January 23, 2004, the second day of the lunar new year, Chinese Lunar Exploration Program was officially initiated.

After nearly 40 years of silence, the world welcomed a new wave of lunar exploration in the 21st century. On January 14, 2004, the United States proposed a plan to return to the Moon; earlier, Europe formulated the "Project Aurora" plan for lunar and Mars exploration, India announced the launch of lunar exploration activities, and Japan was also preparing the SELENE lunar exploration.

The Moon orbits the Earth from 380,000 kilometers away. What is the point that we explore it? What will it bring us?

"The Moon has become the focus of the fight of future aerospace powers for strategic resources," said Luan Enjie, then director of the National Space Administration, "It has a variety of unique resources for human development and utilization. The unique minerals and energy on the Moon are an important supplement to Earth resources, which will have a profound impact on the sustainable development of human society."

Both human curiosity and the urge to outdo others, and the strong social system competition and power show-off gave rise to the peak of the first round of lunar exploration during the cold war. Instead, today's lunar exploration competition is more out of the consideration of practical interests.

"Although the exploitation of lunar resources is too far away to be expected, scientists never doubt that this day will come." At present, there are more than 100 lunar minerals that are made known, five of which are unavailable on Earth. Different rocks on lunar surface are abundant in silicon, aluminum, potassium, phosphorus, uranium, thorium and rare earth elements. Preliminary estimates say the amount of rare earth elements in lunar rocks ranges from 22.5 to 45 billion tons, and the amount of uranium is up to 5 billion tons.

In the thick dust on the lunar surface is stored a very significant element: helium-3, one of the main materials for nuclear fusion, and the total reserves on Earth are only 15 tons. Ouyang Ziyuan, an academician of the Chinese Academy of Sciences and chief scientist of Chinese Lunar Exploration Program, said at the 36th World Space Science Conference: "It is estimated that the helium-3 reserves on the Moon reach 1 to 5 million tons, which can meet the needs of human beings for tens of thousands of years."

The environment of the Moon has advantages that the earth cannot duplicate: high cleanliness, microgravity, zero magnetic field, zero atmosphere, and stable geological state. "It's an excellent environment. Many things that cannot be produced on the earth can be produced on the Moon, such as some particularly expensive biological products, medicines, and special materials." said Ouyang.

The lunar exploration project, reflecting a country's comprehensive strength, will promote new breakthroughs in basic science and applied technology. It is required to break through a series of key new technologies, including tele-data transmission, artificial intelligence, automatic processing, and space nuclear power supply. And it involves many new fields. These new breakthroughs will promote the development of a large number of basic science and applied technology.

According to statistics, the Apollo Moon landing program, which cost US$25.5 billion, has established and improved a gigantic space industry system in the United States. About 3,000 kinds of applied technology achievements have been derived from the program, including those in aerospace, military, communications, materials, medical and health care, computers, etc. For every $1 invested in the program, there was a $4 to $5 output, which has led to the overall development of high-tech in the 1960s and 1970s in the United States.

"The lunar exploration, like the locomotive, has a strong pulling effect. First of all, high requirements are made for the carrying capacity. The need for the improvement of carrying capacity immediately makes the task urgent." Luan Enjie, the director of the National Space Administration said, "The Moon exploration will not only promote the further development of the space industry, but also the development of science. With this lunar exploration, space chemistry, space biology, space environment and space astronomy will all start accordingly. It is a gathering, where experts in geology, minerals, meteorology, astronomy, and the origin of the earth came together."

According to Agreement Governing the Activities of States on the Moon and Other Celestial Bodies adopted by the United Nations in 1984, the Moon is a common treasury of all mankind.

An early entry to the Moon club concerns the long-term interests of every country. The same as a science expedition to the South Pole on earth, whoever arrives first takes the initiative. Academician Sun Jiadong, chief designer of Chinese Lunar Exploration Program said "If we wait until others have found the way to exploit Moon resources, it will be too late for us to start the lunar exploration program. We were absent in the first round of lunar exploration. In the new wave, China must not lag behind."

Finally, the lunar exploration is not only a science project, but also, in a sense, a symbol of national spirit and the power of a country. A nation depends on some things to lift the national spirit. The lunar exploration will provide a new platform for that and for enhancing national cohesion. Chang'e ascends to the Moon is an ancient Chinese myth. Turning it into reality will be one of the important symbols of China's transformation from an aerospace participant to an aerospace leader and will also be another soaring monument on the road of the great rejuvenation of the Chinese nation, thus greatly uplifting the national spirit and enhancing national cohesion.

What can give more courage to explore the unknown than a dream?

What can give more strength to move forward than exploration?

Like a small leaf adrift on an ocean, so this ushered in the era of great navigation. With the vigorous development of the world's aerospace industry, there is no way that China will be absent in the era of great space exploration.

The 1.4 Billion RMB Road to the Moon

On February 25, 2004, the leading group of the lunar exploration project passed the General Requirements for the Development of Lunar Exploration Project. It was given a beautiful name: the Chang'e project.

Subsequently, major domestic newspapers printed out news of great concern, the Determination of the Leading Figures of the Chang'e project: Luan Enjie, director general of the National Space Administration, chief designer of the project, academician Sun Jiadong, and academician Ouyang Ziyuan, chief scientist of applied lunar science.

Hereafter, the fate of the three men are intertwined because of the lunar exploration. And they are called China's lunar exploration Troika.

That year, Luan Enjie was 64, Sun Jiadong 75, and Ouyang Ziyuan 69.

Age did not impede their ambitious goals for the Moon.

On the night of the project launch, Luan Enjie was so excited that he composed a poem: "Sixty thousand years of the earth's existence, five thousand years of Chang'e's homesickness. Ancestors' legacy motivates diligence, individual last years are not to be wasted."

The lunar exploration project faces a great risk. Few people understand why Sun Jiadong, who has already achieved great success, accepted the risky job. In case of failure, his brilliant aerospace career may be overshadowed. To others' incomprehension, Sun responded: "What the country shall need, I shall do."

Ouyang was only 43 years old when he first touched the Moon stone. Over the years, he has been waiting for China to start lunar exploration. When it finally was launched, he held up his glass of wine and toasted: "The day has finally arrived. This is an opportunity and responsibility that history has entrusted me with."

When China launched its first lunar exploration project, it had been nearly four decades since Apollo 11 made the first human Moon landing.

"The United States has already landed on the Moon. What's the point of us going there?" For a while, there was a buzz on the Internet.

To this end, Luan Enjie, general director of China's lunar exploration project, commented: "To walk the path others have walked is not to eat the food others have chewed. This is a Chinese lunar exploration. And it shall be with Chinese characteristics. What the U.S. has done, we will do. What the U.S. hasn't, we will do too."

Ouyang Ziyuan, chief scientist of ground application system of China's first lunar exploration project, also commented: "While ballet is danced abroad, the lion dance is performed here. In short, we should do what has not been done abroad, with new ideas and achievements."

A high starting point, characteristics, and innovation is the blueprint of Chang'e lunar exploration that Chinese scientists have drawn.

—Unmanned lunar exploration has basically gone through five stages: gazing at the Moon, colliding with the Moon, circling the Moon, landing on the Moon, and sampling lunar rocks and bringing them back to earth. The Soviet probe has used all five ways, and the United States the first four. In contrast, the first step of China's lunar exploration was to skip the first two and start from the third.

—The exploration target of Chang'e-1 was basically to fill in the blank of human lunar exploration, a three-dimensional picture of the Moon, which has not been made abroad. The United States has explored five kinds of resources on the Moon. The number that Chang'e-1 aimed at is 14 and the thickness of lunar soil was to be explored for the first time.

—In China's first lunar exploration project, all five systems adopted domestic mature technology. The first phase of the project used entirely domestic products. The project investment was mere 1.4 billion RMB, equivalent to the cost of building a 2-kilometer subway in a city.

Great projects come with great responsibility. The lunar exploration project was composed of five systems, the malfunction of any detail of which could have it fail.

Wei Suping, academician Sun Jiadong's wife, recalled that when Chang'e-1 was in progress, Sun often walked to the veranda at midnight, carefully observed the Moon move slowly in the sky, and silently pondered over the engineering and technical solutions. "Sometimes he stood there for hours, which made me sleepless as well."

"He worked every day. Leather shoes are too exhausting. I used to prepare him four or five pairs of cloth shoes a year," Wei Suping said heartily.

As chief designer, Sun was constantly busy, so were the researchers. Now we often refer to the phrase "China speed." The so-called idea is the result of Chinese aerospace workers' relentless contribution.

To ensure the reliability of Chang'e-1 satellite equipment, academician Ye Peijian led the designers to run experiments everywhere. He once traveled from Xinjiang, the westernmost part of the country, to the very south of Yangtze River in one day.

One afternoon, at the power test of the satellite, the pointer of the power controller suddenly shook, and the voltage rose from 5 V to 10 V. The small change immediately aroused Ye's alert: excessive voltage would instantly burn down the assembled satellite.

"Cut the power supply now," ordered Ye. Then, he immediately reported the malfunction to his superiors, contacted the power module manufacturer, and informed the computer control department to modify the corresponding sample products and software. At once, dozens of science research and production departments were mobilized.

With every circuit and wire checked, the cause finally surfaced. It was a socket with poor conductivity on the ground equipment. Eventually, everyone breathed a sigh of relief. Looking around, many were surprised to see the 62-year-old academician Ye, who didn't close his eyes all night, falling asleep on the ground beside the test stand.

To launch Chang'e, the old launch tower had to be dismantled to make way for a new one.

"At first, I was worried that many comrades had not participated in high-altitude work." Looking at the 74-meter-tall tower, Li Jianping, senior engineer of the general assembly special installation team, could hear his heart pounding loudly.

Therefore, he prioritized safety. After entering the site, the team did not immediately embark on the dismantle. Instead, he first personally organized officers and soldiers to visit the tower, introduced the structure of the tower to them, and secretly inspected who was suitable for high-altitude operations. Two days later, soldiers with the psychological quality of working at heights were selected. A week later, they were able to walk freely on the tower.

The dismantle was difficult. Most of the bolts, exposed too long on the tower, were so rusty that they had to be soaked with diesel oil before they could be unscrewed. For those unable to be unscrewed, gas cutting was needed. Some bolts were hidden in the groove of components. One bolt left behind would fail the dismantle. An attempt to lift the component regardless would face the danger of a broken wire rope.

Moreover, the dismantle took place in mid-air, meaning a component, once lifted, could not be placed back to where it was. The 74-meter, 1,100-ton tower had tens of thousands of steel components. Officers and soldiers had just checked all of them inch-by-inch and found all the connection points.

In only 56 days, led by Li Jianping, they finished dismantling the old tower. Then, they spent 382 consecutive days to build a perfect new ladder to the Moon for China.

Under a strong organization and management system, the design, development, production, and other systems of China's first lunar exploration project began to enter a fast-paced and efficient operation. In just three years, three strides have been made—

2004 was called the first year of the Moon circling project. In this year, there was the successive completion of the formulation of the overall technical scheme of the launch site system, the formulation of the overall design scheme of the measurement and control system, the formulation of the design scheme of the ground application system, the scheme design of the satellite system and the preliminary research of the launch vehicle system. And the cracking of key technical problems began.

The Moon is 380,000 kilometers away. How can we choose a precise road to it? Satellite orbit design was the first problem.

The problem would not be a problem as long as China had a large launch vehicle, such as Saturn 5 of the United States and Proton of the Soviet Union. However, China's large thrust rocket was still quietly on the drawings.

What next? "Let's throw it." Chinese scientists have independently and creatively set up satellite phase modulation orbit. First, the satellite flew and accelerated in the earth orbit, and then a proper timing was chosen to enable it to break free of the earth's gravity and go to the Moon dependent on its own thrust.

Chang'e-1 was the first satellite in the world to go to the Moon this new way.

2005 was the crucial year of the lunar exploration project. In this year, the focus was on the quality and reliability. The preliminary sample development and tests were comprehensively carried out, the interface coordination and technical breakthrough between systems were made, technical risk of the project was gradually solved, the interface relationship between systems was clarified, and technical status of each system was confirmed.

"Climbing Xiangshan Mountain and Everest are two different stories." In the process of developing the Chang'e-1 satellite, scientists were repeatedly frustrated.

In the past, Chinese satellites only involved two bodies' orientation: one was the sun, the other the earth. When Chang'e-1 was in lunar orbit for science exploration, it was necessary to aim the instrument at the Moon, the solar collector at the sun, and the antenna transmitting data at the earth. Three-body orientation tormented scientists. Eventually, an innovative UV lunar sensor gave the satellite the "eyes" to aim at the Moon.

To obtain the three-dimensional image of the Moon, three angles are needed. Generally, three cameras are necessary. But satellites have a strict requirement on weight. Designers created a new method. One camera and one Charged Coupled Device can get three images at the same time. In the detection of the minerals on the Moon's surface, different substances with different wave band reflections were a problem. If the dispersive principle of sunlight on the prism is applied, the spectrometer would be very complex. For the first time, scientists applied the principle of interference to retrieve the spectrum of matter, and consequently the spectrometer was small in size, light in weight, and high in resolution.

The temperature of the satellite side facing the sun was as high as 130°C and -180°C on the back. With the huge temperature difference, which part of it should be covered with a "cotton padded jacket?" Which part should only wear a "T-shirt?" Designers applied 32 "heat pipes" to realize the uniform heat conduction within the satellite.

Chang'e-1 satellite, a cube with a side length of 2 meters, embodies the wisdom of Chinese aerospace workers.

2006 was the decisive year of the lunar exploration project. In this year, Chang'e-1 completed its final assembly, test, and trial. At the same time, the general assembly and test of the Long March 3A rocket were completed, and so were the development and system construction of TT&C, launch site, and ground application system.

Long March 3A, which shouldered the responsibility of lifting Chang'e-1, is the pride of the glorious family of Long March rockets in China. It has been successfully launched 14 times. However, to ensure that Chang'e-1 was infallible, engineers and technicians made a lot of technical improvements and innovations, including the use of remote measurement and control technology, and the application of system level redundancy technology of control system.

In 2007, when the launch of Chang'e-1 was announced to the world, the world was amazed—

In three years, China was ready for the first lunar exploration. What an amazing speed. Internationally, in addition to the quick success of the two superpowers during the cold war, the lunar exploration projects in various countries generally take more than 10 years to implement. Japan launched the SELENE exploration satellite, which was initiated in 1991, in September 2007.

The statue of Hou Yi shooting the sun stands at the launch site. (Photographed by Qin Xian'an)

The development and construction of the entire project was at the expense of only 1.4 billion RMB, about the cost to build 14 kilometers of highway. This amount of money would only suffice to build 2 kilometers of subway. But we managed to build a 380,000-kilometer road to the Moon.

On October 24, China's gold-medal Long March 3A launched Chang'e-1 into space.

That day, Professor Yang Zhenning, a Nobel laureate, told the media in Hong Kong that this was an incredible achievement of China.

On the voyage to the Moon, developed countries have had spacecraft launched one after another. There were Apollo in the United States, Lunar in the Soviet Union, SMART 01 in ESA, and SELENE in Japan. Why did Yang Zhenning describe Chang'e as an incredible success?

He explained that China has caught up with the science achievements accumulated in Europe for 400 years in just 30 years and achieved remarkable results. It took Chinese scientists 3 years to lay the road to Moon exploration, which took 40 years in developed countries. The progress was amazing.

The two sets of data, 30:400 and 3:40, were a wonderful coincidence. The growth ring of China's space industry moving towards the deep universe coincides with the road of independent innovation with Chinese characteristics.

First Contact with the Moon

On October 24, 2007, Chang'e ascending to the Moon, an ancient myth, came true.

This day, I witnessed the magnificent launch of Chang'e-1, China's first lunar exploration satellite, in the Xichang Satellite Launch Center.

At 16:05, the revolving platform of the rocket launch tower slowly opened, and Chang'e-1 lunar exploration satellite stood high on the top of Long March 3A. In the sunset, it was as if a girl about to marry would lift the red bridal veil and show her charming face.

The words Chinese Lunar Exploration on the rocket's fairing were particularly conspicuous from afar. On top of that was drawn the symbol of the project, a crescent Moon, which, shaped like the Chinese character Moon, also resembled a soaring Chinese dragon. Two horizontal strokes of the character were written in the shape of two footprints of human beings. The dragon tail looked like a row of doves flying in peace, expressing the beautiful yearning of Chinese astronauts to go to the universe and make peaceful use of outer space.

Excitement was brewing in the audience, and from time to time they gave out bursts of cheers. For the first time, tourists were allowed to enter the launch site to watch the Chang'e take off, and over 800 tourists were lucky enough to get tickets. However, there were a large number of tourists who went to Xichang. In the mountains and canyons, countless heads were looking up at the sky together.

A thriving China lifted Chang'e. At this moment, countless people were sitting in front of the television, watching the launch in silence with great anticipation.

At this moment, the aerospace workers at Xichang Satellite Launch Center could not have been more nervous. Due to the location of the earth and Moon, it had to be on time for Chang'e to take off. The window where launch conditions were met had only 35 minutes. Once missed, the launch must be cancelled, and the next could not happen until at least one month later.

"Attention, one hour away." At 17:05, the voice of Number Zero commander Li Benqi came from the loudspeaker.

"Thirty minutes!" Low-temperature fuel was slowly injected into the chest of the third stage rocket through an artery-shaped pipeline.

"One minute" The rocket's swing bars were all quickly opened, the magnificent "dragon" stood tall, and the Daliang Mountain Canyon held its breath.

"Ten, nine, eight, seven, six..." Tens of thousands of viewers could not help but count the countdown in one voice, "5, 4, 3, 2, 1, ignition!" The sound of the commander came from the loudspeaker. Pi Shuibing, the operator of the control panel, resolutely pressed the red ignition button. A rumble echoed through the valley. At 18:05, roaring, the nearly 300 tons of giant "dragon" slowly left the launch base with a long tangerine flame from its tail and flew straight into space.

This moment belonged to China.

The successful launch of Chang'e-1 has inspired the Chinese and left the world in awe. The Hong Kong Newspaper *Wenhui* published an editorial. Chang'e's Moon flight was another major breakthrough in space science and technology in China subsequent to the successful breakthrough in manned space technology of Shenzhou 5 and 6. It was a landmark project to achieve the first breakthrough in deep space exploration. Lunar exploration technology was not only necessary for human beings to expand their living space, but also a significant symbol to measure national wisdom and comprehensive national strength, as well as an important factor to determine the international status of a country.

To keep enough power supply for Chang'e to fly to the Moon, researchers

At 18:05 on October 24, 2007, Chang'e-1 satellite was successfully launched at Xichang Satellite Launch Center. (Photographed by Qin Xian'an)

installed 13 engines on the satellite. Meanwhile, 1,200 kilograms of fuel were carried on the satellite, which weighed a total of 2,350 kilograms.

On October 31, 2007, seven days after flying around the earth, Chang'e-1, China's first lunar exploration satellite, began the real journey to the Moon. It embarked on the road to the Moon that scientists carefully designed, the Earth-Moon transfer orbit. Thereon, Chang'e-1 was creating a new height of China's space flight every second.

At 10:40 on November 5, 2007, Chang'e-1, China's first lunar exploration satellite, traversed 1.87 million kilometers to and around the Moon.

At that moment, the satellite was operating at 2.4 kilometers per second, 10 times the speed of a jetliner. Without a brake and slow-down in time, it would pass by the Moon. With excessive braking, it would hit the Moon head on.

Success was at stake. So far, 120 lunar exploration activities have been carried out at a success rate of less than 50%. The first lunar exploration satellites of the United States and the Soviet Union both failed at the last minute.

Commander in chief Luan Enjie and chief designer Sun Jiadong observed the operation of Chang'e-1. (Photographed by Qin Xian'an)

At this moment, the world was watching whether Chang'e-1 could make it?

At 10:40:37, a spectacular scene appeared on the large screen of Beijing Aerospace Flight Control Center: in the dark space, two orange sparks burst out on one side of the satellite, and the satellite attitude adjustment engine ignited on time.

In an instant, Chang'e-1, weighing 2,350 kg, lifted the solar boards, and rotated 180 degrees. It was a beautiful space ballet. Chang'e was ready to brake. However, it was still flying at a high speed.

At 11:15, Ou Yujun, a software expert at the Beijing Flight Control Center, pressed the brake command key. A series of commands, like invisible reins, flew from the heart of the motherland to the satellite 380,000 kilometers away.

Immediately, the main engine of the satellite spewed out flames to the direction the satellite was headed. The speed decreased while the gravity of the Moon increased. The Moon that has shone upon China for centuries opened its arms to embrace Chang'e for the first time. On the big screen, Chang'e-1 pulled a green track, saying goodbye to the large elliptical orbit extending 380,000 kilometers from the earth, and like a curled ivy, approached the silver white Moon.

"Chang'e-1 has successfully entered orbit." Applause roared, and the researchers in the flight control hall of the Beijing Aerospace City exchanged hugs and congratulations with tears sparkling in their eyes.

China's Chang'e successfully held hands with the Moon. On November 5, China's National Space Administration officially announced that the Chang'e-1 satellite had successfully arrived at the Moon. It entered a polar elliptical orbit with a cycle of 12 hours at 210 kilometers away from the perilune and 8,600 kilometers away from the apolune and began to circle the Moon, becoming the first Chinese lunar satellite.

This was the first time a Chinese spacecraft orbited around a celestial body other than the earth. This was the first time that this country with countless romantic legends had made a footprint in the remote Moon kingdom.

"Once we get to the Moon, China's space industry will aim at the next goal: deep space." said Luan Enjie, general director of the lunar exploration project. From the launch on October 24 to the entry to the Moon, Chang'e-1 flew 11 days, 17 hours and 10 minutes. It took Chinese scientists three years, nine months and 13 days from the lunar exploration project initiation. It took China six years and 11 months from when the white paper of China's space industry first revealed the idea of lunar exploration.

"We owe the success to the hard and meticulous preparation and efforts of the vast number of science and technical personnel over the years. There nearly 200 units with more than 17,000 personnel directly involved in the five systems of the project, and countless units and individuals indirectly provided support. Although they don't know one another, they shared the same goal." Luan Enjie told me that this project was like preparing for the final exam. He was glad to say that we passed it with flying colors while none of the 84 contingency plans were activated. After over three years, the tense nerves were finally able to have a relief. The project did not disgrace the country. Instead, it made the country proud.

Chang'e-1

Chang'e-1's arrival at the Moon for the mission was only half of the success.

Next, in this circular orbit 200 kilometers away from the surface of the Moon, Chang'e-1 worked relentlessly for a year, constantly surprising the Chinese people:

On November 2, 2007, China set up a special committee of 122 experts on the science application of lunar exploration engineering, including eight from Hong Kong and Macao to maximize the use of Chang'e-1 in data collection.

On November 19, Chang'e-1 gradually activated science detection instruments and carried out an in-orbit payload test after 135 laps around the Moon.

On November 20, the CCD camera on the satellite started to work. On November 26, a historic day, the first lunar image of China's first lunar exploration project was officially released.

The CPC Central Committee, the State Council, and the Central Military Commission delivered congratulatory messages: the success of China's first lunar exploration project was a landmark achievement of China's independent innovation in science and technology, and another significant leap of the Chinese nation in the climb to the peak of science and technology.

NASA Official Michael Brooks told Xinhua News reporters that the success of China's lunar exploration project was a very important achievement, and scientists around the world showed great interest in the new data obtained by Chang'e-1 from the Moon.

The exploration picked up the pace:

On November 27, the interference imaging spectrometer, the solar high-energy particle detector and the solar wind ion detector on Chang'e-1 satellite were activated one after another. The spectrometer was used to detect the mineral composition and distribution on the Moon, while the two detectors were mainly to detect the space environment around the Moon.

On November 28, the laser altimeter on Chang'e-1 was officially activated, sending out the first laser and receiving the first echo. The combination of laser altimeter with CCD stereo camera obtained three-dimensional images and terrain height data of the Moon's surface.

On January 31, 2008, the first image of the Moon's polar region taken by Chang'e-1 was released.

At 15:05 on November 12, 2008, the first panorama of the Moon in China, which was made on the basis of the data obtained by Chang'e-1, China's first lunar exploration satellite, made the official appearance.

This real panorama covered all the territory of the mysterious Moon. 589 frames of images taken by Chang'e-1 contributed to the completion of the panorama, covering the range of 180 degrees of the Moon's west longitude to 180 degrees of its east longitude and 90 degrees of its north and south latitude.

"Chang'e-1 has looked at every inch of the Moon and transmitted data back in full. The completed Moon panorama was the result of radiometric calibration, geometric calibration, and photometric calibration of the 589 frames of images selected. It was clear and richly layered. Experts of the evaluation group believed

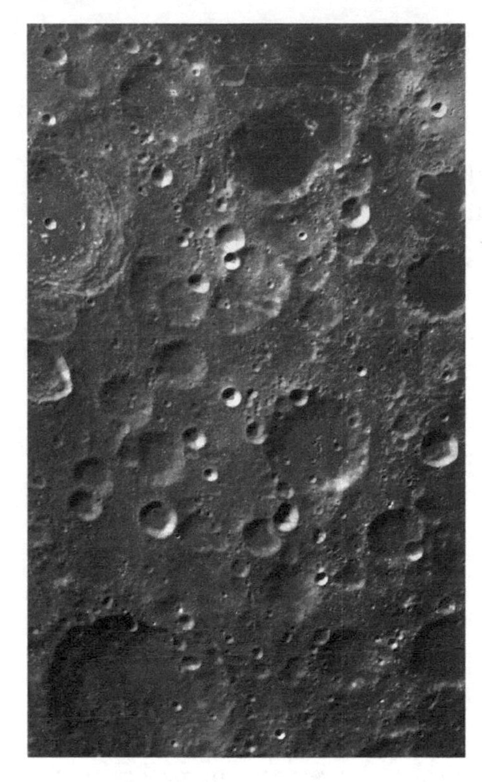

On November 26, 2017, the first Moon picture by Chang'e-1 was released. (Photographed by Qin Xian'an)

that it was internationally advanced," said the head of the ground application system of the lunar exploration project.

"A whole year of flight and exploration has allowed Chang'e-1 to obtain 1.37TB of effective science data, thus successfully completing the science exploration mission," said Chen Qiufa, head of the leading group of China's lunar exploration project and director of the State Administration of Science, Technology, and Industry for National Defense.

Images are silent yet eternal. This is the first panorama of the Moon presented to the world by the nation with countless romantic legends of the Moon and firmly marching into the deep space. It has been also the most complete picture of the Moon published so far.

At 16:13:10 on March 1, 2009, the precise control of science and technical personnel had Chang'e-1 descend into the designated area of the Moon's surface with 52.36 degrees east longitude and 1.50 degrees south latitude. A hard landing was pulled off. China's first lunar explorer wrote the last touch of splendor in life at the moment of impact on the Moon's surface. "The successful collision of Chang'e-1 with the Moon has laid a foundation for the soft landing of China's space probe in the future," said Wang Sichao, a researcher at Zijinshan Observatory, Chinese Academy of Sciences.

Three years later, the national Project 863 and Project Application and Research of Science Data of Lunar Exploration Engineering passed the acceptance on September 28, 2012. In the coming of the Mid-Autumn Festival, Chinese scientists brewed the sweet "osmanthus wine" – a large number of technological innovations and a series of scientific achievements made through two years of hard work to study the data by Chang'e-1.

"It was not until now that I dare to call Chang'e-1 a complete success. The greatest wish of the researchers is that the data obtained by the lunar exploration satellite would play a role in the science research on the Moon," said Ouyang Ziyuan, academician of the Chinese Academy of Sciences and senior consultant of the leading group of the Chinese lunar exploration project.

Chang'e-1 has created many firsts in China's aerospace history:

It was the first to explore the Moon; the first to break through the earth's low earth orbit; the first to photograph the Moon by obtaining three-dimensional images with a stereo camera; the first to explore the elements of the Moon's surface; the first time to explore the thickness and distribution of the lunar soil with a microwave radiometer; the first to introduce astronomical measurement approaches in the measurement and control of the spacecraft; and the first to use the international network to conduct a tighter control over the spacecraft.

"Chinese astronauts' journey to space announced that China has become a major aerospace country while the success of Chang'e-1 project has opened a new era of China's aerospace industry," said Georgy Polischuk, chief designer of Russia's lunar exploration project, to Xinhua News Agency when congratulating Chang'e-1's successful transmission of Moon images.

Chang'e-2: Faster, Closer, and Clearer

In another three years, Chinese aerospace workers had taken big leap again. On September 30, 2010, the Long March 3C carrier rocket, holding up the Chang'e-2 satellite, was ready for the journey on the launch tower of the Xichang Satellite Launch Center.

"Chang'e-2, transformed from a backup satellite as it was, has a mission that is not a simple repetition of Chang'e-1," said Chen Qiufa, who succeeded Luan Enjie as the commander-in-chief of China's lunar exploration project, "According to the three-step plan of China's lunar exploration project, it is actually a bridge and link between "circling the Moon" and "falling on the Moon." The purpose is to carry out key technical verification for the realization of the soft Moon landing of Chang'e-3 in the future."

In February 2008, the second phase of the lunar exploration project aiming at "falling on the Moon" was officially approved. At the time, the Chang'e-1 command team, once known as the troika: General Director Luan Enjie, chief designer Sun Jiadong, and chief scientist of lunar application science Ouyang Ziyuan, three academicians, took the initiative to retire to the second tier. Young Chen Qiufa, Wu Weiren, and Yan Jun took over the baton of the lunar exploration project from the older generation of aerospace experts.

As pioneer of the second phase of the lunar exploration project, Chang'e-2 was required to verify six key technologies, including direct orbit entry, Moon acquisition, X-band measurement and control, orbit maneuver, data transmission and high-resolution imaging. At the same time, it was required to conduct more accurate exploration of the alternative landing area of Chang'e-3 to prepare for its safe landing.

Wu Weiren, chief designer of the lunar exploration project, summed up that compared with Chang'e-1, Chang'e-2 had new requirements: fly faster, arrive at the destination in 5 days, a week earlier than Chang'e-1; get closer, work in an orbit 100 kilometers away from the Moon, 100 kilometers closer than Chang'e-1; shoot more accurately, descend into an orbit of 100 × 15 kilometers to perform high resolution imaging test on Bay of Rainbows, an alternative landing area.

Who was paving the way of Chang'e series to space? In the past three years, Chinese aerospace workers independently innovated a way to the Moon with unique Chinese characteristics.

The first hurdle Chang'e-2 faced was that the carrier rocket would send the satellite directly into the Earth-Moon transfer orbit, enabling a straight, faster way to the Moon. Li Dan, chief designer of the Long March 3C, and his colleagues successively consulted and studied hundreds of documents, deduced dozens of pages of trajectory design models, programmed nearly 10,000 software and thousands of trajectory simulation calculation and optimization designs, and finally accurately figured out a new Moon orbit.

The GNC subsystem of Chang'e-2 was the key to the satellite attitude control and on orbit stability. In an electrical survey of Chang'e-2, one cable communication of the UV sensor was abnormal. To thoroughly solve the problem, Wang Xiaolei, deputy chief engineer of the GNC subsystem and relevant personnel devoted themselves to exclusion tests involving hundreds of possibilities in laboratories.

The first combination did not lead to the fault. Next. The second combination either. Next. Finally, the fault was eliminated. Excited, they realized that it had been days in the laboratory.

To ensure total safety, 111 contingency plans were formulated for the satellite and rocket system; the measurement and control system prepared 108 emergency plans and formulated 153 troubleshooting countermeasures; the key equipment to be focused on was sorted out for the launch site system, and contingency plans were improved to ensure that each piece of equipment was in normal operation.

In those three years, Chinese aerospace workers strived to make the most of every second and paved a safe and fast road to space.

Chang'e-2 had a pair of sharp eyes: a high-resolution CCD stereo camera. To deliver the camera as soon as possible, Gao Wei, deputy director of the Chinese Academy of Sciences, worked tirelessly. When his son had a high fever, Gao did not make time to take care of him but fought with everyone in the lab. All-nighters and overload of work led to premature beatings of his heart. But to ensure the smooth progress of the task, Gao stayed on the frontline of production, regardless of illness.

Liu Yong, a young expert at Beijing flight control center, returned home from work late every night. He would check his daughter, who was already asleep. To launch Chang'e-2 successfully, working overtime was common for Liu and his colleagues. Sometimes he sat in front of the computer searching for answers all day until the problem was solved. "We are young. We have more to give." Liu, who participated in the Chang'e mission for the second time, often encouraged himself this way.

As a huge and complex system engineering, Chang'e-2 truly interpreted the meaning of "unity is strength." In the Chang'e-2 rocket and satellite system, female designers took up 40%. These women silently escorted the journey of Chang'e.

"When the task is finished, I will go home, get together with my family and spend time with my little girl." Before the launch of Chang'e-2, Yan Liqing, commander of Xichang Satellite Launch Center 01, talked about his expectations. It was almost a rare luxury for the aerospace expert to enjoy some family time who was occupied with missions all the year round.

For the success of Chang'e-2 lunar exploration, countless Chinese aerospace workers have sacrificed rest, recreation, and family life, as well as youth, sweat, and health as Yan has.

Three years witnessed the hardship and fruit. The world had the eyes fixed on Chang'e from China flying to the Moon again.

Chang'e-2: New China Height

The night fell on October 1, 2010,

In Tian'anmen Square, in shining neon and blooming flowers, Chinese people celebrated the 61st birthday of the country.

Meanwhile, the atmosphere was tense in the flight control hall of Beijing Aerospace Flight Control Center. Hundreds of researchers in blue anti-static gowns shuttled back and forth among the computers.

The soaring launch tower held the beige Long March 3C carrier rocket at Xichang Satellite Launch Site thousands of miles away. At the top of the rocket, the Chinese lunar exploration project logo, composed of a bright Moon and a pair of footprints, was particularly eye-catching.

Three years ago, Chang'e-1 took off from here. The thousand-year-old dream of Moon travel among Chinese people became a reality. Three years later, Chang'e-2 embarked on the journey with a new agenda: to capture high-definition three-dimensional images of Hong Wan, the designated landing area of Chang'e-3.

"Ten, nine, eight..." The final countdown came from the loudspeaker. Everyone in the hall held their breath.

"Ignition!" "Take off!" At 18:59:57, the rocket, spitting tangerine flames, lifted up Chang'e-2 from the Xichang Launch Site, and flew into the vastness of space. The long tail flame left a brilliant track across the night sky.

People on the ground took no break while Chang'e-2 was flying up there. At that moment, the sound of keyboard striking in terminal room at the back of flight control hall was relentless. Technicians were fully engaged in the processing of satellite telemetry data and orbit data. They closely monitored the flight status of the satellite-rocket complex. "To lead Chang'e-2 to the Moon, some colleagues haven't closed their eyes for two days," said Ma Yongping, deputy commander of the flight control center.

However, they were not the only busy people. The flight of Chang'e to the Moon took great efforts, every ounce of which came from the scientists and technicians.

At 19:35, the large screen read that Chang'e-2 entered the Earth-Moon transfer orbit with a perigee altitude of over 200 kilometers and an orbit inclination of 28.5 degrees. Hereafter, the satellite began the 112-hour journey to the Moon.

Tonight, the aerospace workers lit the most splendid fireworks in the night sky to celebrate the motherland's birthday.

China has launched a second lunar probe; another step toward China's ambitious plan to become the second country successful in manned Moon landing, AFP reported. The launch of Chang'e-2 on October 1 was symbolic as it is China's national day.

At 11:06 on October 7, 2010, the precise control of the ground scientists and technicians enabled the high-speed Chang'e-2 to achieve a perfect "brake" in space, thus becoming the second Moon satellite in China.

At this very moment, Wu Weiren, chief designer of China's lunar exploration project, aged 57, wept in front of live cameras; in the spotlight of the media, young faces were cheering.

At 18:59:57 on October 1, 2010, Chang'e-2 lunar exploration satellite was successfully launched in the Xichang Satellite Launch Center by Long March 3C, which was the 131st flight of the Long March series rockets. (Photographed by Qin Xian'an)

At this very moment, not far behind, academician Sun Jiadong, 82-year-old chief designer of Chang'e-1 and distinguished scientist of atomic bombs, hydrogen bombs and man-made satellites in China, sat quietly enjoying every bit of it.

Ordinary as the moment seemed, it was of great significance. A kind of inheritance happened in silence: a new generation of astronauts took the baton of China's aerospace industry.

The more important a thing, the more silent it gets. The elder's eyes were filled with joy, praise, and expectation; these young people were becoming the next Qian Xuesen and Sun Jiadong. They are the future of China's aerospace industry.

From this moment on, the stage belongs to the young generation. Energetically, China's aerospace industry will continue to march forward with the speed of New China.

The great era has chosen the young generation. The great cause has trained the young generation.

I witnessed that Han Liangbin, the 28-year-old chief controller, calm and confident, gave nearly 400 dispatch orders in just 18 minutes in the Beijing aerospace flight control hall.

Looking at him giving orders, people rarely knew:

In 1982, when China successfully launched the first launch vehicle underwater, he had just been born; in 1991, when China first proposed to build a Moon satellite, he was a second grader; in 2004, when China first set up the lunar exploration project, he was just a young man who graduated from college.

Whereas, positioned as the general controller of flight, he has completed the work nearly perfectly in the first project.

Han's young resume is a specimen of the rise of young talents in China's aerospace industry in recent years. Meng Linzhi, chief designer of satellite system, is 30 years old; Jiang Mengzhe, telemetry software designer of the TT&C system, is 27; Hao Jun, chief operator of launch site system, is 28.

The 380,000-kilometer road to the Moon is a beautiful Concerto of youth. From development and production to launch test, Chang'e-2 has involved five systems, thousands of units, and tens of thousands of test personnel. Experts born in the 70s are the main technical force, and those born in the 80s pillared the front-line operation. Qian Weiping, chief designer of the measurement and control system, told reporters: "The average age of our chief designers is only 32.5, about 15 years younger than those in Europe and the United States."

Fifteen years. What does it mean?

A look into the dusty history of the PRC will help us have a deeper understanding.

In the late 1950s, when the Soviet experts who came to offer technical support stepped off the plane at the Jiuquan base, the base leader, who had been waiting in the cold wind for long, was taken by surprise: "You are so young?"

At that time, the leaders were all generals with great achievements in the war. However, they were nobody in the aerospace battlefield. What an embarrassing image of a country lacking high-tech talents!

Time flies.

50 years later, there was dramatic irony when our aerospace science and technology delegation went to Russia for exchange. Mostly the elderly with grey sideburns made up the Russian delegation, while China presented all young faces. Russian experts exclaimed, "I envy that you have such a young team!"

In years of vicissitudes, history says that with the world's youngest aerospace talent promises China a better tomorrow for the aerospace industry.

At 21:45 on October 26, 2010, Chang'e-2 successfully completed the orbit reduction control and entered the 15 km × 100 km lunar Bay of Rainbows imaging orbit.

At about 10:00 on November 8, 2010, the regional image of Bay of Rainbows taken by Chang'e-2 was unveiled.

The time of imaging was 18:25 on October 28. The satellite was about 18.7 kilometers away from the Moon and the resolution was about 1.3 meters. The central position of the image was 31° 3 west longitude and 43° 4 north latitude, corresponding to the 8 km east-west width of the Moon and about 15.9 km north-south length. The image showed that the surface of the area was relatively flat, covered by basaltic lunar soil, with craters and stones of different sizes, the largest of which was about 2 km in diameter.

The Bay of Rainbows, a romantic and poetic name for a place, lies 380,000 kilometers away from us.

Never have so many eyes been focused on this remote land. And never has there been so much to expect on this isolated land.

This was because the Bay of Rainbows was about to become where the Chinese dream takes off; here is the preferred landing area for Chinese lunar exploration.

Once upon a time, we gazed at it with our naked eyes. The Bay of Rainbows was a small spot in the shadow of the Moon. Today, it is so clearly presented in front of us – resolution of 1.3 meters of high-definition picture hanging on the wall. The 1.3-meter resolution means that every "black mole" on the charming face of the Bay of Rainbows is clearly seen.

There's a good reason to the choice of the Bay of Rainbows:

Wu Weiren, chief designer of China lunar exploration project, explained that the geological structure in the area is complex, typical and of high scientific exploration value. For the sake of science, the Bay of Rainbows was selected.

Ouyang Ziyuan, senior consultant of China lunar exploration project, said that the Bay of Rainbows area is relatively flat so the detector landing would be relatively safe; the lighting is good so the energy supply is guaranteed; and it is located on the side facing the earth, which is convenient for operation control management and science data transmission for the ground.

What's more, the Bay of Rainbows is a virgin land that human beings have not yet exploited. The exploration of this land would surprise us with more discoveries beyond expectation.

The transmission of the partial image of the Moon's Bay of Rainbows marks the success of Chang'e-2 mission.

From April 1, 2011, Chang'e-2 safely operated for 180 days, reaching half a year's design life, and fully achieved the established engineering objectives and exploration tasks.

To give full play to the role of the satellite, Chang'e-2, which served beyond the expected term, flew to deeper space.

Since June 9, 2011 it was pulled away from the Moon, Chang'e-2 satellite successfully entered the Lagrangian L2 orbit from the Moon orbit for the first time in the world after a 77-day-long journey on August 25, 2011. On September 21, the first batch of precious deep space data was sent back, 1.72 million meters away from the earth.

"This is the first time that China has explored about 1.7 million kilometers away from earth, which is equivalent to over 200 earth radiuses. The exploration data is precious to China's space science," said Sun Huixian, deputy chief designer of lunar exploration project.

On February 6, 2012, the State Administration of Science, Technology, and Industry for National Defense released the 7-meter resolution panorama of the Moon obtained by Chang'e-2 lunar probe, which was another major scientific and technological achievement of China's lunar exploration project.

"At present, no other country in the world than China has released a panorama of the Moon with a resolution better than 7 meters and 100% coverage of the entire Moon surface." said Liu Dongkui, deputy commander-in-chief of the lunar exploration project. "The resolution of the image by Chang'e-2 was 17 times better than the 120-meter resolution of Chang'e-1." For example, while Chang'e-1 can see airports and ports, Chang'e-2 can discover airplanes and ships there."

On July 15, 2013, the distance between Chang'e-2, which had become China's first man-made solar system asteroid, and the earth exceeded 50 million kilometers, once again making a new China height.

Since Chang'e-2 flew over and explored the Toutatis asteroid and successfully carried out the re-expansion test on December 13, 2012, it had covered the longest flight distance of China's spacecraft, continuously refreshing the China height record. The distance between the satellite and the earth exceeded 10 million kilometers on January 5, 2013, 20 million on February 28, 30 million on April 11, 40 million on May 24, and 50 million on July 14.

At present, the satellite, in good condition, is continuing to probe further into deep space. According to calculations by the Beijing Aerospace Flight Control Center, Chang'e-2 is expected to fly as far as 300 million kilometers from the earth.

"Unlike Chang'e-1, which collided with the Moon, Chang'e-2 would work there until the last moment," said academician Ouyang Ziyuan.

Chang'e-3

It is a long road to space, and a long dream for China.

The sky in 2013 was destined to leave more unforgettable memories for the Chinese.

In the early summer of 2013, Shenzhou 10 docked with Tiangong-1 twice; China's first space lecture won the world's acclaim. In early winter, Chang'e-3 would implement China's first soft Moon landing, which would be another new start for China's deep space exploration.

The success of Chang'e-1 and Chang'e-2 missions benefited China's lunar exploration with more confidence and calm:

In March 2012, the State Administration of Science, Technology, and Industry of National Defense announced that the Chang'e-3 probe should plan to be launched in 2013. It would carry out China's first soft Moon landing and lunar surface patrol survey.

The Chang'e-3 prototype was successfully developed, the flight products were basically prepared, and the launch vehicle system was completely assembled. Since then, progress for Chang'e-3 has been made public on newspapers.

On September 25, 2013, Wu Weiren, chief designer of China's lunar exploration project, announced that China's first lunar rover carried by Chang'e-3 would have a global poll for the name. Two months later, the poll ended after 650,000 opinions were collected. And the final decision was it would be named Yutu ("jade rabbit").

Li Benzheng, deputy commander-in-chief of China's lunar exploration project, said that in Chinese myths and legends, Chang'e flew to the Moon in the companion of a rabbit. The kind, pure, and agile image of the rabbit corresponds to the structure and mission of the lunar rover, which reflects the position of China's peaceful use of space.

The world's first unmanned lunar rover was launched by the Soviet Union on November 17, 1970. After landing in the designated area on the lunar surface, having traveled 10.5 km, it carried out ten and a half months of scientific exploration and inspected 80,000 square meters of the lunar surface, until the power supply was exhausted. And it stopped working on October 4, 1971.

The first manned lunar rover in the world was sent to the Sea of Rains area of the Moon by Apollo 15 on July 31, 1971. Astronauts Dave Scott and Jim Irwin were the first to steer a lunar rover in space history. Before returning to the lunar module, two astronauts drove the 4-wheel rover 27.9 kilometers across the rugged surface of the Moon, collecting 77 kilograms of lunar rock samples.

A close look at the Yutu, China's first lunar rover enables a better sight of her "golden armor" and "wind fire wheel." Wei Ran, deputy chief designer, said that the purpose of the "golden armor" was not a good appearance, but to reflect the strong light of the Moon in the daytime, reduce the temperature difference between day and night, and block the radiation of various high-energy particles in the universe. All was to support and protect the 10 sets of precious science exploration instruments, such as infrared imaging spectrometer and laser lattice device.

In addition, it was also equipped with a long "ear," (the Earth-Moon communication antenna); with four sharp "eyes" – "(overhead navigation camera and front and back obstacle avoidance camera); an agile "arm" (a mechanical arm responsible for drilling, grinding, and sampling).

Regarding technical difficulty, Yutu was the most difficult "vehicle" in the history of China, as almost every link involved a technical innovation. In terms of manufacturing technology, it was the most rigorous "vehicle" in Chinese history because every component was greatly polished. An overweight of 5 g must be reported to the chief designer for approval.

The rover was built from scratch. In the past decade, a large number of researchers have made great efforts to complete this masterpiece, which consisted of eight subsystems, i.e. mobile, power, thermal control, etc., on a lunar rover capable of withstanding extreme environments such as surface vacuum, intense radiation, and high temperature differences.

For example, to adapt to the harsh conditions of the Moon, Yutu had a high requirement for wheels. In this case alone, researchers racked their brains out. In the development, they have come up with dozens of schemes, such as four wheels, six wheels, eight wheels, and track-wheel. Finally, the scheme of six-wheel independent drive with four-wheel independent steering made it as the perfect combination of high reliability and low weight.

Wu Weiren, chief designer of China's lunar exploration project, said that Yutu was an all-Chinese product.

The design life of Yutu was three months, meaning that it was planned to endure three months of lunar days and nights. All work should be done in the morning and afternoon of lunar time. It was introduced that a lunar radar was installed at the bottom of Yutu, which could launch radar waves to detect the Moon's soil structure as thick as 20–30 meters, and the places as deep as 100 meters. "This is what the rest of the world has never done" said academician Ouyang Ziyuan.

On November 26, 2013, Wu Zhijian, spokesman of the State Administration of Science, Technology, and Industry of National Defense, claimed that Chang'e-3 was by far one of the most complex and difficult missions in China's aerospace industry. The implementation of the project involved seven barriers to break, including multi-window narrow-width on-time launch, soft landing on the Moon, separation of the two devices, remote operation between the Moon and the earth, Moon survival, measurement and control communication, and ground test verification.

This time, Chang'e-3 shouldered the duty of the first soft landing on the Moon and the inspection and survey of the lunar surface. From a stare at the face of the Moon to close contact, Chang'e-3 landing on the Moon was a qualitative leap for China's lunar exploration project.

"Chang'e 1 and 2 have verified the technology of circling the Moon, but there was no experience to land on the Moon from 15 km high." Ye Peijian, academician of the Chinese Academy of Sciences and chief scientist of the Chang'e-3 detector system, explained that Chang'e-3 would first search for a landing point and then hover for some time like a helicopter at 100 m to determine the final landing point due to the uneven surface of the Moon.

From 15 kilometers to the surface of the Moon, the whole process was about 12 minutes.

The breathtaking 12 minutes was called the "Black 12 minutes." Chang'e-3 would completely rely on its own control to complete altitude reduction, ranging,

At 1:30 a.m. on December 2, China successfully launched Chang'e-3 with the Long March-3B at Xichang Satellite Launch Center. (Photographed by Qin Xian'an)

speed measurement, landing site selection, and free fall landing.

No human intervention was possible during the landing. This is because every action in the process was very short. Once the sensor installed on the detector receives the information, an immediate response must be made. The 380,000 km distance between the earth and the Moon together with signal delay kept the information from reaching the ground on time and the corresponding instructions from the ground reaching there.

Ye Peijian elaborated that to achieve a soft landing while preventing the detector from colliding with the Moon surface, it was necessary to lift it and deactivate the engine when it was close to the Moon surface. So, raising Moon dust and polluting the camera lens or affecting the work of other equipment would be avoided. And to reduce the impact, each leg of the lander was designed to absorb the impact, and each foot equipped with a large "insole."

Space experts introduced that in addition to the probe's soft landing on the Moon and the rover's lunar surface exploration, Chang'e-3 would also set a number of records. It was the first to have put an "observatory" on the Moon, and the first to have looked back from the Moon to the earth's plasma layer.

In this cold winter, the crescent Moon ignited a fire within the Chinese.

At 1:30 on December 2, 2013, tangerine flames billowed from the bottom of the rocket. Chang'e-3, carrying China's dream of exploring the Moon, set out with Yutu to the Moon 380,000 kilometers away.

This was the 186th flight of the Long March rockets, the 130th lunar exploration in human history, and the third launch of lunar exploration satellites in China.

An article published on the evening of December first on the French News Radio pointed out that the take-off of China's Yutu lunar rover would be followed up by the exploration of the Moon's Bay of Rainbows, an area none set foot in during the next few months. China was going through a crucial stage in realizing the ambition of conquering space.

Sorae JP, Japan's largest space development website, published an article reading that Chang'e-3, the third lunar probe launched by China, undertook China's first lunar landing mission. It was ready to launch an unmanned lunar rover to challenge the lunar walk. Should the soft landing succeed, it would be another feat since the Soviet "Luna 24" probe which landed on the Moon in 1976.

An article on UOL, a Brazilian national website, predicted that by the 2020s, Chinese astronauts would be seen walking on the Moon, or even staying there for a long time.

Chang'e and Yutu

"Someone asked you to pay attention to a fair maiden with a rabbit on the Moon. In an old legend, a Chinese beauty named Chang'e has lived there for 4,000 years. You can also seek her companion, a big Chinese rabbit, which should be easy to find because it always stands under the laurel tree."

"Well, we'll definitely pay close attention to the rabbit girl."

This was an interesting conversation between Apollo 11 and the ground command center in 1969. The recording, which was recorded on the day of the first human landing on the Moon and stored on NASA's website, went viral on the Internet because of China's Chang'e-3 mission to the Moon.

In 1976, the Soviet Luna 24 performed the world's last soft landing on the Moon. After 37 years, it was exciting for another to happen. Chang'e took the Chinese "rabbit" up to the Moon:

At 21:11 on December 15, 2013, Chang'e-3 successfully landed in the Bay of Rainbows, a pre-selected landing area on the Moon's surface, marking that China has become the third country in the world to achieve soft landing of extraterrestrial objects. The 130th Moon exploration of mankind starred China.

After the landing, the Yutu lunar rover was detached from the satellite and took the first step on the Moon.

From as far as 380,000 kilometers away, who would guide the way for Yutu? The answer is the young researchers in the teleoperation hall on the top floor of the Beijing aerospace flight control center.

In the late night of December 15, the remote-control operation hall, with an area of more than 200 square meters, began the busy work to the applause for the successful landing of Chang'e-3. Keyboard strikes echoed. They were about to perform a gripping "puppet show" in space: the first remote control operation of China's spacecraft in extraterrestrial motion.

"Should the general space TT&C be compared to steering a rail locomotive, remote operation is driving an off-road vehicle on unknown terrain." Standing in front of the huge arc-shaped screen in the hall, Song Jun, deputy leader of the flight control team, was both excited and nervous. Despite his participation in all the measurement and control tasks of manned spaceflight and lunar exploration projects, the remote operation was unprecedented because of the challenges brought by real-time and large amount of information exchange.

Guiding Yutu to move on the Moon was a highly difficult space drive involving hundreds of operators. I have noticed that in the teleoperation hall, there were nearly 10 kinds of work positions, such as command and dispatching, terrain establishment, visual positioning, path planning, and over 40 employees work shifts 24/7. Song Jun pointed to the large crowd of operators, telling me that the lunar surface environment is complex and unpredictable. Slope, light, and obstacles would affect the lunar rover's progress. Before starting the lunar rover, the movement path of all movable parts on the rover must be carefully planned and control instructions be written and sent.

Certainly, the driving team was not limited to this hall. There were more of them; at this moment, in the mountains, forests, and seas.

At 4:35 a.m. on December 15, 2013, Yutu left Chang'e and set foot on the Moon for the first time. It went on and off, observing along the way. Every step was creating a new history of China's aerospace industry.

Chang'e successfully landed on the Moon. Yutu embarked on the Moon walk. Both were fascinating enough to receive waves of praise overseas:

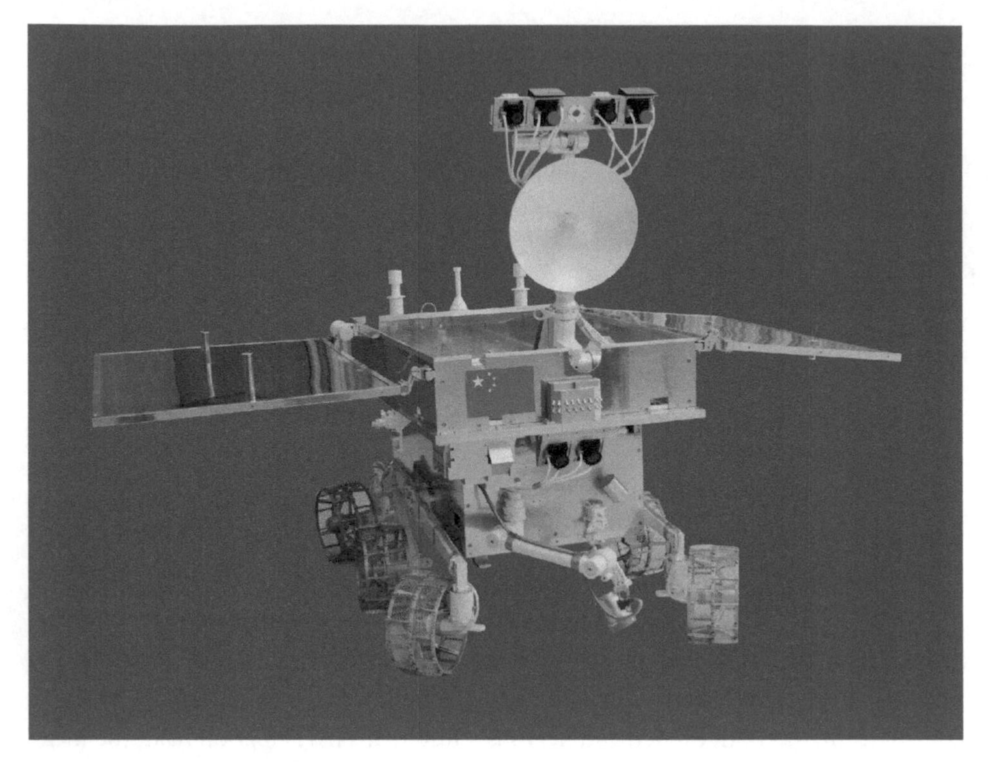

Yutu, the Chinese lunar rover. (Photographed by Qin Xian'an)

"Go out and have a look at the Moon. There's a something new and it's glimmering," wrote someone whose cyber alias was Astroo. An American civil society organization named Operation Asteroid announced that the solar wing had been launched 37 years earlier, and now there was another detector on the Moon.

"This is the first time in 37 years that a detector made a visit to the Moon without smashing into it. I guess the Moon had forgot what it was like to have a visitor," read a post on twitter from the University of Manchester in England. "Congratulations to China, it's great!" praised a U.S. netizen. "37 years later, the human probe finally came back to the Moon. Thank you, China. Thank you, Chang'e-3," said one Peter of Sweden.

At 23:35 on December 15, 2013, the Chang'e-3 mission reached a climax: with the order of the control commander issued, Chang'e-3 and Yutu widened eyes for an affectionate gaze.

Tick, tick, tick: every second ticked clearly.

Boom, boom, boom: every heart beat sounded passionately.

Ten meters was the distance for mutual picture shooting. In the flight control Hall of the Beijing Aerospace City, everyone held the breath waiting for the exciting moment.

In the rhythmic command sound on the ground, Chang'e-3 and Yutu successively pressed the shutter and recorded the most beautiful moment of each other.

The one minute of mutual shooting was more than a simple photo moment. A few hours before approaching the shooting spot, the lunar rover successfully verified the remote operation technologies, including lunar surface walking, terrain establishment, visual positioning, perception planning, and the operation modes such as vehicle body control, in-situ turning, camera use, etc.

Silent as the image is, the moment is eternal. On the large screen of the Beijing Aerospace Flight Control Center, the desolate and lonely surface of the Moon 380,000 kilometers away presented a bright Chinese national flag.

This was the first picture of the five-star red flag taken by China on the Moon. At this very moment, it solemnly declared to the world that the Chang'e-3 mission was a complete success and the second-step strategic goal of China's lunar exploration project was fully achieved.

The five-star red flag on the Moon voiced for the aspiring Chinese spacemen.

Applause roared in the China Astronaut Center of Beijing Aerospace City. Astronaut Zhai Zhigang said: "Five years ago, I contributed to the success of Shenzhou 7 mission bringing the five-star red flag to space. This time, Chang'e-3 took it to the Moon. The dream of the great rejuvenation of the Chinese nation will surely come true in the continuous efforts of the Chinese people."

Wang Yaping, a female astronaut known as the real Chang'e, said: "I am proud of another great feat of Chinese spacemen and our great motherland. I don't see myself as Chang'e. I'm looking forward to the day when Chang'e lands on the Moon. How high the national flag is raised, how high the eyes can reach."

The five-star red flag shone on the Moon like a flame burning in the hearts of the Chinese people.

The Tibetan mountaineer Gongbu in front of the TV watching the live broadcast, sighed: "The great motherland is booming, and the five-star red flag goes higher and higher!" Fifty-three years ago, Gongbu became the first Chinese who successfully climbed Mount Everest from the north slope and put the flag there.

Wang Yue, a volunteer who participated in the Mars-500 experiment, said: "No one can predict how far Chinese will go in exploring space now that Chang'e-3 successfully landed on the Moon. I hope that the national flag will be on Mars one day."

Chang'e-3 successfully landed on the Moon, and Yutu ambled in the Bay of Rainbows. Immediately, praise came in abundance from international media.

An article published on the website of the Wall Street Journal said that China has become the third country to achieve a soft landing on the Moon after the Soviet Union and the United States. It was the latest major victory of China's space program and another milestone for the rising power to achieve space ambition.

The French News Radio website reported that China's lunar exploration plan was fulfilled as promised and this achievement was an important step forward in China's larger space program.

Yutu, the first lunar rover in China, carried the millennium dream of the Chinese people to the Moon, and every move of hers concerned hundreds of millions of Chinese.

On December 16, 2013, Yutu met the high temperature test of the Moon and turned to the "lunch break" mode: the mobile subsystem stopped working, and the break ended on December 20, and work resumed.

On January 12, 2014, Chang'e-3 lander and Yutu lunar rover realized independent wake-up. China successfully broke through the technology of night survival of the probe.

On January 25, 2014, Yutu malfunctioned. As the news spread, blessings and encouragement were widely received. One netizen wrote: "You have experienced the trek that we have never experienced, endured the cold and heat that we cannot bear, and brought us the scenery that we have never seen. Yutu, you have done well. We pray that you will get better, but no matter what the result is, it is you that allowed the red flag to shine brightly over 300,000 kilometers away."

Those days was a roller coaster ride for the Yutu fans. When the sun rose on the Moon, Chinese people were anxious to wait for it to wake up as scheduled; when the situation improved, hope was rekindled; when they learn that it has survived the extreme cold and woken up again, there was a great relief.

This spring was especially warm, because of the perseverance of science researchers, and because of the patient waiting of hundreds of millions of Chinese, which supported Yutu, a high-end robot.

It was over five months since Yutu landed on the Moon on May 28, 2014. It endured one cold night after another under abnormal conditions. Over 100 meters of tracks was left on the Moon surface, while a large amount of engineering and scientific data was obtained.

Wu Weiren, chief designer of China's lunar exploration project, disclosed that except for the mobile system, four major science instruments on the lunar rover were in normal operation on September 4, 2014, when Yutu started its 10th month on the Moon.

"The design life of Yutu is three months. But it has been working to the 10th month, which is definitely overdue service. It is undeniable that its function will gradually fail over time." Wu said the Yutu lunar rover and lander would remain on the Moon forever.

Ten years of efforts had the Chinese Moon dream come true. It was a small step for Yutu, a giant leap for China.

In the Xichang Satellite Launch Center, the place where Chang'e-3 set out, there was a road named Tanyue (meaning Moon exploration). Three boulders beside the road were engraved with: Chinese road, Chinese spirit, and Chinese power.

At the very moment, countless Chinese looked up at the bright Moon in the sky. We wondered if Chang'e and Yutu 380,000 kilometers away were also looking down at us, homesick.

Return Ticket

The pace of Chinese aerospace workers has always been unexpectedly fast.

In the autumn of 2014, when the pride in the success of Chang'e-3 landing on the Moon lingered, Chang'e-5 T1 (re-entry and return-flight tester) set out—

At 02:00 on October 24, the third phase re-entry and return-flight tester of the lunar exploration project developed independently in China was launched at the Xichang Satellite Launch Center with the Long March 3C carrier rocket. It entered the Earth-Moon transfer orbit precisely at the perigee altitude of 209 km and the apogee altitude of 413,000 km.

This test, although not named after Chang'e, was closely related to the lunar exploration. Real flight of the spacecraft would generate test data to verify the key technologies of orbit design, aerodynamics, thermal protection, guidance, navigation, and control for Chang'e-5's re-entry to the earth at a speed close to the speed of escape (11.2 km/s). This would explore the way for Chang'e-5's mission in 2017.

As the pathfinder of Chang'e-5 mission, the spacecraft was composed of a service cabin and a return device. It appeared glamorously armored from a distance. Researchers had customized it various steel bones. A petite shape yet strong physique was cleverly designed, and full efforts were made in new products, new technology, and new crafts. A series of technological innovations armed it with a lighter weight yet more stable and heat resistant suit.

Deep space exploration has never been simple, but a road full of barriers and unknown challenges. This journey would not cover the whole journey of Chang'e-5. Instead, it would focus on the way home from the Moon, 380,000 kilometers away from the earth.

This was a road no Chinese has walked on. And the conditions are strange, complex, and changeable. As a newbie, Chinese aerospace workers were faced with unprecedented risks and challenges. Yang Mengfei, commander-in-chief and chief designer of the flight tester, explained that it was like throwing stones from high; the higher the distance, the faster the landing speed. T1 would return from the Moon 380,000 kilometers away, and the velocity when it reaches the earth would be very terrifying. Specifically, it would be about 3 km/s faster than the return speed of Shenzhou spacecrafts. The friction heat would pose a serious threat to the spacecraft.

How to ensure a safe way home? Researchers specially designed the special return mode: half ballistic jumping.

Hao Xifan, deputy chief designer of the third phase of China's lunar exploration project, made an analogy. "It's a bit like playing ducks and drakes. When the spacecraft returns to the atmosphere for the first time, it jumps up first, then flies for a period of time, and then enters the atmosphere for the second time to return to the earth. The main purpose of this design is to reduce energy consumption and speed, so as to ensure the smooth return."

On November 1, 2014, I was in the Flight Control Hall of the Beijing Aerospace Flight Control Center, witnessing the safe return of the pathfinder.

At about 6:10, the general dispatcher Dai Kun loudly ordered: "Establish a return and re-entry posture."

At the time, the spacecraft, about 120 kilometers from the earth's atmosphere, flew towards it like a meteor. Technicians in the flight control hall all looked up, watching the progress on the large screen.

In the next three minutes, it entered the black barrier by friction with the atmosphere at high speed. In the black barrier area, it bounced up due to atmospheric lift, just like playing ducks and drakes, in which a stone would bounce up and fall again under water buoyancy.

"This jump is critical," said Zhou Jianliang, chief engineer of Beijing flight control center. "Should the jump up fail, she will fall into pieces. Should the jump-up be overdone, she will escape from the atmosphere and cannot enter again." This thrilling motion took place in the earth's atmosphere. My heart could not help racing when experts elaborated it in a concerning tone, although I could not see it with my own eyes.

Jump up and fall down. "The first lift control was complete, and next is the free flight." At 6:15, there was thunderous applause in the flight control hall, and technicians cheered for the perfect jump.

In this jump, it broke through the black barrier and the atmosphere; in this jump, she crossed the most dangerous part of the return journey.

At 6:42, after eight days and nights of space flight, the test lander landed on the Amugulang grassland in Siziwang Banner, Inner Mongolia. At this moment, the dark brown returner was lying in the deep grassland, shining with charming light in the morning ray. It seemed to smile at the people who came to greet, saying, "It was a smooth flight, and I feel great."

This is the very spacecraft returning from the Moon since Apollo 18 made it back from there 42 years ago in 1972.

This meant the success of the third phase of China's lunar exploration project. Chinese aerospace workers have obtained the first return ticket for lunar exploration.

The return ticket was the fruit of independence and continuous self-improvement. The success of the self-developed "pathfinder" marked the end to the era of one-way trip to the Moon and laid a technical foundation for the future third phase of the main mission of Chang'e-5 to carry out more complex return tasks. Behind the success of Chang'e series and T1 was the Chinese independence and innovation.

The foreign press was more than willing to acclaim the success of Chang'e-5 T1.

Igor Marinin, editor in chief of Russia's aerospace news magazine, told reporters that while Russia, the United States, and the European Union were currently formulating their own Moon development plans, China has taken a leading position in the world; China had a complete roadmap for the development, ranging from the launch of spacecraft to circle the Moon to the landing of the probe, and even the establishment of a lunar base in the future.

Dr. Fraser, a professor at the U.S. Naval Academy of Military Affairs, said the test was of great significance; it demonstrated China's aerospace technological capabilities and willingness to achieve long-term space goals in the future, which were what some old aerospace countries lack.

"I would like to see men walking on the Moon again. Thank you to China. Will that be by 2020?" wrote a netizen on the US space network.

Three Barriers to a Manned Moon Landing

The furthest distance in the world is the distance between dream and reality.

The closest distance in the world is the same.

Regarding the realization of the dream of lunar exploration, Luan Enjie, the first commander in chief of China's lunar exploration project, once sighed with deep emotions, "At the crucial moment when human activities were expanding into space again, China has not lagged years and efforts behind as the previous two expansions to the ocean and sky, but was finally admitted to the Moon club."

Before China, the four members of the Moon club were the United States, Russia, Europe, and Japan.

To earn the admission, Chinese spacemen relied on strength and action; in 2004, the lunar exploration project was officially approved. In 2007, Chang'e-1 successfully orbited the Moon. In three years, China accomplished what western lunar exploration achieved in 40 years. It was a quick admission to the Moon club.

One step at a time. Chinese aerospace workers have at an incredible speed turned the millennium myth of Chang'e ascending to the Moon into reality.

In 2010, Chang'e-2 made it to the Moon, in 2013, Chang'e-3 succeeded a soft landing, and in 2014, Chang'e-5's T1 effectively explored the way.

Nowadays, in the face of the new upsurge of deep space exploration in the world, Chinese spacemen have stepped up the pace.

In November 2017, Chang'e-5 achieved lunar sampling and returned, meaning the completion of the last step of China's lunar exploration project.

In 2018, Chang'e-4 was launched to implement the world's first soft landing on the back of the Moon and patrol exploration. Therefore, the fourth phase of China's lunar exploration project also started.

How far is China's manned lunar landing? Faced with the rapid progress of China's lunar exploration, global media have made predictions.

Forbes commented that should China's lunar exploration project be carried out as scheduled, China would realize manned lunar landing sometime in 2025.

"In 15 years, the next person to walk on the Moon will probably be Chinese," said astrophysicist David Whitehouse on the *Independent*.

Compared with the optimism of the international public opinion, Chinese spacemen exhibited more caution. According to the white paper *China's Space 2016*, China would lay the foundation for manned exploration and development of Earth-

Moon space in the next five years.

It seems that no manned lunar landing plan has been formally approved. The precise time of China's manned landing on the Moon remains undetermined.

At present, China's lunar exploration stays in the stage of unmanned lunar exploration. But unmanned lunar exploration is the prelude to manned lunar landing, which has been the goal of Chinese spacemen for many years. Not long ago, academician Ye Peijian said in an interview: "Human beings must go out of the earth, and the Moon is the closest destination. I don't believe Chinese spacemen will give up manned landing on the Moon."

Regarding the rapid development of China's space industry, Joan Johnson, an American space policy expert, once evaluated, "Obviously, China has two very meaningful and well-organized space exploration projects: the Chang'e project and the Shenzhou project. At present, separated as two projects are, they will eventually be closely linked, and China will eventually launch a manned lunar landing project."

Those words spoke the truth. A crucial prerequisite for the realization of manned landing on the Moon is the capability of both manned spaceflight and lunar exploration.

That is to say, the progress of manned space flight will determine the schedule of China's manned landing on the Moon. A man with both legs cannot walk with only one.

To ordinary folks, there is little difference in technical principles between manned spaceflight and manned Moon landing. In fact, numerous technical problems, such as the space walk and space docking, must be overcome to evolve from manned spaceflight to a manned Moon landing.

Take the spacesuit for example. In the Shenzhou 7 mission in 2008, Zhai Zhigang, a Chinese astronaut, wore a domestically made extravehicular spacesuit. It was more than a simple suit. It was actually a little space vehicle, which was developed after four years of research and development.

Compared with the orbit environment as high as 300 kilometers, the living environment on the Moon is worse at 380,000 kilometers away. The temperature at night goes as low as −180°C and in the daytime as high as 150°C. Even the unmanned Chang'e-3 detector had to be "armored" to enhance viability.

In the future of manned lunar landing, a Moon suit is an indispensable equipment. Science and technology and developmental difficulty will far exceed that of a simple extravehicular spacesuit. In July 1969, U.S. astronaut Neil Armstrong made the first Moon walk in a 19.69 kg Moon suit, with a life support system on his back that could guarantee an astronaut to work on the Moon for 8 hours.

The overall development plan of China's manned space project scheduled the construction of China's manned space station to be completed around 2020. Experts envisioned that the manned space station in operation would serve as both a transit station for lunar exploration and an important science research and test base to accelerate the breakthrough of some core technologies in the manned lunar landing project.

In the bigger picture, three barriers must be taken down to achieve manned lunar landing in China. The first is to develop a heavy-capacity launch vehicle, which can send people and the lunar module to the Moon; the second is to secure life support, safety, and working conditions of astronauts to and from the earth and the Moon; the third is to construct more ground support systems to ensure that all tests are fully verified.

"I hope to see Chinese astronauts on the Moon and Chinese deep space probes flying to Mars in my lifetime." Looking up at the starry sky, academician Sun Jiadong's words echoed. "The lunar exploration project is the beginning of China's science exploration in deep space. It is not limited to the Moon, but will irreversibly extend to Mars, Venus, Saturn, and all the other planets in the solar system and beyond."

CHAPTER 7

Beidou Shines Bright

Have you ever ridden a shared bicycle?

Should you answer no, you have to keep up.

All of a sudden, in the spring of 2017, shared bicycles gained massive popularity in Beijing overnight. Orange, yellow, and blue shared bicycles are seen on the streets and lanes of Beijing, forming a unique scenery.

"Jingling, jingling." Bathed in the gentle spring breeze riding a shared bicycle, are you aware that the convenience brought to you is closely related to the Chinese Beidou Navigation Satellite in the vastness of space?

On April 8, 2017, a strategic cooperation was reached among Beidou Navigation Beijing, the public platform of the location service industry, and the shared bicycle industry. The spatial information positioning technology of China's satellite navigation system has become an important strategic weapon for the future products and services of the shared bicycle platforms. It set a new direction for the shared bicycle industry to adopt the positioning technology. (Update: by 2020, the broken funding chain left the company in great debt and bankruptcy.)

It is reported that ofo shared bicycles will be equipped with Beidou Smart Lock with global satellite navigation and positioning in Beijing, Tianjin, and Hebei in the future. At the same time, Beidou navigation and positioning further optimizes its electronic fence positioning to achieve more refined vehicle operation management.

At the foot of Tianshan Mountain thousands of miles away, cotton farmers in Bole City, Xinjiang Autonomous Region, also benefited greatly from Beidou.

Ofo shared bicycles. (Photographed by Qin Xian'an)

In spring ploughing, a seeder with antenna on the top of the cab was guided via satellite signal to accurately pave hundreds of straight and equally spaced cotton mulches on the open farmland of Dalete Town. It was quite a spectacular scene.

"As long as the navigation system is activated and the route is set, seeders can run automatically day and night" said Weng Lianzhong, a seeder driver in Dutuaobo village. "The efficiency has been increased by more than 20% compared with the manual operation. Now about 150 mu is sown every day. The cost of seeding per mu is 30 RMB, which will be soon earned back."

Bole Agricultural Machinery Department has begun to apply the satellite positioning navigation automatic seeding technology to accurately plant cotton, helping cotton farmers reduce costs and increase income since 2015.

"It's amazing!" Li Bichun, a big cotton farmer does not understand the perplexing technical principles, but the immediate benefits had her acclaim Beidou relentless. Satellite navigation Precision Planting improved the utilization rate of cultivated land. According to the average yield of 400 kg per *mu* of her cotton field last year, the income per *mu* was increased by about 200 RMB.

In this spring, the Beidou trend swept cotton fields at the foot of Tianshan Mountain, and then the entirety of China.

In Beijing, the safety of the gas pipeline network leading to residents' homes is monitored 24 hours/day by Beidou Intelligent Housekeeper. Beijing Gas Group is the first municipal public enterprise to apply the national Beidou Precision Service

Network. Since the pilot application in 2013, it has been used in intelligent inspection, pipeline retest, emergency repair, leakage, and anti-corrosion coating detection.

In Shishi, Fujian Province, many fishermen have installed Beidou system terminals on fishing boats, which integrate the functions of positioning and navigation, SMS sending, and voice calling, improving their ability to cope with complex sea conditions.

In recent years, Beidou's application in the civil field has been continuously expanded from improving the operational efficiency of urban lifelines such as water, electricity, gas, and heat, to providing convenient services for personal care for the elderly, intelligent transportation, and emergency rescue. Up to now, the National Beidou Precision Service Network has provided service for various industry applications in over 300 cities. "By the end of 2016, Beidou civil users had reached tens of millions," said Miao Qianjun, Secretary General of China Satellite Navigation and Positioning Association.

In the spring, Beidou stepped out to the world.

On March 30, 2017, Wuhan Optics Valley Beidou Holding Group Co., Ltd. built Beidou international cooperation prototype rooms in Thailand and Sri Lanka. The construction of a Beidou base station can extend the coverage radius of Beidou to Southeast Asia and South Asia by at least 3,000 kilometers. Therefore, Beidou will benefit 10 ASEAN countries.

On April 18, 2017, the Beidou Navigation System Seminar, jointly organized by the China Satellite Navigation System Management Office and King Abdullah Aziz Science and Technology City of Saudi Arabia, was held in Riyadh, Saudi Arabia. This was the first time that China's satellite navigation system has entered the Saudi market, with a huge demand for navigation products and applications on a large scale.

During this year's National People's Congress and the Chinese Political Consultative Conference, Yang Yuanxi, deputy chief designer of Beidou Navigation System, revealed to the media that China's Beidou navigation satellite system would make a big move. It planned to launch 6 to 8 satellites. The first launch would take place in July in a way that a rocket could lift up two satellites.

According to the three-steps development strategy of the Beidou system, it is expected to provide basic service for countries of the One-Belt-One-Road project and become a first-class navigation satellite system in 2020.

According to Medium- and Long-Term Development Plan of the National Satellite Navigation Industry issued by the State Council, it is predicted that by 2020, the scale of China's satellite navigation industry will have exceeded 400 billion

RMB, and the industry scale of Beidou will have reached 240 billion RMB.

Beidou is another new legend that Chinese spacemen created with wisdom and sweat.

It took them only 17 years from the launch of the first Beidou navigation satellite in October 2000 to multiple Beidou satellites navigating from above.

China's Beidou satellites have changed the lives of ordinary people and affected all walks of life. The world benefits from them.

Unintentional Positive Outcome

"Where am I?"

Simple as the question seems, it is difficult to answer. For thousands of years, people around the world have been searching for the answer in various ways.

First, people invented the compass; then, radio and radar were created.

It was not until June 26, 1993, when the global positioning system of the United States completed the in-orbit networking of 24 satellites that the question became really easy. People can accurately determine real-time location coordinates (including longitude and latitude, altitude, etc.) via GPS receivers.

This is satellite positioning.

Like many great inventions in the history of science and technology, satellite positioning is also a work of an unintentional positive outcome.

On October 4, 1957, the Soviet Union launched Sputnik 1, the world's first man-made satellite, marking the beginning of the era of humans in space. The world paid close attention. In the Johns Hopkins University Applied Physics Laboratory, young mathematician Bill Guy and physicist George Weiffenbach received the Sputnik-1 signals, which they jointly studied.

Soon, they discovered that the specific position of the satellite could be calculated according to the frequency differences of the received signals by setting up multiple receivers on the ground. This is the Doppler Effect resulting from relative motion.

Two young scientists excitedly came to Frank McClure, director of the laboratory. They had succeeded in the Doppler tracking of satellites. As director of the laboratory, McClure was frying a big fish for the U.S. Navy: how to locate warships in a vast sea. Faced with the unprecedented problem, he was deeply puzzled.

Hearing the description of two young scientists, he was inspired and came to a great deduction: since we can locate the satellites, the satellites can locate us if the approach is reversed. And that would make it possible to locate the naval ships.

Thus, the first satellite navigation plan, the transit navigation system of the U.S. Navy, was born. Johns Hopkins University Applied Physics Laboratory took full charge. Starting from the original concept in 1958, the first Transit satellite test was carried out in 1960. In January 1964, the system was put into operation.

The U.S. Navy used the system for submarine navigation with intercontinental missiles, and also for sea navigation of ordinary ships. As the first real satellite navigation system put into operation in the world, it was not until 1996 that Transit was replaced by GPS system (global positioning system) and it took a bow on stage.

On the basis of the successful application of Transit, the United States Ministry of Defense put forward the overall vision of GPS in 1973 24 medium-orbit satellites were launched to cover the world.

GPS system launched the first satellite in 1978 and put it into full operation in 1995. There are about 30 satellites in the system, which can be divided into military use and civil use, of which the latter is globally open. Nowadays, GPS is the most widely used global satellite navigation system in the world.

During the Cold War, to counteract the transit of the U.S. Navy, the Soviet Union also formed Tsikada, a navigation system for the Soviet Navy.

In the 1970s, the Soviet Union, keeping up with the U.S., launched the GLONASS satellite navigation system. In the planning and design, the system required 18 satellites to provide positioning and navigation services for the whole territory of Russia, and 24 satellites in orbit to provide global services.

GLONASS was officially put to use in 1993. Since the collapse of the Soviet Union, the number of GLONASS satellites decreased dramatically with only six satellites in orbit. In recent years, Russia has increased efforts to build GLONASS. At present, all 24 satellites rated by the system are in orbit operating, and another four satellites are in orbit for backup. It is understood that the positioning service it provided to civilians can only be accurate to 30 meters, while the accuracy of GPS civil positioning and navigation error of the United States is within 10 meters.

In the future, Russia plans to increase the number of in-orbit satellites of GLONASS to 30, which will greatly improve the navigation accuracy of GLONASS. And the efforts are made to accelerate the progress. According to Vladimir Popovkin, the Deputy Minister of Defense of Russia, the navigation accuracy of GLONASS will have reached 0.6 m by 2020.

"Whoever grasps the advantage of satellite navigation has the initiative of the war," commented the *US Military Review*. In recent years, the operation platform and guidance weapons supported by GPS contributed greatly to the U.S. victory of multiple local wars at a very small cost. No wonder some American generals have expressed that "Without GPS, we can't fight at all."

To prevent countries from using the system against the United States, GPS only provides low-precision satellite navigation signals to foreign countries. In the event of a military conflict threatening its own security, the U.S. will immediately cut off satellite navigation services. Europeans found it unsafe. Therefore, they launched the Galileo project to establish a global satellite navigation system for civil use controlled by Europe itself.

The proposal of the Galileo satellite navigation system was first put forward in a report of the European Commission in 1999 and was formally launched in March 2002 after rounds of arguments. The initial completion was scheduled in 2008, but technical problems have badly delayed the progress.

The Galileo satellite navigation system is a new generation of global satellite navigation system for civil use led by the European Union. It cost more than three billion euros. The system is composed of two ground control centers and 30 satellites, of which 27 are working satellites and three back-up satellites. It is expected to realize all-satellite networking by 2020.

As a European Union led project, Galileo system did not turn down foreign participation. China, South Korea, Japan, Argentina, Australia, and Russia were involved in the plan with provisions of financial and technical support. After the completion of the Galileo satellite navigation system, it will form four global satellite navigation systems together with GPS of the U.S., GLONASS of Russia, and Beidou of China.

European Space Agency experts believe that the Galileo system is more advanced and reliable than the U.S. GPS system. Regarding the satellite signals provided by GPS of the U.S. to other countries, objects no less than 10 meters long can be discovered on the ground, while Galileo's satellite can find objects as small as one meter long. To put it figuratively, the GPS system can only see the streets, while Galileo sees the doors.

Dual-satellite Positioning

History will always remember that on June 9, 2012, China presented the Beidou navigation satellite model to the United Nations as a gift.

At the presentation ceremony, Cheng Jingye, resident ambassador of China to the United Nations and other international organizations in Vienna, introduced Beidou as the name of a constellation (the Big Dipper) in China, and people around the world have relied on it to guide directions for thousands of years. Chinese named their own navigation satellite system after it in the hope that the system can provide global service.

Beidou impressed representatives of all countries attending the 55th session of the United Nations Committee on the peaceful uses of outer space.

What took the world by surprise was that merely six months later Beidou-2, with 16 Beidou satellites operating above, was officially put into use, covering the Asia Pacific region.

History says that it was not too late when the research of China's satellite navigation started.

As early as the 1960s and 70s, China began to study the use of satellites for ground positioning services, but poor national strength left this project Lighthouse stranded.

In a flash, 10 years have passed.

The turning point of the development of Beidou appeared at the joint meeting of the Chinese Academy of Sciences and the Surveying and Mapping Bureau of the Former Headquarters of the General Staff of the PLA in October 1985.

At the meeting, academician Chen Fangyun, a famous scientist in China and one of the advocates of Project 863, proposed that only two geosynchronous fixed-point satellites were able to cover a large area, locating and navigating ground targets and mobile objects on the sea, while providing communication service.

This theory, later summarized as dual-satellite positioning, became the theoretical foundation of China's Beidou plan in the future.

In fact, two years ago, Chen Fangyun had already discovered and put forward this theory. The same theory in the United States was made public five years later. Unfortunately, the idea was too far ahead of its time, which led to very little acceptance.

Hardship makes greatness. At the age of 67, Chen Fangyun, a legendary scientist who enlisted the army in his early years and has achieved much in China's aerospace cause, put forward one of the most important and brilliant theories in his career.

The dual-satellite positioning communication system was accepted by the Surveying and Mapping Bureau of the Former Headquarters of the General Staff of the PLA and initiated in 1986. After unremitting efforts, on September 25, 1989, the function demonstration of the first dual-satellite positioning communication system came out a success in a temporary computer room less than 30 square meters in Beijing. In the moment, the excitement had academician Chen welling tears in his eyes.

The next day, Xinhua News Agency announced that China had integrated rapid positioning, communication, and timing with two satellites and obtained ideal test data. This was the first time in the world that the accuracy of rapid positioning has reached an international advanced level.

Some experts said that GPS, manned space station construction, and lunar exploration were the three most influential space projects across the globe. To complete each of these projects, a strong national strength is necessary. Large expense is expected.

Data shows that the GPS system development in the United States cost about US$10 billion, and the annual cost of operation and maintenance of 24 satellites is about US$500 million. Since the development in the 1970s, at least US$40 billion has been spent.

At the end of last century, China was still in economic reform and opening up. There was no economic strength back then.

The successful practice of the theory of dual-satellite positioning has paved a research and development way in line with China's national conditions.

Two navigation satellites can basically meet the needs of national satellite navigation with little money spent. This cutting-edge technological breakthrough has greatly incentivized China to launch the Beidou navigation system project.

Chief Designer of Beidou

A person has been present in every historical juncture of China's aerospace development.

He is academician Sun Jiadong.

In increasingly fierce competition, a scientist's strategic thinking often outweighs technical background. Sun Jiadong is such an outstanding strategic scientist.

In the early 1990s, the great success GPS of the U.S. achieved in the Gulf War attracted global attention. The war thousands of miles away enlightened the scientists that the satellite navigation and positioning system with independent intellectual property rights and control is essential for national security, national defense construction, and military combat effectiveness. "The value and benefit cannot be measured and reflected in numbers. And the economic value in civil application market cannot be accurately estimated either," said Sun.

At that time, Chen Fangyun's dual-satellite positioning theory had been proved in practice. This was exactly what the launch of China's Beidou satellite navigation system needed.

"It is imperative to develop satellite navigation." Sun Jiadong came to Shen Rongjun, then-deputy director of Commission of Science Technology and Industry for National Defense. Coincidentally, they shared the same idea. Jointly, they wrote a letter to suggest the launch of the Beidou satellite navigation system.

There was the voice of doubt. GPS is free and highly accurate. Why should we bother to develop Beidou?

Sun Jiadong replied, "GPS is available at peace time. What if a war breaks out? As Russia is engaged in GLONASS, Europe in Galileo, Beidou is a major strategy related to our national security."

There were the sincere words of the veteran spaceman. "Practice has proved that the core technology of military industry cannot be bought, and neither can cutting-edge aerospace products. We must rely on our own strength to develop space technology."

I remember seeing academician Sun for the first time on the eve of the launch of Chang'e-1. Faced with questions from me and his peers, Sun smiled, saying: "I've made an indissoluble bond with satellites in my life, and my greatest wish is to build Chinese satellites all my life."

So, he did.

On a sunny winter day in December 1994, the Beidou navigation test satellite system project was approved by the state. And the Beidou satellite navigation system construction was officially a go.

Sun Jiadong, 65, once again took charge as the chief designer of Beidou navigation test satellite project.

In the history of China's aerospace development, no one shares the same experience as Sun Jiadong, chief designer of Dongfanghong 1, chief designer of China's lunar exploration project, and chief designer of China's Beidou navigation project.

Beidou navigation test satellite was a huge and complex system engineering. Sun's great wisdom was embodied in excellence at coordinating various complex technical problems and finding the most economical and reasonable solutions.

"Space engineering emphasizes on the whole, that is to use the most reliable technology, the least cost, the shortest time, and the most favorable cooperation to work out the most feasible scheme and ensure the best results," said Sun."When the table is high and the chair is short, you can raise the chair or lower the height of the table, depending on which is more convenient."

The construction of Beidou satellite navigation system started much later than Europe and the United States, but the Chinese aerospace workers were determined: "We will catch up."

According to the national deployment, independence, openness, compatibility, and gradual progress made up the development principles. And the Beidou system followed the three-step development strategy. First, Beidou satellite navigation test system was built in 2000, making China the third country in the world with an independent satellite navigation system. The second step was to build the Beidou satellite navigation system and form a service capacity covering most of the Asia Pacific region around 2012. Step three involves global coverage of Beidou satellite navigation system in 2020.

A road was planned. However, it was full of challenges and difficulties, which required researchers to overcome them step by step.

For example, how to pull off a test satellite launch? How to improve the launch site system? Whether the dual-satellite positioning test system can form regional navigation service capability? In addition, there was no precedent in China and zero references to certain technical conundrums abroad.

Xie Jun, then the overall designer of Beidou navigation satellite, said: "The pressure is huge, but the era has chosen me and the responsibility too, so I must not snub it, but give it my all."

This was the common belief of Xie and all the researchers participating in the Beidou mission.

To make Beidou happen is a national mission that must succeed.

The Creation of Beidou

"Ignition!"

"Take off!"

In an instant, the earth trembled, and the fire rose to the sky.

At 00:02 on October 31, 2000, the nearly 300-ton Long March 3A carrier rocket spat a blazing flame, lifting up China's first Beidou navigation test satellite and heading straight to space at the Xichang Satellite Launch Center.

Only 50 days later, China's second Beidou navigation test satellite was successfully launched into geosynchronous orbit by Long March 3A at 00:20 on December 21, 2000.

Two satellites shone upon the land of China. On December 21, the Xinhua News Agency announced to the world that this was the start of the first generation of satellite navigation and positioning system independently developed by China.

At 00:00 on January 17, China's third Beidou navigation satellite was successfully launched in Xichang Satellite Launch Center with the Long March 3 This was the first satellite launch that year and the 122nd flight of the Long March series. (Photographed by Qin Xian'an)

The system is a regional navigation system that provides satellite navigation information 24/7. Upon completion, it will mainly provide navigation services for road transportation, railway transportation, and marine operation, which will play a positive role in promoting the national economy.

Now, China became the third country to possess its own satellite navigation system, after the United States and Russia.

At that time, it had been only six years since the Beidou project was launched.

Starting from scratch, Beidou made it to space at an amazing speed in six years. But little knew how much sweat and wisdom of thousands of Chinese aerospace workers were behind the feat; in how many sleepless nights they buried their heads tackling problems; and how many times they rose from where they fell.

When Beidou was newly launched, it encountered the double restrictions of foreign technology blockade and a relatively backward domestic industrial base. Wang Shufang, a senior engineer involved in the Beidou project, believed it was extremely difficult to build Beidou up to international standards.

What to do? "There is only one way ahead, that is to walk the path of innovation with Chinese characteristics," said Wang Shufang.

Wang was the first batch of college students recruited in the construction of the Beidou navigation system in China. At the age of 22, she participated in the research and development of the Beidou system immediately after her graduation. Since then, Beidou has been her sole focus for 23 years. She witnessed the birth of the Beidou system.

Wang has a sacred sense of mission for the research and development of the Beidou System. "The GPS of the United States will interfere with the satellite signal. It was made available according to America's preference. Before 2010, GPS was open to civil use with an error of 100 meters, and now it is about 15 meters after the cancellation of human interference. But the control lies in the hands of Americans. What if they adjust the error to 1 km or even several kilometers?"

"The most shocking thing in my memory is the scene of a simulated satellite ground joint commissioning in an aerospace factory in 1998. In the simulated vacuum liquid nitrogen tank, the satellite was spitting out white smoke. R&D personnel were tirelessly debugging in the closed large factory building." They gathered from all over the country and joined forces to coordinate the satellite and the ground control, which left an unforgettable impression in Wang. Up to now, she keeps the *ci* poem she composed:

"Sunny in the capital as it is in May, little good of spring that I was made aware. Researchers came to coordinate from all directions, grey-haired and black-haired, both smiled. Strangers as they remained to one another, the satellite was well-behaved up above."

"No one knew whether it would work or not, and technology verification was like walking through a maze." Wang Shufang and her team went through trails and errors and eventually found the solution.

When the first generation Beidou user end was developed, it weighed more than five kilograms, very heavy on the back. Wang, weighing only 40 kilograms, was crouching in the snow in the northeast with a user end on her back. One test lasted over ten hours. She had fainted several times.

In the warehouse, when Wang discovered the Beidou user end she designed, she burst into tears. "That is my youth."

On Beidou's journey to space, every step is engraved with the name Wang Feixue.

Wang Feixue, director of satellite navigation and positioning at the Engineering Research Center of National University of Defense Science and Technology, has been dedicated to Beidou for 21 years.

In 1995, the development of a key equipment of Beidou satellite navigation had the researchers stuck. The national technical elites joined talents, but still failed to achieve a breakthrough.

Anxiety led to aimless wandering. Yet, at this time, three young doctors, Wang Feixue, Yong Shaowei, and Ou Gang, from the University of National Defense Science and Technology, came forward regardless of the great difficulty presented to them. A few thin pages convinced academician Chen Fangyun to give them a try, allocating 50,000 RMB for the experiment to build a simulation model first.

The laboratory was temporarily borrowed, and so was the equipment. In spite of hardships, the three young scientists worked tirelessly. This was a road that none had walked. Looking at the three immature faces, peers believed they would fail.

Who would have thought that three years later, they would break through the bottleneck of Beidou satellite navigation and positioning project at one stroke, namely, digital fast precise capturing and tracking? This exciting news attracted experts involved in the development of the Beidou navigation system to fly from all over the country to watch the testing process.

At that moment, the air froze, and dozens of eyes were fixed on the tiny experimental platform. In the dark, as pulse flashed, the signal was captured successfully. There was thunderous applause. At the celebration banquet, the experienced predecessors took the initiative to toast the three young talents in turn to show their respect. The bottleneck technology that had not been tackled in 10 years has been perfectly tackled by several young minds in only three.

This year, Wang Feixue was as young as 27 years old.

Those who think forward made a difference at the turning points in history. Wang Feixue said: "If you see people enjoy a nice lunch and prepare yours with the same recipe, you would only find you are one step behind at dinner. Why not prepare tomorrow's lunch earlier?"

The first generation Beidou user machine, bulky and heavy-weighted, was problematic. The fact that it must be carried in a backpack made it inconvenient to use. Wang Feixue led the team to work on the miniaturization and portability of the machine. In a mere two years, they successfully developed the first small hand-held Beidou user machine with fixed equipment in China.

In April 2007, shortly after the entry into orbit, China's first Beidou-2 test satellite encountered high-power complex electromagnetic interference, and the success rate of signal reception was less than 50%.

In the vastness of space, should the interference stay unsolved in time, the plan of over 10 satellites launched to be networked would be postponed indefinitely, and it would be difficult for the launched satellites to meet the expectations.

How to solve the problem? There were two schemes. One was to hide by altering the signal frequency, thus avoiding interference. The technical difficulty is small, but it would require a re-construction of the ground system that had been built in the early stage. The country would suffer billions of RMB of in losses. Once the electromagnetic environment should change, the hiding would have to take place again. The other was anti-jamming by adding anti-jamming devices and building an electromagnetic shield, thus protecting the satellite from interference, which would cost less money. However, the technology was perplexing, the risk high, the time urgent, and whether it could be achieved hard to predict.

How tough was it? "It's like trying to put an elephant in a small refrigerator." Professor Ou Gang in the team recalled that the satellite was limited in size while the anti-jamming equipment was usually the opposite and high in power consumption; to have a satellite in small size yet with strong capability to avoid interference was no piece of cake at all.

Who had the courage to take the hot potato?

When relevant departments were lost in tremendous anxiety and hesitation, Wang Feixue came forward, "We will do it. And I promise to have it done within three months."

Where was the courage from?

"Soldiers are born to win. The time is fleeting. Once we slack today, tomorrow it would take us more efforts to catch up," said Wang Feixue.

In May 2008, they won the battle with the courage and wisdom of soldiers. The satellite payload with super anti-jamming capability, 1,000 times better, was presented. Academician Sun Jiadong, then the chief designer of China's satellite navigation engineering, exclaimed, "You are a research team as Li Yunlong's squad. There is the audacity to shine, and victory ensues."

A dispute over frequency happened between the Galileo satellite navigation system in Europe and Beidou in China. At the negotiation, EU experts were aggressive: there was no priority unless the European and American patented technology was bypassed. Wang Feixue, as one of the academic leaders of the research of satellite-based navigation and positioning technology of the university, the top-level design expert, and special expert group leader of China's second generation satellite navigation system construction, had participated in international negotiations between China and the United States, China and Russia, China and Europe many times. She understood that to safeguard national security interests, real skills and mastery of technology allowed one to speak up at the international negotiation.

Bravery triumphs, and why is that?

"Strength. Strength and skills must be shown to convince the world." Wang Feixue and her team solved a series of core problems, such as systems and coding, and successively cracked dozens of key technologies in satellite navigation, with arduous spirit and unimaginable perseverance and speed.

The average age of the five members headed by Wang Feixue was less than 29 at the beginning of the establishment of the satellite navigation and positioning technology team in 1996. The team has grown to a size with over 200 members after 16 years, and the average age is still below 30. However, their achievements are astonishing: one was awarded the second prize of national science and technology progress, one was the first prize winner and one the second prize winner of military science and technology progress, one the first prize of the Central Military Commission, one the National Outstanding Youth Fund Award, one won the helium-3, two the Truth Seeking Engineering Award, and one the May Fourth Medal of Chinese Youth.

Behind every award was a major technological breakthrough. This young research team was a veritable national team of Beidou satellite navigation system construction.

Beidou, a national treasure, was the harvest of great wisdom and efforts. There were many more units that had contributed to the development of Beidou project, and many more researchers worth lyrical praise. It is impossible to elaborate on them one by one.

"In the high-tech field concerning national security interests, core technology cannot be bought or imported. We have to rely on independent innovation. To have the core and key technology with independent intellectual property rights, a great deal of passion and an unimaginable amount of efforts have to be devoted."

History will always remember their courage and wisdom, persistence and tenacity, and their figures bold enough to innovate, challenge, and overcome difficulties. It was their loyal wisdom and fiery youth that let Beidou shine in the vast starry sky.

On May 25, 2003 and February 3, 2007, the third and fourth Beidou navigation test satellites were launched successively as the replacement and backups of the first generation.

On May 12, 2008, a catastrophic earthquake struck Wenchuan, Sichuan Province. Houses collapsed, people were injured, traffic was congested, and communications broke down.

When aftershocks kept coming and communications remained cut, an army took the lead in entering Wenchuan and issued a disaster report in the first minute.

What served as their beacon? What delivered the information of survival in the disaster area?

It was the Beidou satellite navigation and positioning system. Compared with foreign satellite navigation systems, Beidou has uniquely combined location reporting and SMS services, which enabled the knowledge of both where I am and where you are.

"This is the advantage of Beidou. It reflects the excellent system top-level design and future expansion." According to academician Tan Shusen, the passive positioning method of GPS only answers, "Where am I," while the active positioning method of Beidou answers both "Where am I" and "Where are you" at the same time. It can be said that "one Beidou user machine allows one to travel around the world."

When carrying out emergency rescues in the desert, mountains, oceans, and other sparsely populated areas, the Beidou satellite navigation system can navigate and position, and use the satellite navigation terminal equipment to report the location and disaster situation in time via SMS, thus effectively shortening the rescue search time and improving effectiveness.

It can be said that Wenchuan earthquake relief was a practical test of the first generation of the Beidou satellite navigation in China. In the disaster rescue and relief, the stable performance of Beidou solved the most critical positioning and communication problems and greatly improved the efficiency of earthquake relief and was highly recognized by all sectors of society.

Three months later, the Beidou satellite navigation system once again worked magic during the Beijing Olympic Games, making Chinese aware of the importance of the possession of a Chinese satellite navigation system.

Needless to say, compared with the GPS system of the United States, the first generation of Beidou satellite navigation system in China still had a long way ahead in positioning accuracy, range, and timeliness. But it matters that it has taken a historic leap from zero satellite navigation system to one.

The Coverage of Asia-Pacific

Behind every success are unexpected thrills.

At 4:00 a.m. on April 14, 2007, Long March 3A with China's first Beidou navigation satellite (Compass-M1) stood at the launch pad, waiting for ignition.

Five minutes before launch, testers reported that one air supply connector on the rocket did not fall off as required. It was only four minutes away from ignition. If the fall-off did not succeed in three minutes, there would be serious consequences.

At the critical moment, the on-site launch commander did not panic in the face of danger. Within one minute, seven orders were made one after another. And personnel of relevant posts cooperated closely and operated accurately to eliminate faults.

At 4:11, the rocket took off on time and China's first Beidou navigation satellite was successfully launched.

This was the first networking satellite in the construction of the second-generation Beidou satellite navigation system in China. The construction has entered a new stage.

Both the improvement of comprehensive national strength and the rapid development of aerospace technology have led to the acceleration of the construction of Beidou navigation since 2007.

One year later, China's second Beidou navigation satellite was successfully launched on April 15, 2009; five Beidou navigation satellites entered the orbit accurately in 2010; three Beidou navigation satellites settled down in space in 2011, including the ninth, which was successfully launched in pouring rain and lightning; six Beidou navigation satellites were sent up in four launches in eight months, both of which witnessed one rocket lifting up double satellites.

In five years, Chinese aerospace workers had 16 Beidou satellites hanging in space.

At 23:33 on October 25, 2012, China's 16th Beidou navigation satellite blasted off.

At this moment, Li Benqi, head of the launch station of the Xichang Satellite Launch Center, eyed the beautiful curve the rocket left and laughed with utmost joy.

"In five years I saw 16 Beidou satellite launched effectively. And I was present at every launch," said Li Beiqi with pride. "In the competition for space resources, Beidou started two decades later than GPS, but it is catching up with strong momentum."

On October 27, 2012, a historic milestone was set up in the development of Beidou.

Ran Chengqi, spokesman of the Beidou satellite navigation system, announced that day at a press conference held by the Information Office of the State Council that the system would now officially provide regional services to most of the Asia Pacific region. The coverage extended from 84 to 160 degrees East Longitude to 55 to 180 degrees East Longitude. The positioning accuracy was enhanced from 25 meters in the plane surface and 30 meters in the elevation to 10 meters in the plane surface and 10 meters in the elevation. The velocity measurement accuracy was increased from 0.4 meters per second to 0.2 meters per second.

"Since today, Beidou system promises the provision of continuous passive positioning, navigation, and timing services to most parts of the Asia Pacific region on the basis of retaining the active positioning, two-way time service and SMS service of Beidou satellite navigation test system." Ran Chengqi said, "Beidou's current ability to serve has been greatly improved compared with that of last year's trial operation. It is basically as capable as GPS. And it has a unique skill."

The unique skill referred to Beidou's SMS function. Other global satellite navigation systems tell users when and where only. In addition, Beidou system can send out the location of users, so that others who users intend to need to know the location can know it.

There were unimaginable difficulties and dangers at every major leap.

It took five years for the Beidou system to realize the evolution from the first generation to the second. It has been five extraordinary years, from being buried in pursuit to leading the way, from others in control to us in control, and from dual satellite positioning to covering the Asia Pacific. The hardship and bitterness were only known to the researchers, who worked tirelessly.

At the celebration of the great leap forward of the Beidou satellite navigation system, domestic major media disclosed to the public a mysterious unit with outstanding achievements in the construction of Beidou: the former general staff satellite navigation and positioning terminal of the people's Liberation Army.

On December 27, 2012, the Beidou-2 satellite navigation system was officially in operation. "It's like changing the engine for a high-speed car to have users of Beidou 1 change to Beidou 2." Tan Shusen, academician of the terminal, disclosed that when designing the system of Beidou No.2, they made sure that users of Beidou 1 would complete the upgrade without noticing it in less than two minutes.

This terminal came into being with the development of the Beidou cause in China. In recent years, they have been advocating the Beidou spirit of independent innovation, unified cooperation, overcoming difficulties, and pursuing excellence. Fourteen years of hard work with hundreds of cooperative units led to the Chinese navigating beacon shining in the vastness of space.

Hardship at the beginning was deeply engraved in Academician Tan Shusen's mind. He never forgot for a second that without holidays, a dozen experts including himself started their journey with Beidou in a tin room of less than 20 square meters.

At the time, countries had divided up the frequencies. Beidou was faced with the unavailability of frequency.

As chief expert of Beidou navigation frequency design and international coordination, academician Tan Shusen, took the initiative to actively mediate and negotiate after careful preparations. He creatively put forward the evaluation criteria of satellite navigation signal compatibility. And that proved that Beidou and other satellite navigation systems did not affect each other when frequencies overlapped. A consensus of frequency sharing was reached, and he earned China valuable frequency resources.

In the construction of Beidou 1, the terminal has organized and designed a technical system integrating positioning, position report, SMS, and time service based on the theoretical assumption and practical application requirements of dual satellite positioning. "GPS does not have the function of position report and SMS. We do not have to have them. "Voices of disagreement immediately ensued. Academician Tan Shusen firmly said: "We started late, but our ideas must be ahead of our time and we must break the pattern of being the one to catch up with the others."

In the argumentation of the Beidou 2 development scheme, he led the team to conceive an overall framework integrating two technical systems of position reporting and continuous navigation. The problems of "Where am I" and "Where are you" were solved at the same time. It went beyond the conventional satellite navigation development, led the direction of satellite navigation development, and became the characteristic and advantage of Beidou.

Later, they and the manufacturers jointly carried out chip technology research. Consequently, Beidou user ends were all equipped with self-developed chips, the bulk was downsized to one-fifth of the original, and the weight reduced to one-twentieth.

In order to improve the ability of Beidou equipment to adapt to the special environment, they made trips to the Gobi Desert and snowy plateaus for more than 150 times, testing equipment performance. Technical indicators were optimized, low-temperature working time was prolonged by five times, and the corrosion and high-temperature resistance was greatly enhanced.

Over the years, they have nailed more than 50 key technologies, such as digital fast acquisition, precise orbit determination, two-way time service, multi-path interference, and digital multi beam. It made Beidou one of the brightest stars in the hearts of Chinese.

On December 29, 2008, two researchers in the terminal called to report to Beijing after six hours of continuous testing at a testing ground in Qinling, where it was -30 degrees Celsius cold: "data testing is complete." Wen Rihong, director of the technical support center on duty, managed to hold the tears in his eyes and replied calmly, "Happy Chinese New Year."

A strong old man has great ambition. On January 30, 2016 was the 74th birthday of Tan Shusen. Staring at the sky, he said: "My star age is only 22, pretty young. I can still do a lot of things for Beidou global networking." As the main pioneer and builder of the system, academician Tan has postponed his retirement three times. He is still at the front line of Beidou in his late years. And he said "What my country requires illuminates the direction I head to. I have the responsibility and obligation to build China's Beidou into the world's Beidou."

There were no regrets. Zhou Jianhua, a female doctor in her prime years, was elated when she learned that her superiors wanted her to work on Beidou. She gave up her superior life and resolutely joined the cause.

"People with hot blood have a low boiling point, and the passionate are easy to burn." On the eve of the opening of Beidou 2, she was too occupied to take care of her mother. As a result, a minor illness that missed the best treatment period turned into a tragedy and a permanent pain in her heart.

"Sometimes life is like war. In order to win, sacrifices must be made," said Zhou Jianhua.

Beidou is the tomorrow. The average age of the chief engineer of Beidou team has dropped from 46 to 41 and to 38 from Beidou 1 and 2 back then to the current Beidou global test system.

"We spent a third of the time and one-fortieth of the funds of the foreign navigation system to build Beidou-1." The leader of the terminal explained that in recent years, researchers of terminal have rewritten the world satellite navigation technology record six times and achieved a group breakthrough in the core technology of satellite navigation. The stable operation rate of the system of Beidou-2 has remained above 99.98%.

On December 28, 2012, the Central Committee of the Communist Party of China, the State Council, and the Central Military Commission congratulated the Beidou-2 satellite navigation system. "The development and construction of Beidou-2 satellite navigation system has gathered the wisdom of the vast number of engineers and technicians, and embodies the spirit of Beidou, which is independent innovation, unified cooperation, overcoming difficulties, and pursuing excellence. The completion and use of the system is an major milestone in the informatization of the country and the military, a major contribution to the economic and social development of China, a comprehensive victory in the second step of the three-step strategy for the development of China's satellite navigation, and a significant and far-reaching step in the process of building an independent and controllable satellite navigation system."

Beidou shines amongst the stars, illuminating the road of Chinese people. This is another comprehensive improvement of China's aerospace science and technology industry capacity since the manned space project.

Beidou's accession to the global satellite navigation family will not only provide more users with higher service accuracy, richer navigation functions, and better user experience, but also further enhances China's voice and initiative in international navigation and China's core competitiveness.

The Last Distance to Mass Application

"Human imagination is the sole limitation to the application of global satellite navigation system." The changes in the world are proving every word of the saying to be true.

The breadth and depth of satellite navigation applications far exceeded the imagination of the designers. The chain reaction generated by the satellite navigation system has brought huge economic and social benefits, thus massively changing the whole planet.

2013 was the first year of the Beidou system application.

"As a strategic emerging industry, Beidou navigation industry has taken shape in the past year. And a relatively complete industrial system of basic industry, application port, system application, and operation service has been formed." Ran Chengqi, spokesman of the Beidou satellite navigation system, summarized at the end of 2013 that Beidou products had been applied in transportation, marine fishery, hydrological monitoring, meteorological forecast, geodesy, intelligent driving testing, communication timing, and disaster relief and prevention.

This year, the application of Beidou satellite navigation system picked up the pace.

In January, the Beidou navigation service began service in nine provinces and cities of road transportation in China; in March, the first Beidou satellite navigation ground enhancement network, Beidou ground-based enhancement system in Hubei Province was completed and put into trial operation, whose position accuracy is up to centimeter level; in May, the National Defense University of Science and Technology successfully developed a high-performance satellite navigation receiver, which can receive three signal systems, Beidou, GPS, and GLONASS, with positioning accuracy within 10 meters; in August, the national Beidou satellite navigation product testing agency was established; in November, the first independently developed 3D high-precision positioning Bohai Bay demonstration system in the northern sea was put into operation, and China's maritime positioning system entered the era with centimeter-level precision.

Specially, the general office of the State Council issued the Notice on the Medium and Long-term Development Plan of the National Satellite Navigation Industry, which clearly defined the medium and long-term development goals. Beidou satellite navigation system and its compatible products will be widely used in important industries and key areas of the national economy, and gradually popularized in the mass consumer market. By 2020, the innovative development pattern of China's satellite navigation industry shall have been basically formed, and the industrial scale shall have exceeded 400 billion RMB.

According to the plan, with a focus on the overall objectives and main tasks of the development of satellite navigation industry, a number of major projects will be organized, including basic projects to enhance the performance of satellite navigation, innovation projects to enhance the core technical capabilities, safety projects to promote the application of important fields, and mass projects to promote the development of industrial scale. As a result, the cultivation and development of satellite navigation industry shall be accelerated, the industrial base be improved, technology in key fields be innovated, and large-scale application, and international development be promoted.

On December 27, 2013, *Open Service Performance Specification of the Beidou Satellite Navigation System (version 1.0)* and *Spatial Signal Interface Control of the Beidou Satellite Navigation System (version 2.0)* were released in Chinese and English. It symbolized the Beidou navigation system's entry to the era of multi frequency application, which means that it can provide higher precision navigation services.

In 2014, the Beidou satellite navigation system entered the mass application market.

In April, Xihe, a high-precision positioning service system independently developed in China, was officially applied. In order to kill the last distance from satellite navigation service to mobile phone users, the Xihe system was a deepening and refinement work of satellite navigation system service. Xihe is the sun god in ancient Chinese mythology. The name means to provide a seamless navigation and positioning service system in the whole airspace and time domain. Xihe is capable of a seamless and high-precision positioning service with sub-meter level outdoors and three-meter level indoors. And satisfactory application demonstration results have been obtained in Beijing, Tianjin, and Shanghai.

In September, the first China Beidou application summit was held in Urumqi, Xinjiang Autonomous Region. With the focus on Beidou's global service, over 600 representatives from national ministries and commissions, local governments, military and local units, and science research institutes, as well as specially invited academicians, experts, and overseas guests, conducted in-depth discussions on the application and development, industrial policies, industry outlook, and international cooperation of Beidou.

To further regulate the Beidou application market and promote the fair and orderly competition and healthy development of the industry, China Satellite Navigation Application Management Center released the List of the Qualified Units of Beidou Navigation Civil Service and List of Beidou Navigation Product Quality Inspection Organizations. Also, it cooperated with National Development and Reform Commission, Ministry of Industry and Information Technology, Ministry of Housing and Urban-Rural Development, and Ministry of Transportation, Ministry of Science and Technology of People's Republic of China, China National Space Administration, National Bureau of Surveying, and Mapping Geographic Information to issue the proposal to jointly promote the application and development of Beidou to the whole society. The acceleration of the improvement of Beidou application service system was demanded, and so were the vigorous promotion of the large-scale application of Beidou in key industries, the popularization of Beidou in all aspects, the active expansion of the international application of Beidou, and wide benefit of Beidou to the world.

In November, a Beidou radio-frequency chip Hangxin I only 5 mm square, attracted massive attention at the *Third Shanghai Military-civilian Technology Promotion and Project Exchange Conference 2014*. This newly developed 40-nanometer-level China core has been installed into domestic electronic devices, such as smartphones, to provide accurate navigation during people's daily life. This indicates that the key equipment of Beidou navigation system has penetrated consumer electronics.

In 2015, the Beidou satellite navigation system picked up the pace to benefit every citizen.

In May, Beidou Satellite Communication Co., Ltd. and Xinxing Communication Co., Ltd. released a new generation of high-precision multi-mode and multi-frequency satellite navigation system-level chip. It is also the world's first all system multi-core high-precision navigation and positioning chip, marking China's satellite navigation chip in the lead around the globe.

In June, the Beidou "100 cities, 100 links, and 100 uses" campaign initiated by the China satellite navigation and positioning association made new progress. Hundreds of cities in China used the Beidou navigation system to carry out the construction, management, and maintenance of underground gas pipelines. At present, the Beidou Precise Service Network has been established in over 10 urban gas industries nationwide, which can provide more precise location information for integrity management, pipeline leakage monitoring, pipeline anti-corrosion monitoring, emergency repair of pipeline networks, intelligent inspection of pipeline networks, construction management of pipeline networks, etc. The informatization and intelligence of gas management was therefore promoted.

In November, China's first mass-produced 40-nanometer high-precision Beidou navigation application chip was released, whose first positioning would take less than 2 seconds.

According to the *White Paper on the Development of China's Satellite Navigation and Location Service Industry* issued by the China Satellite Navigation and Positioning Association in 2015, China's satellite navigation and location service industry has maintained a rapid development, with the total output reaching 173.5 billion RMB in 2015, an increase of 29.2% over the previous year. Among them, the proportion of Beidou application has been further improved, and the market contribution rate was close to 20%.

The white paper shows that in the industry market, the traditional demand of transportation, surveying and mapping, and logistics is relatively stable; in the mass market, the demand of vehicle navigation, smart phones, and location services, is growing rapidly. At the same time, the demand for UAV, indoor and outdoor integrated positioning, wearable equipment, and safety emergency, has increased significantly.

Since 2016, the Beidou satellite navigation system has been widely used to accurately serve people's lives.

Beidou satellite technology benefits the Qinghai Tibet Plateau with intelligent grazing; Beidou Safety Information Broadcasting System is open to the public free of charge; Beidou Disaster Prevention and Monitoring System is located in Bashan, Ankang, Shaanxi Province; China began the construction of Kuilong, the first Beidou global centimeter-level positioning; the trial operation of Beidou precision positioning service system in Bohai Bay area of China realized real-time centimeter-level positioning service; in the Beidou/GPS dual-mode era, the positioning accuracy of differential navigation in the North China Sea area can upgrade to the sub-meter level.

"At present, Beidou has achieved cross-border integration in urban gas, urban heating, electric power grids, water supply and drainage, intelligent transportation, and care for the elderly. It fundamentally improved the informatization capacity of urban operation and management, enhanced people's livelihoods, and brought technological innovation and breakthroughs to the infrastructure construction and management of smart cities," said Wang Yanyan, Secretary General of the Precision Application Committee of the China Satellite Navigation and Positioning Association, at the summary meeting of Beidou's "100 cities, 100 links, and 100 uses" campaign.

In 2017, the application results of Beidou satellite navigation system were reported positive, and the pace of promotion of the application accelerated.

In January, the first phase of the Beidou foundation enhancement system in China has been completed, including all the development and construction of 150 frame network base stations and 1,269 regional encryption network base stations. "One network in China" and "One platform in China" were officially put in use, capable of providing real-time dynamic centimeter-level, post-processing millimeter-level, and rapid auxiliary positioning in major economic regions of the country.

In February, a Beidou chip for domestic high-precision vehicle use was released with meter-level positioning accuracy and three-second fast positioning. China's vehicles would be equipped with meter-level navigation.

"Beidou has made great achievements in spatial use. We should work hard on ground use." Sun Jiadong, known as the Father of Beidou, said that in order to produce benefits, Beidou satellite navigation service must pay attention to the uses. Victory in the global competition of satellite navigation system ultimately depends on whose ground application is deeper, more extensive, and successful.

In the age of information, data is in the cloud and people on the Internet. As an information system of integration of space and earth, the trend of cross-border integration of the Beidou system and emerging information technology is more inevitable.

—To integrate the network, the focus is on promoting the integration of Beidou system with Bluetooth, broadband mobile Internet, wired internet, and satellite communications networks, and realizing the interconnection among Beidou foundation enhancement networks, high-precision location service networks, and broadband mobile Internet. This will enable Beidou to have faster information transmission, more accurate location, clearer images, and more skillful usage;

—To integrate data, big data, cloud computing, and navigation grid code are adopted to realize the integration and development of the Beidou System, high-precision remote-sensing digital maps, navigation grid codes, and cloud computing platforms. This would make the isolated and static Beidou spatiotemporal data dynamic, and improve its application value;

—To integrate terminals, mobile phones, tablet computers, and wearable products are the largest market for Beidou applications in the future. It concentrates on promoting the integrated chip research and development of Beidou navigation, wireless LAN, mobile communication, and satellite communication as a whole. This will create a multi-functional integrated information terminal product.

"Many new development concepts have emerged. For example, in the car network, Beidou has achieved an effective application, which report the vehicle safety and reduce the zero-load rate; in the shipping network, it can realize emergency rescues and disaster predictions; in the Internet of Things, we have also achieved product traceability and food safety supervision." Ran Chengqi, director of the China satellite navigation system management office, said, "After Beidou is built, the precision can reach to the centimeter-level at most. High precision application is what characterizes Beidou. In the future, the new business form of the combination of Beidou, the Internet, and other industries will promote the transformation of production and lifestyles. And imagination will be the only limit to what can be achieved."

According to *Medium- and Long-Term Development Plan of the National Satellite Navigation Industry* issued by the State Council, it is predicted that by 2020, the scale of China's satellite navigation industry will have exceeded 400 billion RMB, and the scale of Beidou industry will have reached 240 billion RMB.

At the second Beidou Civil Promotion Conference, Fu Yong, director of the China satellite navigation and positioning application management center, said that since Beidou system was put into use, it has been in stable operation and constantly good condition. Currently, Beidou civil users have reached tens of millions. Regarding passive service, the monitoring data shows that the overall performance of the system meets the design requirements, especially the key indicators such as positioning accuracy and timing accuracy, which are obviously better than expected. It has provided 1.2 billion active positioning services, 6.1 billion SMS services, and more than 90 million two-way time services. In 2016, the number of users and services increased significantly, with over 70,000 registered users, 250 million active positioning services, 3.08 billion SMS services, and 8.33 million two-way time services.

Where is the Big Dipper? According to the folk songs, "The river flows eastward, and the stars in the sky point to the Big Dipper."

Where is the Beidou? In the boundless starry sky patrols the Beidou navigation satellite.

International vision allows the Beidou system to be qualified enough. Independent innovation allows the Beidou system to be down-to-earth enough.

From an experimental article to convenient and easy-to-use mass products, Beidou has gone through a decade of a bumpy journey. Today, the application contains disaster prediction, precision agriculture, ocean fishing, Internet of things, and all aspects of people's lives. With the continuous acceleration of industrialization,

Beidou products will accumulate more popularity and go international. The performance-price ratio keeps getting better and more ordinary folks are benefiting from the creation.

Global Coverage by 2020

"China's Beidou shall benefit the world too."This is what is often mentioned amongst Chinese Beidou developers, who share both national feelings and international perspectives.

In recent years, Beidou satellite navigation system has been expanding its application in both China and the world.

In October 2013, Thai officials revealed in Bangkok that Beidou, one of the world's four major satellite navigation systems and the independent creation of China, would be put to use in Thailand at the beginning of next year, taking the first step into ASEAN. A spokesman for Thailand's geospatial technology bureau said that the construction of the disaster prediction system based on Beidou's satellite remote sensing technology would be included in the economic development of the Thai government. And it would mainly manage early warning of agricultural disasters and also provide services in transportation, electricity, and environment in Thailand.

In May 2014, the first overseas networking project of the Beidou satellite navigation system, Pakistan National Location Service Network Phase I, jointly constructed by China Great Wall Industry Group Co., Ltd. and Beijing Hezhong Sizhuang Technology Co., Ltd., was completed in Pakistan.

Experts explained that, at present, five base stations and a processing center have been established in Karachi City, Pakistan, to form a regional Beidou positioning enhancement network. And it would cover the whole Karachi region, with real-time positioning accuracy of 2 cm and post-processing accuracy of 5 m. Pakistan would be provided with real-time and reliable Beidou high-precision positioning services.

Pakistan's National Location Service Network project is a national key infrastructure. Compared with previous single station projects that China led overseas, this is the first networking project of Beidou overseas, whose broad application in the future is rather promising. Pakistan's effective use of the high-precision services of the Beidou system shall meet the social needs of various modern information management in urban planning, land surveying and mapping, cadastral management, environmental monitoring, disaster prevention and mitigation, traffic monitoring, etc.

In November 2014, the Maritime Safety Committee of the International Maritime Organization (IMO) reviewed and approved the navigation safety letter of confirmation for the Beidou satellite navigation system. The Beidou satellite navigation system has officially become the third global satellite navigation system, following the United States GPS and Russia GLONASS satellite navigation system. It serves global marine navigations.

The person in charge of the Ministry of Transport pointed out that as a class A member of the IMO, the official recognition of the Beidou satellite navigation system will push forward the internationalization and industrialization of the Beidou system in marine sailing. And it has, as an integral part of the global radio navigation system, gained a legitimate seat in maritime applications. This is the first time that international organizations accepted China's Beidou satellite navigation system standard.

Since the approval of the IMO, China will has continued to comprehensively promote the formulation and revision of standards, specifications, and guidance documents of the International Electrotechnical Commission, The International Association of Marine Aids to Navigation and Lighthouse Authorities, the International Maritime Radio Technical Commission, and the International Telecommunication Union, and further realize the all-round application of the Beidou System in the international maritime field.

In August 2016, Laos began to receive the China Beidou satellite navigation system service. Underpinned by China National Bureau of Surveying and Mapping of Geographic Information, Yunnan Provincial Bureau of Surveying and Mapping of Geographic Information, and Laos National Bureau of Surveying and Mapping, the Laos satellite positioning integrated service system based on the Beidou system construction was officially launched. It can be applied to land planning, urban construction, transportation, water conservancy, and electric power, can provide real-time, accurate, and reliable data sources for various engineering construction, three-dimensional spatial position and time monitoring and management in transportation, public security, finance, etc., mapping and geodetic application for various industries; and can be used for monitoring natural disasters, and more.

According to the introduction, the application of the Beidou System in Southeast Asia has better time availability, space availability, high-precision navigation, high-intensity encryption, etc., over other systems.

In March 2017, Chinese Beidou-related enterprises, planned to jointly develop and build at least 10 base stations (CORS base stations) for Sri Lanka's specific geographical conditions and use the environment in cooperation with Sri Lanka's National Bureau of Surveying and Mapping. Beidou served as the main base station compatible with U.S. GPS, Russia GLONASS, and other satellite navigation systems. They would carry out land mapping, marine fishery, and disaster warning in Sri Lanka as demonstrative applications.

Before that, China built the first three overseas foundation enhancement system base stations (CORS base stations) of the Beidou System in Chon Buri, Thailand, all of which were networked. In Thailand, it has successively carried out Beidou intelligent transportation, marine fishing boats, intelligent industrial parks, etc. as demonstrative applications. At present, the China ASEAN Beidou Science and Technology City, which promotes Beidou's application in key areas including communication, transportation, and the Internet of things in the ASEAN region, is also being planned for construction.

China's Beidou will soon 'light up' ASEAN. Experts introduced that the construction of Beidou base stations in Thailand and Sri Lanka could extend the coverage radius to Southeast Asia and South Asia by at least 3,000 kilometers. As a result, the Beidou high-precision navigation and positioning network can reach more users.

In April 2017, the Beidou Navigation System Seminar, jointly organized by the China satellite navigation system management office and King Abdullah Aziz Science and Technology City, was held in Riyadh, Saudi Arabia. This is the first time that China's satellite navigation system entered the Saudi market, with huge demand for navigation products and applications on a large scale.

The Sino-Saudi satellite navigation cooperation will play an important role in promoting China's One Belt One Road initiative and Vision 2030 of Saudi Arabia. It is expected to become a new value and new engine for pragmatic cooperation between the two countries.

As stated in the *White Paper of China Beidou Satellite Navigation System,* published in 2016, China would continue to promote the international development of the Beidou system, actively and practically carry out international cooperation and exchanges, serve the "One Belt and One Road" initiative, propel the development of the global satellite navigation business, and enable Beidou to better serve the world and benefit mankind.

In the future, Beidou's pace towards the world will continue to accelerate with the development of the Beidou network. It is expected to have covered the "One Belt and One Road" countries first by 2018, and the whole world by 2020.

Great Responsibility

Einstein's theory of relativity revealed the nature of the universe.

The universe is boundless space and endless time. It is made up of matters that are in perpetual motion.

Space and time are the two largest parameters in the world. Satellite navigation perfectly realizes the integration of space and time, thus becoming the core of new space-time technology and an important drive to promote the world's new information technology revolution.

Today, satellite navigation systems, as a ubiquitous national infrastructure, are a major technical support for national defense security, economic security, social progress, and people's livelihood improvement.

Because of the increasing importance of satellite navigation system, the competition is increasingly fierce among the countries.

To ensure a leading position, the United States is upgrading GPS technology on the largest scale ever. The transmission capacity of the upgraded third generation GPS system is 500 times that of the current one. In addition, the anti-jamming ability and positioning accuracy will be greatly improved.

To compete against the U.S. GPS system, Russia will launch 13 GLONASS-M satellites and 22 of its upgraded GLONASS K satellites. According to the plan, by 2020, the GLONASS system is expected to have 30 satellites in orbit, six of which will serve as back-up, and the navigation accuracy will reach 0.6 m. It is reported that the "GLONASS" system currently operates more than 30 satellites in orbit.

The construction of the Galileo navigation system in Europe has also picked up the pace. In November 2016, the 15th, 16th, 17th and 18th satellites of the European Galileo satellite navigation system were successfully launched by an Ariane rocket from Kourou of Centre Spatial Guyanais. The previous 14 satellites were launched by Russian Union rockets, two at a time. According to the plan, another eight satellites will be launched by Ariane in 2017 and 2018, and satellite networking is scheduled to be realized by 2020.

The competition is so fierce that Chinese spacemen are not slacking for a second.

—Global networking is launched. Key technologies are cracked, new generation of special satellite platforms for navigation satellites, direct launch of vehicles into orbit, high-precision atomic clocks, inter satellite links, and autonomous navigation. And a large number of high-end scientific and technological achievements with independent intellectual property rights were made.

—The new generation of Beidou navigation satellite for global networking has achieved many breakthroughs, the first time to directly launch into orbit medium and high orbit satellites, the first time to use a new navigation satellite special platform, the first time to adopt a newly designed navigation signal system, etc.

At 21:52 on March 31, 2015, China's first new generation Beidou navigation satellite was launched at the Xichang Satellite Launch Center. The successful launch started the new journey of global networking around 2020 for the Beidou satellite navigation system.

"In terms of performance, the new generation of global coverage navigation satellite is one to two times better than the system currently in orbit, and the highest accuracy will reach two-to-three meters." Yang Changfeng, chief designer of China's Beidou satellite navigation system, said that in 2015, three more Beidou navigation satellites were launched, forming a complete test system. This laid a basis for global networking of Beidou.

The first new generation of Beidou navigation satellites independently adopted the general CPU named Dragon Core and the homemade rubidium atomic clock developed by the Institute of Computer Science of the Chinese Academy of Sciences. In addition to the lifespan extending from eight years to 10 to 12, function and performance will be greatly upgraded. It will be able provide higher precision satellite navigation services for global users.

At 15:29 on February 1, 2016, the fifth new generation Beidou navigation satellite was successfully launched. This is the last satellite involved in the test verification of inter-satellite links and the new navigation signal system. The completion of test verification will help them consolidate the global networking and invest in new products.

At present, the Beidou engineering team has completed the testing and verification of the new generation Beidou navigation satellite, conquered all core technologies of Beidou system satellite networking, and basically consolidated the status of new generation Beidou navigation satellites.

According to Ran Chengqi, director of China's satellite navigation system management office, it is expected that around 2020, five geostationary orbit and 30 earth geostationary orbit satellites will be networked around the world as the Beidou satellite navigation system global network construction scheme plans. In 2018, it will provide the basic services for the "One Belt One Road" countries. in 2020 the global service capacity will be formed, and the world's first-class global satellite navigation system be built.

In the next three years, China's Beidou navigation satellite will enter the intensive launch period again, China will continue to launch 18 Beidou navigation satellites before 2018, and nearly 40 satellites be launched in the next five years."

The goal of 2020 is clear. Chinese spacemen are sprinting at maximum efforts.

It can be predicted that in the near future, the global satellite navigation system will have four dominating systems. However, we should be aware that compared with other countries, Beidou still has a long way to go. Only by maintaining its own advantages while constantly improving innovation, positioning, and navigation abilities, can it stay invincible in the fierce competition.

EPILOGUE

The Heart Unchanged, the Dream in Pursuit

Time ages people. And time makes greatness.

Sixty years, a cycle of life. Generations of Chinese aerospace workers have gone through thick and thin with little sleep taken but great efforts. The wisps of dark hair have turned silver and the enthusiasm has become everlasting loyalty. Generations of Chinese aerospace workers have devoted themselves one after another, making the Chinese's space dream come true.

Sixty years, a moment in history. Generations of aerospace workers have been exploring the vast space with youth and even life, creating China miracles one after another and writing legendary stories one after another in the history of world spaceflight from Dongfanghong satellites to manned spaceflight, from Chang'e lunar exploration to Beidou navigation system.

The wisdom of a nation and the creativity of a country are often seen in remarkable achievements. They symbolize the prosperity of a nation. There is no doubt that aerospace has become the most presentable national business card of China today. As the crystallization of the extraordinary wisdom and great creativity of the Chinese nation, the footprints that Chinese aerospace workers left in space announce to the whole world that the Chinese nation, fully capable and self-reliant, deserves to stand tall and proud in the world.

The space dream is a Chinese dream. A dream can make a nation strong and confident and stay vital and vigorous. The Chinese space dream is the dream of a strong country in science and technology and of national rejuvenation. At present, China's space technological capability has been admitted to the advanced club of the world. China has nearly 150 satellites in orbit, second only to the United States and Russia, ranking third in the world. From near earth to lunar exploration, Chinese look up at the stars, down to earth.

In the past six decades, China's aerospace industry was able to achieve Chinese-style leapfrog development at an incredible speed owing to the "Chinese road": the superiority of the socialist system to concentrate efforts on major issues, and hundreds of thousands of scientific and technological forces aiming at one direction; owing to the "Chinese spirit": self-reliance, hard work, dedication, and perseverance, which are the deepest endowment of the Chinese nation, passing down and shining from generation to generation, and becoming the inexhaustible power to explore space; and owing to the "Chinese power": independent innovation is the core of national development strategy, the soul of national progress, and the powerful engine to realize the space dream.

All for the country, and all for the success. The sixty years' glorious experience of China catching up has condensed sixty years of remarkable leaps. Only being in the coordinates of history can we understand the difficulties along the way. The historical epic of China's aerospace industry is a brilliant national memory and a proud national album. China Aerospace engraves China's status as a great power and the confidence and glory of the Chinese nation on the boundlessness of space.

Today, looking back on this history is not to indulge in glory, but to find the original heart of China's aerospace industry: where do we come from and where will we go?

Why did we still insist on making satellites in the age of starvation? Why was the light still on in the lab in the darkest hours? Why could we still concentrate on tackling problems in the face of the embarrassment that missiles were considered no better than feeding empty stomachs? Why could generations of aerospace workers regret nothing to devote their youth? Because, no matter what, they never forget why they set out, never forget the original heart: for the country's prosperity and national rejuvenation. On the development path of China's aerospace industry, tens of thousands of aerospace workers have linked their personal ideals with the fate of their motherland and their personal choices with the needs of their country. The incomparable cohesion of the Chinese nation erupted like a volcano.

When dreams become glory, glory breeds new dreams.

Brilliant achievements are yesterday. New dreams are tomorrow.

The heart remains unchanged, and the dream in pursuit continues. There is no end to space exploration. And there is a long way to realize dreams. China's aerospace industry has a long journey to explore.

From now on, it will take five to seven years to complete major on-going projects, such as manned space flight, lunar exploration project, Beidou navigation, a high-resolution earth observation system, etc.; around 2025, it will finish the national civil space infrastructure, and promote the scale, business, and industrial development of space information application; in 2030, it will realize the overall leap and become one of the aerospace powers.

Now, dark-matter satellite Wukong is on an adventure in space; hard X-ray Modulation Telescope will fly there; Mozi, the world's first quantum science experimental satellite, is being tested; China's Mars exploration mission has just been officially approved and the Mars probe will be launched in 2020.

The footprint reaches as far as the dream takes. As Robert Goddard, an American scientist, said in 1914, "It is hard to say what is impossible, because the dream of the past is the hope of today and the reality of tomorrow."

Looking to the future, we believe and expect: China aerospace will write new highlights.

References

60 Events in the 60th Anniversary of the Founding of China's Space Industry. Beijing: Press and Publicity Office of the State Administration of Science, Technology, and Industry of National Defense, 2016.

Bei Fang. *Five-star Flag in Space*. Beijing: People's Publishing House, 2011.

Bian Dongzi. *The Long Memory: Stories in Zhongguancun's Special Building*. Shanghai: Shanghai Education Press, 2008.

Cao Chong et al. *Beidou with us in the World*. Beijing: China Aerospace Press, 2011.

Chen Shanguang. *Flying Heroes: Tracking the Footprints of Astronauts*. Beijing: China Aerospace Press, 2011.

China Academy of Space Technology. *Beacon in Space: The Legend of Chinese Satellite Spacecraft*. Beijing: China Aerospace Press, 2009.

China Aerospace White Paper. 2006, 2011, 2016.

Commission of Science Technology and Industry for National Defense, Lunar Exploration Engineering Center. China Lunar Exploration. Beijing: Science Press, 2007.

Dong Renhai et al. *Satellites in Space*. Beijing: China Overseas Chinese Press, 1993.

Editorial Board. *The Soul of Space: A Chronicle of Aerospace Spirit*. Beijing: China Aerospace Press, 2011.

Gong Xiaohua. *Inside Story of China's Aerospace Decision Making*. Beijing: China Literature and History Press, 2006.

Hu Shihong. *The Military Career of General Zhang Aiping in the Founding of the People's Republic of China*. Beijing: People's Publishing House, 2006.

Jia Yong et al. *Interpretation of China's Manned Space Flight*. Beijing: Xinhua Press, 2008.

Jin Chongji. *Biography of Zhou Enlai*. Beijing: Central Literature Press, 1998.

Lan Tao and Chen Xin. *Autobiography of Ouyang Ziyuan*. Nanjing: Jiangsu People's Publishing House, 2008.

Li Mingsheng. *A Dream for Thousands of Years: The Story of the First Time Chinese Left the Earth*. Chengdu: Tiandi Publishing House, 2016.

Li Mingsheng. *A long Way to Space: A Record of China's Rocket and Satellite Launch*. Beijing: Party School Press of the CPC Central Committee, 1995.

Li Mingsheng. *China Long March Series*. Chengdu: Tiandi Publishing House, 2016.

Li Mingsheng. *General of Launches*. Beijing: Beijing October literature and Art Press, 2010.

Li Mingsheng. *Going out of the Earth Village—Journey to the Space of the First Manmade Satellite in China*. Beijing: People's Publishing House, 2009.

Li Mingsheng. *Risky launch of AusSat-1*. Chengdu: Tiandi Publishing House, 2016.

Li Mingsheng. *Thirty-Six Thousand Kilometers of Expeditions*. Sichuan: Tiandi Publishing House, 2016.

Li Peicai. *Space Tracking: A Record of China's Space TT &C*. Beijing: Party School Press of the CPC Central Committee, 1995.

Li Xuanqing. *Looking Back on the Historical Moment*. Beijing: Long March Press, 2012.

Liang Dongyuan. *Secret Travel to Space: The Legend of China's First Manned Space System*. Beijing: Contemporary China Press, 2010.

Liu Huaqing. *Memoirs of Liu Huaqing*. Beijing: PLA Press, 2004.

Liu Jianguo. *The Spiritual Track of the Coming True of the Space Dream—Interpreting the Spirit of Manned Space Flight*. Beijing: Military Science Press, 2004.

Liu Ping. *Biography of Qi Faren*. Nanjing: Phoenix Publishing and Media Group, 2006.

Ma Jingsheng. *Biography of Chen Fangyun*. Beijing: China Youth Press, 2016.

PLA General Equipment Department. *Two Bombs and One Satellite—Monument of the PRC*. Beijing: Jiuzhou Book Press, 2000.

Propaganda Department of General Equipment and Political Department. *Dream in Space: Selected News Works of the 10th Anniversary of China's Manned Space Flight*. Beijing: National Defense Industry Press, 2014.

Propaganda Department, Political Department, and General Equipment Department. *Shenzhou Chronicle: Selected News Works of China's First Manned Space Flight*. Beijing: PLA Press, 2005.

Space Shepard: A Collection of Li Jisheng's Advanced Stories and New Works. Beijing: National Defense Industry Press, 2002.

Tao Chun and Chen Huaiguo. *National Destiny: The Secret Course of "Two Bombs and One Satellite" in China*. Shanghai: Shanghai Literature and Art Press, 2011.

To the Space (5 episodes of TV documentary script). Beijing: Published by Founder Press of China.

Tu Yuanji. *Stories of Qian Xuesen*. Beijing: PLA Press, 2011.

Wang Jianmeng. *Biography of Sun Jiadong*. Beijing: China Youth Press, 2015.

Wang Jianmeng. *Satellite in the Heart: Sun Jiadong, a Famous Aerospace Engineering and Technology Expert*. Beijing: China Aerospace Press, 2009.

Wu Chuansheng. *Expedition Log*. Chengdu: Sichuan Literature and Art Press, 2004.

Wu Lilin. *World Famous: The Successful Launch of the First Man-made Satellite in China*. Jilin: Jilin Publishing Group, 2010.

Wu Shuhua and Zhu Yuhua. *Passion for the Space: Ren Xinmin, Pioneer of Chinese Missiles, Satellites, Launch Vehicles and Spacecraft*. Beijing: People's Publishing House, 2013.

Wu Zhuo. *The Story of Chinese Aerospace Workers*. Beijing: China Aerospace Press, 2015.

Xin Hua. *China's Three-level Lunar Exploration Leap*. Beijing: Xinhua Press, 2013.

Xu Xiaoyan. *Panoramic Documentary Series of China's Aerospace Development*. Shenyang: Baishan Publishing House, 2015.

Yang Bing. *Far Towards the Space: Report from the Beijing Aerospace Flight Control Center*. Beijing: People's Daily Press, 2006.

Yang Liwei. *Cloud Nine*. Beijing: PLA Press, 2010.

Yao Kunlun. *Biography of Wang Yongzhi*. Beijing: AVIC Industrial Press, 2015.

Ye Yonglie. *A Closer Look at Qian Xuesen*. Shanghai: Shanghai Jiaotong University Press, 2009.

Zhao Jibao. *The Road of Rocket Flight*. Beijing: PLA Press, 1997.

Zhu Zengquan. *Space Dream Comes True*. Beijing: Huayi Press, 2003.

Index

ABOUT THE AUTHORS

Li Xuanqing, Vice-General Editor of the People's Liberation Army Daily and Major General in the People's Liberation Army (PLA). He planned and carried out the report on the major events including the launches from Shenzhou 1 to Shenzhou 7 and Chang'e-1's lunar exploration. He interviewed a great number of hero-models such as astronaut Yang Liwei, test pilot Li Zhonghua of the Air Force, and well-respected military doctor Hua Yiwei. His news works have successively won China News Award and PLA News Award.

Liu Gang, Vice-Director of the Ministry of Propaganda of the PLA Daily and Senior Colonel in the PLA. He performed the interviews about the manned space programs and lunar explorations after Shenzhou 7. He covered the story of the aircraft carrier Liaoning, the air force test pilot group of China. His news reports have successively won the China News Award and the PLA News Award.

PHOTO CREDIT

QIN XIAN'AN: former senior reporter of the Propaganda Department of the General Armament Department, was engaged in the photo shoot of Shenzhou 1–Shenzhou 10, Chang'e lunar exploration project, missiles, satellites and other major national defense science research and test missions.